Wireless Networks

Series Editor

Xuemin Sherman Shen, University of Waterloo, Waterloo, Canada

The purpose of Springer's Wireless Networks book series is to establish the state of the art and set the course for future research and development in wireless communication networks. The scope of this series includes not only all aspects of wireless networks (including cellular networks, WiFi, sensor networks, and vehicular networks), but related areas such as cloud computing and big data. The series serves as a central source of references for wireless networks research and development. It aims to publish thorough and cohesive overviews on specific topics in wireless networks, as well as works that are larger in scope than survey articles and that contain more detailed background information. The series also provides coverage of advanced and timely topics worthy of monographs, contributed volumes, textbooks and handbooks.

Xiangjun Xin

Advanced Optical Fiber Transmission Systems

Springer

Xiangjun Xin
School of Information and Electronics
Beijing Institute of Technology
Beijing, China

ISSN 2366-1186 ISSN 2366-1445 (electronic)
Wireless Networks
ISBN 978-3-032-09515-2 ISBN 978-3-032-09516-9 (eBook)
https://doi.org/10.1007/978-3-032-09516-9

This Springer imprint is published by the registered company Springer Nature Switzerland AG
The registered company address is: Gewerbestrasse 11, 6330 Cham, Switzerland

Declarations

Competing Interests The author has no competing interests to declare that are relevant to the content of this manuscript.

Contents

Chapter 1
Introduction

1.1 Background

We find ourselves in the era of information, enabling communication with anyone, anytime, and anywhere on diverse topics. Information dissemination occurs through communication systems, including hybrid coaxial cables and a range of Radio Frequency (RF)-based millimeter-wave (mmW) and terahertz (THz) wireless communication technologies. With the rapid advancement of information technology, a plethora of emerging data services such as Internet Protocol Television (IPTV), Mixed Reality (MR), Internet of Things (IoT), big data, and cloud computing continuously emerge. While these services enhance the convenience of users' lives, they also pose new requirements and challenges to the capacity of communication networks. According to the Mobile Market Report released by Ericsson in November 2022, the global number of contracted users for the Fifth Generation mobile communication technology (5G) has already reached approximately 870 million, with an estimated increase to 5 billion by 2028, achieving an 85% population coverage [1]. Predictions indicate that the widespread deployment and extensive commercial use of 5G will lead to a rapid growth in global mobile data traffic.

However, with the accelerating global demand for bandwidth, copper/coaxial cables and RF cellular/microwave technologies face limitations in bandwidth and high-speed information transfer, rendering them inadequate to meet future requirements. This inadequacy becomes particularly apparent due to crowded spectra and limited access bandwidth. Moreover, these technologies incur expensive licensing fees and are susceptible to security issues, leading to higher installation costs. To address these challenges, optical communication has been proposed as an alternative technology, which utilizing various mediums and spanning a range of frequencies. Optical communication systems specifically utilize high-frequency electromagnetic waves within the infrared, visible, or ultraviolet spectrum for transmission [2]. It offers fast transmission rates and a large information-carrying capacity.

X. Xin, *Advanced Optical Fiber Transmission Systems*, Wireless Networks,
https://doi.org/10.1007/978-3-032-09516-9_1

A pivotal milestone in the evolution of optical communications was the invention of the laser in 1960, which resolved the longstanding challenge of generating a coherent and stable optical source [3]. However, the practical implementation of optical transmission remained constrained by the prohibitively high attenuation in optical fibers, with early prototypes exhibiting losses exceeding 1000 dB/km. The paradigm shifted in 1966, when British-Chinese physicist Charles Kuen Kao, in collaboration with G. A. Hockham at Standard Telecommunication Laboratories, proposed a visionary solution: utilizing purified glass fiber as a transmission medium [4]. Their seminal analysis demonstrated that, by substantially reducing the concentration of impurities within the fiber core, attenuation levels could be lowered to below 20 dB/km–rendering fiber-optic communication not only viable but transformative. This foundational insight catalyzed the development of modern fiber-optic networks and earned Charles Kuen Kao the 2009 Nobel Prize in Physics, in recognition of his pioneering contributions to the field. In 1970, corning announced a significant milestone–the production of the world's first low-loss fiber, reducing the fiber loss to 17 dB/km. Subsequent relentless efforts by numerous researchers further decreased the optical fiber's transmission loss. Today, the loss of optical fiber has reached its theoretical limit, plummeting to as low as 0.15 dB/km. This remarkable progress underscores the transformative journey of optical communication from ancient techniques to cutting-edge advancements. Table 1.1 reveals the evolution of fiber optic communication prior to 1980. Since the advent of low-loss optical fiber in 1980, these systems have undergone widespread adoption, catalyzing a rapid transformation in global communication network architectures.

Recognizing the significant development prospects of optical communication systems, researchers committed themselves to advancing optical communication technologies. Over the past several decades, the field of optical communications has witnessed a transformative evolution, driven by continuous advances in device technology and system design. These innovations have enabled exponential growth in data transmission capacity, achieving enhancements by orders of magnitude–scaling

Table 1.1 Developmental trajectory of fiber optic communication prior to 1980

Period	Unit/scholars	Research achievements
1966	STC laboratory High Kun and He Ke Han	Quartz-based glass fibers (optical fibers) with an attenuation below 20 dB/km can facilitate long-distance transmission of information.
1970	Corning incorporated	A fiber optic cable with a length of 30 meters and an attenuation of 17 dB/km has been successfully prepared.
1972	Corning incorporated	A fiber optic cable with an attenuation of 4 dB/km has been successfully manufactured.
1976	Bell laboratories	A fiber optic transmission system with a data transfer rate of 44.7 Mbit/s has been established.
1977	Corning, CSELT, and pirelli companies	The first successful implementation of optical signal transmission achieved a total length of 9 km with a speed of 140 Mbit/s.

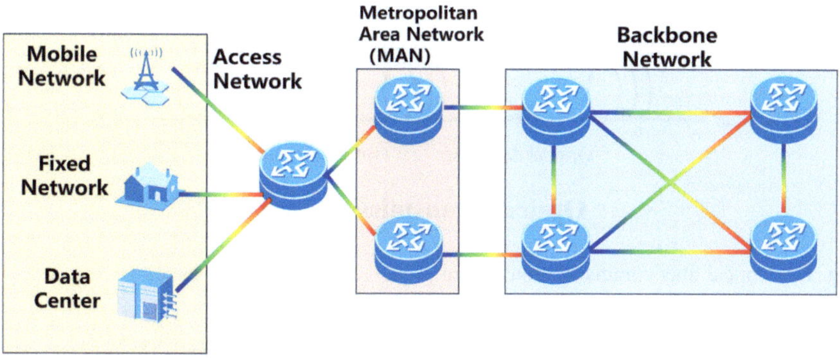

Fig. 1.1 High capacity optical network structure

hundreds to thousands of times over the span of half a century, with no signs of plateauing. Modern optical networks now embody highly sophisticated architectures capable of supporting unprecedented bandwidth demands. As depicted in Fig. 1.1, a representative wide-area optical communication network spans distances exceeding one thousand kilometers and sustains aggregate transmission rates on the order of terabits per second (Tb/s), underscoring the remarkable scalability and efficiency of current photonic infrastructure.

This publication aims to familiarize readers with contemporary optical communication systems, highlighting sophisticated coding techniques, IMDD and coherent communication methodologies, advancements in amplifier technology, and DSP techniques applied in receivers. The primary goal is to offer readers a thorough comprehension of the principles, applications, and innovations in optical communication systems catering to both short-distance and long-distance transmission requirements.

1.2 Overview of Optical Fiber Communications

As illustrated in Fig. 1.2, optical fiber communication represents a cornerstone of contemporary telecommunication infrastructure, enabling ultra-high-speed data transmission across vast distances with exceptionally low signal attenuation. The architecture of such systems integrates a suite of advanced technologies, including intensity modulation with IMDD, coherent detection techniques, high-order modulation formats, forward error correction (FEC) algorithms, sophisticated digital signal processing (DSP), and optical amplification. Together, these components synergistically enhance system capacity, spectral efficiency, and transmission fidelity, underpinning the continued scalability and resilience of global optical communication networks.

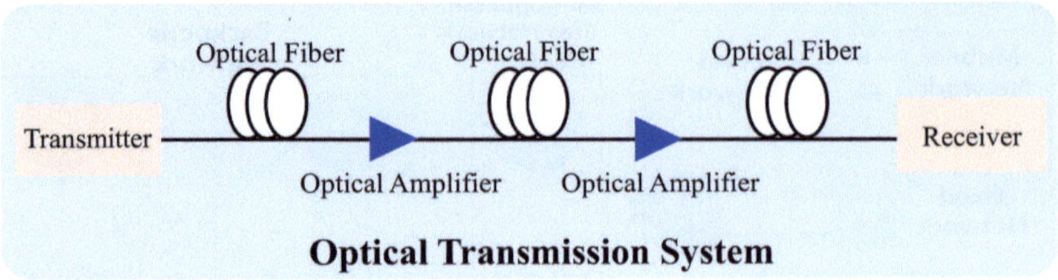

Fig. 1.2 Optical fiber communication system

1.2.1 Optical IM/DD for Short Reach Optical Fiber Systems

Optical IMDD technology stands as a cornerstone in short-reach optical fiber systems, offering a robust and cost-effective solution for data transmission over relatively short distances. This technology, characterized by its simplicity and efficiency, employs intensity modulation to encode digital information onto the optical carrier signal, which is subsequently detected directly at the receiver without the need for phase or polarization-sensitive components. In IM/DD systems, data transmission is achieved by modulating the optical intensity of a laser diode in accordance with the binary information stream. The resulting optical signal is coupled into the fiber and guided through the core via total internal reflection. Upon arrival at the receiver, a photodetector transduces the optical signal into an electrical current, enabling the retrieval of the original data. This modulation scheme, while relatively simple in implementation, forms the basis of many high-speed short- and medium-reach optical links due to its cost-effectiveness and robustness. One of the primary advantages of IM/DD technology lies in its simplicity and cost-effectiveness. By eliminating the need for complex modulation schemes and sophisticated detection techniques, IM/DD systems offer a straightforward and economical solution for short-distance optical communication applications. Moreover, the direct detection approach simplifies receiver design, enabling compact and energy-efficient receiver architectures. However, IM/DD systems are not without their limitations. One notable challenge is their susceptibility to fiber impairments, such as chromatic dispersion and fiber nonlinearities, which can degrade signal quality and limit transmission distance. Additionally, IM/DD systems typically exhibit lower spectral efficiency compared to coherent modulation schemes, constraining their suitability for high-speed, long-distance transmission scenarios.

1.2.2 Optical Amplifier for Long Haul Optical System

As illustrated in Fig. 1.3, optical Raman amplification stands as a pivotal technology in long-haul optical wavelength-division multiplexing (WDM) systems, offering a potent solution for mitigating signal attenuation and enhancing optical transmission

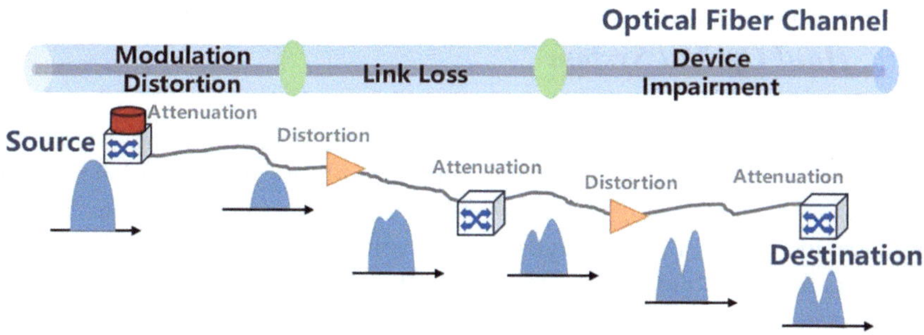

Fig. 1.3 Long-haul optical fiber transmission with Raman amplifier

performance over extensive fiber optic networks. This technology harnesses the nonlinear Raman effect to amplify optical signals directly within the optical fiber, enabling extended transmission distances and enhanced signal quality.

Raman amplification fundamentally relies on the nonlinear optical process of stimulated Raman scattering (SRS) [5], in which incident pump photons interact with the vibrational modes of the silica fiber matrix, facilitating energy transfer from the higher-energy pump photons to the lower-energy signal photons. Unlike discrete amplification schemes that require external gain media or specialized modules, Raman amplification operates intrinsically within the transmission fiber itself. In conventional implementations, one or more high-power pump lasers, tuned to wavelengths aligned with the fiber's Raman gain profile, are co-propagated or counter-propagated with the signal. As the pump light traverses the fiber, it induces Raman scattering interactions that continuously amplify the signal over the transmission span [6]. This distributed gain mechanism enables extended reach, improved noise performance, and enhanced system flexibility, making Raman amplification a key enabler in modern long-haul and ultra-long-haul optical communication networks.

One of the key advantages of Raman amplification lies in its ability to provide distributed amplification throughout the fiber length, effectively compensating for signal attenuation and nonlinear effects that degrade signal quality over long transmission distances. By amplifying the signal directly within the fiber, Raman amplifiers offer superior performance compared to traditional erbium-doped fiber amplifiers (EDFAs), particularly in scenarios requiring ultra-long-haul transmission [7]. Moreover, Raman amplification exhibits excellent compatibility with optical WDM systems, allowing for simultaneous amplification of multiple signal channels across the entire WDM spectrum. This feature enables seamless integration of Raman amplification into existing optical network architectures, facilitating scalable and flexible network deployment.

1.2.3 *Coherent Optical Communication Technology for Long Haul Optical System*

In the realm of long-haul optical fiber systems, coherent technology stands as a cornerstone, heralding a new era of high-capacity, robust communication networks. This advanced technology, characterized by its ability to manipulate both the amplitude and phase of optical signals, has revolutionized the landscape of optical communications, enabling unprecedented levels of spectral efficiency, transmission distance, and signal integrity. Optical coherent technology stands as a cornerstone of modern long-haul optical fiber systems, embodying the convergence of advanced optical modulation, coherent detection, and digital signal processing techniques [8, 9]. With its unparalleled capacity, reach, and reliability, coherent technology continues to drive the evolution of optical communication networks, laying the foundation for the realization of a truly interconnected world.

Figure 1.4 presents a schematic overview of a coherent optical communication system, delineating its fundamental distinction from direct detection architectures–namely, in the modalities of signal modulation and detection. Coherent systems employ external modulators at the transmitter, enabling sophisticated modulation of the optical carrier in terms of amplitude, phase, and frequency. This flexibility facilitates the implementation of advanced modulation formats, which are discussed in greater depth elsewhere. At the receiving end, coherent detection is employed, leveraging a local oscillator (LO) laser source to enable phase-sensitive reception. The incoming signal is optically mixed with the LO in a coherent mixer, effectively down-converting the optical carrier to an intermediate frequency in the microwave domain. This heterodyne or homodyne mixing process allows the photodetectors to extract both the amplitude and phase information of the transmitted signal, significantly enhancing sensitivity and spectral efficiency compared to direct detection systems.

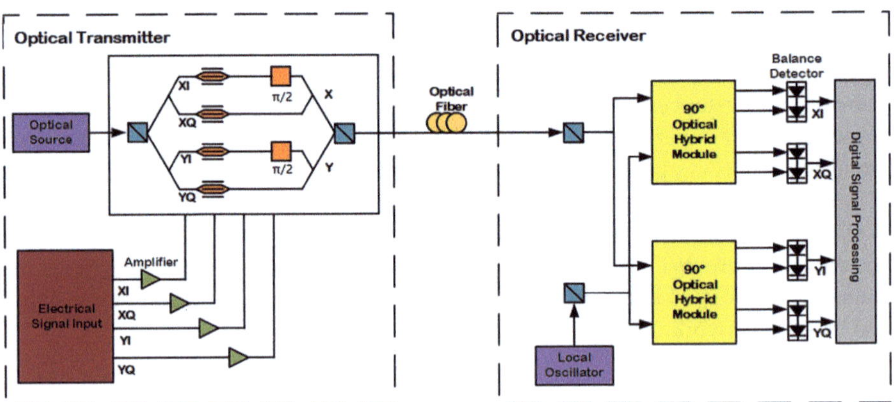

Fig. 1.4 Coherent optical fiber communication system

1.2.4 Advanced Modulation Technology and DSP Applications

In the domain of optical fiber communications, the relentless pursuit of higher data rates and more efficient bandwidth utilization has necessitated the evolution of Forward Error Correction (FEC) technologies. Advanced FEC algorithms, such as Turbo Codes, Low-Density Parity-Check (LDPC) Codes, and Polar Codes, have emerged as pivotal in augmenting the error-correcting capabilities, thereby enhancing the overall system reliability and efficiency. These FEC schemes, characterized by their near-Shannon limit error correction performance, are instrumental in combating the intrinsic impairments of optical channels, including noise, dispersion, and non-linear effects, which tend to exacerbate with increasing data rates and transmission distances [10, 11].

The pursuit of maximizing the data throughput of optical fiber systems has catalyzed the development and deployment of advanced optical modulation formats. While On-Off Keying (OOK) once served as the foundational approach, the field has since evolved toward more spectrally efficient schemes such as Quadrature Phase Shift Keying (QPSK), Quadrature Amplitude Modulation (QAM), and Orthogonal Frequency-Division Multiplexing (OFDM). As illustrated in Figs. 1.5 and 1.6, these formats exploit multiple degrees of freedom–specifically the amplitude, phase, and frequency–of the optical carrier to encode data, thereby substantially enhancing spectral efficiency and overall transmission capacity. In particular, the integration of high-order QAM formats, made feasible by coherent detection techniques and sophisticated DSP algorithms, has marked a transformative advance. This synergy has enabled optical networks to support transmission rates on the order of terabits per second, positioning these technologies as foundational enablers for next-generation high-capacity communication infrastructures.

The convergence of coherent optical communication and advanced DSP has markedly advanced the spectral efficiency and overall performance of modern fiber-optic systems. In coherent architectures, the transmitter encodes information by

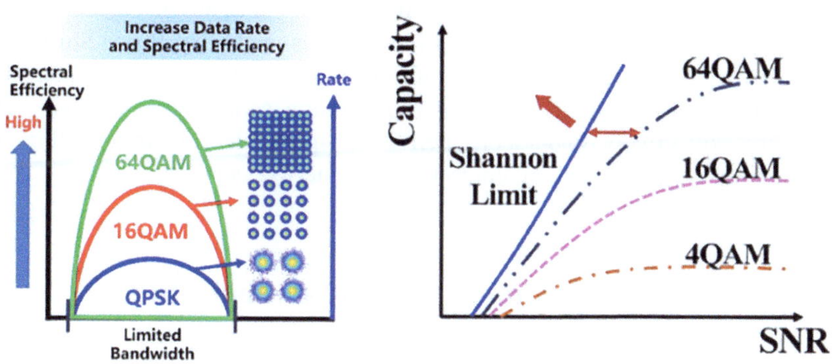

Fig. 1.5 Advanced modulation formats vs transmission rate and capacity

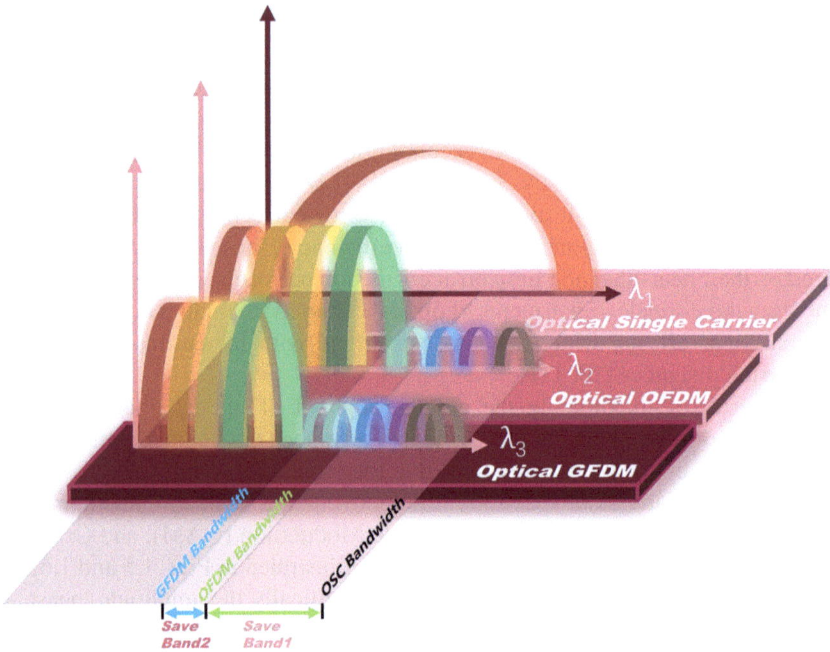

Fig. 1.6 Multicarrier technology for optical fiber communication system

modulating the phase or frequency of the optical carrier, enabling the use of complex modulation formats. On the receiving end, signal demodulation is achieved via homodyne or heterodyne detection schemes, both of which are intricately supported by sophisticated DSP algorithms. DSP is central not only to coherent systems but also to intensity-modulated direct detection systems utilizing non-OOKformats. Techniques such as Pulse Amplitude Modulation (PAM), Discrete Multi-Tone (DMT), and Carrierless Amplitude and Phase Modulation (CAP) rely heavily on DSP for robust signal recovery and impairment mitigation. Furthermore, the implementation of higher-order modulation schemes–including Phase Shift Keying (PSK), QAM, polarization-division multiplexing, and OFDM has substantially elevated the achievable data rates and transmission capacity per channel. These innovations collectively underscore DSP's indispensable role in enabling next-generation high-capacity optical communication networks [12, 13].

1.3 The Evolution of Optical Fiber Communication

The progression of optical communication systems–from their conceptual origins to today's intricately engineered networks–has been marked by a series of transformative technological breakthroughs. Each advancement has contributed decisively

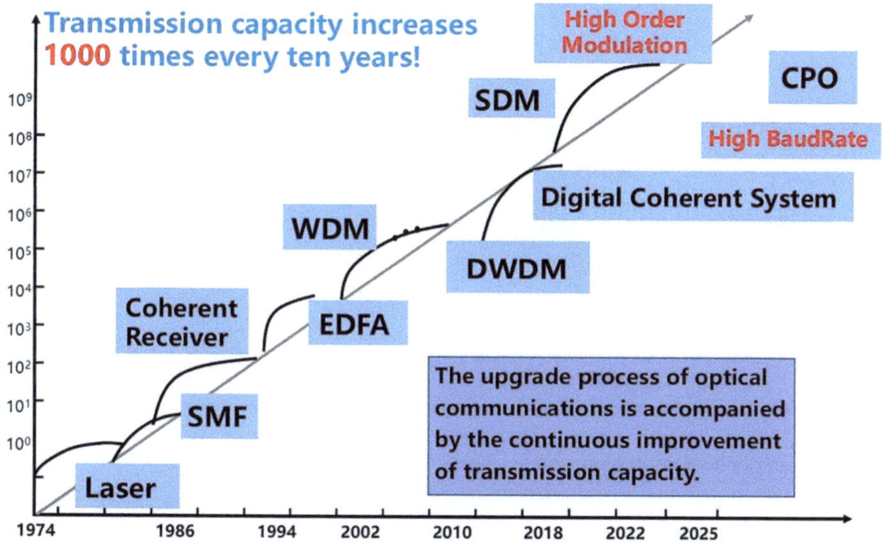

Fig. 1.7 Transmission capacity trends of optical fiber technology

to scaling the capacity and functionality of optical transmission infrastructure. Figures 1.7 and 1.8 chronicle the trajectory of transmission capacity and enabling technologies in both experimental and commercial optical communication systems spanning from 1970 to 2025. This historical overview reveals a striking, sustained exponential growth in system capacity across the past five decades. Notably, within the last twenty years alone, spectral efficiency has experienced an order-of-magnitude increase per decade. By 2025, projections indicate that spectral efficiency will reach a record-breaking level of approximately 150 bit/s/Hz. This trend underscores the relentless pursuit of higher data throughput and the continuous refinement of photonic technologies that define the modern landscape of optical communications.

Coherent optical communication, empowered by DSP, has emerged as the cornerstone of modern high-capacity optical networks. The widespread adoption of coherent detection techniques in experimental and commercial systems underscores their critical role in enabling ultrafast data transmission. Currently, coherent systems operating at 100 Gbit/s and 400 Gbit/s per channel have been extensively deployed worldwide, establishing a robust foundation for global data infrastructure. With the impending commercialization of systems achieving single-channel transmission rates of 1 Tbit/s, coherent optical technology stands at the threshold of a new era [14–16]. This trajectory points toward a future wherein coherent architectures, in conjunction with advanced DSP algorithms, will remain central to the evolution of optical communication networks, continually pushing the boundaries of data rate, spectral efficiency, and system scalability.

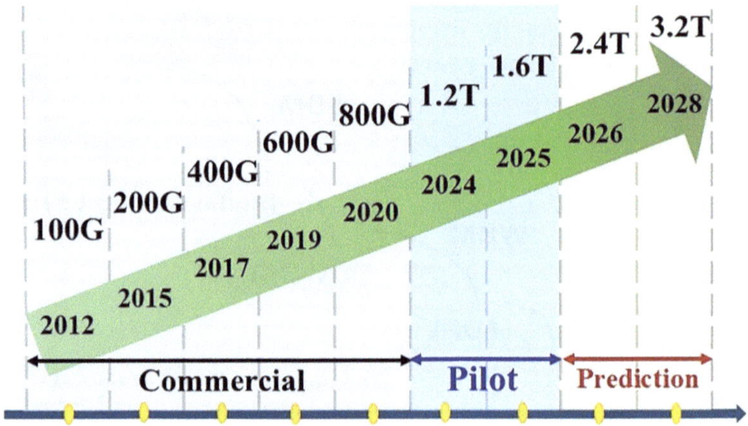

Fig. 1.8 Fiber optic transmission system single channel rate development trend

1.4 Organizations of This Book

In the current era of rapid technological evolution, the domain of optical communications has witnessed significant advancements, catalyzing a revolution in the global information infrastructure. This book meticulously delineates these advancements, providing an in-depth exploration of the foundational and emerging technologies shaping the future of optical fiber communication systems. As shown in Fig. 1.9, this book was divided into seven critical chapters in three ways, the manuscript offers a comprehensive discourse on the principles, applications, and research frontiers of optical communications.

Chapter 1: Introduction sets the stage by elucidating the historical context and the theoretical underpinnings of optical fiber communications. It charts the trajectory of the field from its nascent stages to the present day, highlighting key technological milestones and the underlying scientific principles. This chapter establishes a foundational understanding, facilitating readers' navigation through the subsequent, more specialized discussions.

Chapter 2: Advanced Coding Technology delves into the sophisticated realm of coding techniques that enhance the performance and reliability of optical communication systems. It addresses the evolution of error correction codes from traditional approaches to cutting-edge solutions, including low-density parity-check codes, probabilistic shaping, adaptive RS codes. The discussion emphasizes the role of these technologies in improving the spectral efficiency and transmission distance of optical systems, crucial for meeting the burgeoning demand for data transmission capacity.

Chapter 3: Advanced Modulation Technology explores the innovations in modulation schemes that have been instrumental in optimizing the use of optical bandwidth. This chapter provides a critical analysis of various modulation tech-

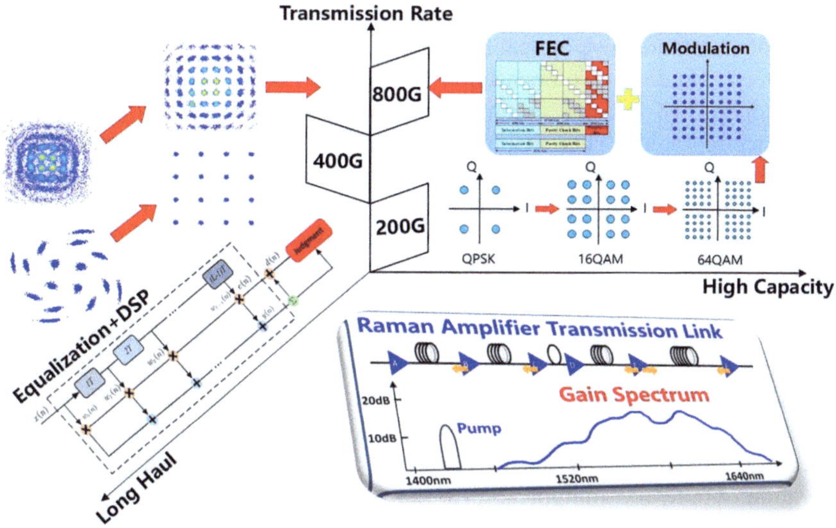

Fig. 1.9 The organization of this book

niques, from the established bandpass delta-sigma modulation to the more complex probabilistic shaping 4D modulation. The narrative underscores the trade-offs between spectral efficiency, system complexity, and robustness against impairments, offering insights into the selection of modulation schemes for different communication scenarios.

Chapter 4: Raman Amplifier examines the pivotal role of Raman amplification in enhancing signal strength over long-haul optical transmission. The chapter provides a rigorous examination of the principles of Raman scattering and the operational dynamics of Raman amplifiers. It also discusses the integration of Raman amplification with other technologies, such as EDFA, to achieve composite amplification schemes, thereby extending the reach and capacity of optical communication systems.

Chapter 5: Advanced Equalization Technology addresses the challenges posed by signal distortion and dispersion in optical fibers, presenting advanced equalization techniques as solutions. This chapter reviews digital signal processing (DSP) strategies, including feed-forward and decision-feedback equalizers, adaptive filters, and machine learning algorithms, highlighting their efficacy in compensating for channel impairments and enhancing signal integrity.

Chapter 6: Nonlinear Compensation Technology delves into the strategies for mitigating the impact of nonlinear optical effects, which become pronounced at high signal powers and long transmission distances. It critically assesses various compensation techniques, from digital backpropagation to Volterra series-based methods, and introduces the emerging paradigm of artificial intelligence (AI)-based nonlinear compensation. The discussion emphasizes the potential of AI and machine learning algorithms in dynamically optimizing compensation strategies, heralding a new era of smart, adaptive optical networks.

Chapter 7: Nonlinear Compensation Technology Based on Artificial Intelligence offers a deep dive into the cutting-edge intersection of AI and optical communication technologies. This chapter elaborates on how AI-driven approaches are redefining the landscape of nonlinear compensation, providing adaptive, efficient solutions that transcend the limitations of conventional methods. Through detailed case studies and theoretical analyses, the narrative showcases the transformative potential of AI in enhancing the performance and reliability of optical communication systems.

Together, these chapters weave a comprehensive tapestry of the current state and future directions of optical fiber communications, articulated with the precision and insight characteristic of scholarly contributions to the Nature Journal. The book stands as an invaluable resource for researchers, practitioners, and students in the field, offering a lucid, detailed exposition of the technologies driving the next generation of optical communication systems.

1.5 Summary

In the rapidly evolving landscape of optical communication, the book presents a comprehensive examination of the latest advancements and technologies shaping the future of high-capacity, long-haul, and short-reach optical fiber systems. With a meticulous approach resonant with the academic rigor, this book encapsulates the theoretical underpinnings, practical implementations, and future directions of optical fiber communication technologies. It delineates the complexities and challenges inherent in the deployment of these systems, while proposing innovative solutions and frameworks to overcome them.

References

1. R.A. Fezeu, C. Fiandrino, E. Ramadan, J. Carpenter, D. Chen, Y. Tan, F. Qian, J. Widmer, Z.L. Zhang, Roaming across the European union in the 5G era: performance, challenges, and opportunities, in *IEEE INFOCOM 2024-IEEE Conference on Computer Communications* (IEEE, Piscataway, 2024), pp. 2378–2387
2. I. Kaminow, T. Li, A.E. Willner, *Optical Fiber Telecommunications Volume VIB: Systems and Networks* (Academic Press, Cambridge, 2013)
3. T.H. Maiman, Stimulated optical radiation in ruby. Nature **187**(4736), 493–494 (1960)
4. K.C. Kao, G.A. Hockham, Dielectric-fibre surface waveguides for optical frequencies, in *Proceedings of the Institution of Electrical Engineers*, vol. 113, no. 7 (IET, London, 1966), pp. 1151–1158
5. G.P. Agrawal, *Fiber-Optic Communication Systems* (Wiley, Hoboken, 2012)
6. S. Babin, I. Vatnik, A.Y. Laptev, M. Bubnov, E. Dianov, High-efficiency cascaded raman fiber laser with random distributed feedback. Opt. Exp. **22**(21), 24929–24934 (2014)

7. J. Simpson, P. Becker, N. Olsson, *Erbium-Doped Fiber Amplifiers: Fundamentals and Technology (Optics and Photonics)* (Academic Press, Cambridge, 1999)
8. E. Ip, J.M. Kahn, Digital equalization of chromatic dispersion and polarization mode dispersion. J. Lightw. Technol. **25**(8), 2033–2043 (2007)
9. P.J. Winzer, R.J. Essiambre, Advanced modulation formats for high-capacity optical transport networks. J. Lightw. Technol. **24**(12), 4711–4728 (2006)
10. M. Mazur, A. Lorences-Riesgo, J. Schröder, P.A. Andrekson, M. Karlsson, 10 tb/s pm-64qam self-homodyne comb-based superchannel transmission with 4% shared pilot tone overhead. J. Lightw. Technol. **36**(16), 3176–3184 (2018)
11. J. Cho, P.J. Winzer, Probabilistic constellation shaping for optical fiber communications. J. Lightw. Technol. **37**(6), 1590–1607 (2019)
12. K. Roberts, D. Beckett, D. Boertjes, J. Berthold, C. Laperle, 100g and beyond with digital coherent signal processing. IEEE Commun. Mag. **48**(7), 62–69 (2010)
13. M. Xiang, S. Fu, O. Xu, J. Li, D. Peng, Z. Gao, Y. Wang, Y. Qin, Advanced DSP enabled C-band 112 gbit/s/λ PAM-4 transmissions with severe bandwidth-constraint. J. Lightw. Technol. **40**(4), 987–996 (2022)
14. R.J. Essiambre, G. Kramer, P.J. Winzer, G.J. Foschini, B. Goebel, Capacity limits of optical fiber networks. J. Lightw. Technol. **28**(4), 662–701 (2010)
15. M. Kong, J. Yu, H.C. Chien, K. Wang, X. Li, J. Shi, L. Zhao, B. Zhu, X. Pan, X. Xin, et al., WDM transmission of 600G carriers over 5,600 km with probabilistically shaped 16QAM at 106 Gbaud, in *Optical Fiber Communication Conference* (Optica Publishing Group, Washington, 2019), pp. M4I–3
16. W.S. Saif, S.K.O. Soman, O.A. Dobre, Deep learning-assisted nonlinearity compensation in subcarrier-multiplexing coherent optical systems. J. Lightw. Technol. **42**, 8162–8172 (2024)

Chapter 2
Advanced Coding Technology

Advanced coding techniques encompass various encoding methods, optimizing signal encoding to enhance data transmission efficiency and reliability. Key technologies include constellation shaping and forward error correction coding, providing higher-quality data transmission and error correction capabilities. First, we propose a novel probabilistic shaping (PS) scheme based on the quasi-Maxwell–Boltzmann (quasi-MB) distribution model in order to solve the incompatibility between PS and carrier phase recovery (CPR) algorithms, such as blind phase search (BPS) and principal component-based phase estimation (PCPE) [1]. Then, in order to enhance the data transmission reliability, we propose an adaptive Reed–Solomon (RS) coding scheme and envelope-detector-based (ED-based) down-conversion for 16-quadrature amplitude modulation (16-QAM) vector signal used in optical transmission system [2]. Finally, we propose a LDPC-coded optical GFDM (LCO-GFDM) IM/DD system with high order modulation which is proved to be an efficient coded modulation technique in optical transmission. Extensive results show that these advanced coding technologies have significant performance advantages [3].

2.1 Carrier Phase Recovery Friendly Probabilistic Shaping Scheme Based on a Quasi-Maxwell-Boltzmann Distribution Model

With the rapid increase of bandwidth demand in optical communication systems, the past half-decade has witnessed a growing interest among both academics and service suppliers in the capacity-approaching techniques, such as probabilistic shaping (PS), geometric shaping, and time-division hybrid modulation [4–7]. Among these emerging technologies, PS stands out due to two distinct benefits, namely the ability to achieve an arbitrarily fine rate/reach trade-off and the capability to bridge the

X. Xin, *Advanced Optical Fiber Transmission Systems*, Wireless Networks,
https://doi.org/10.1007/978-3-032-09516-9_2

shaping gap to closely approach ultimate performance [7]. However, as Mello et al. pointed out in [8], further research is still required with respect to the PS's practical implementations, especially its interplay with carrier phase recovery (CPR) algorithms. To start with, the blind phase search (BPS) algorithm is considered as a benchmark algorithm for high-order modulation formats due to its convenient implementation in hardware [9]. The work in Ref. [8] has already proven that PS strongly impairs the system performance of BPS, especially at a low signal-to-noise ratio (SNR), which may cause potential implementation penalties. On the other hand, a principal component-based phase estimation (PCPE) algorithm was recently proposed by Diniz et al. in Ref. [10]. PCPE is also suitable for realistic hardware implementation and, more importantly, it outperforms the BPS algorithm with respect to mutual information (MI) performance at low SNR values, which makes it a potential solution to the problem of incompatibility between PS and BPS. Unfortunately, based on the results, although the application of PCPE to PS signals can achieve near-ideal results at low SNR values, a serious incompatibility can be observed under a high SNR, especially for highly shaped constellations. Clearly, the existing PS scheme, especially when the degree of shaping is strong, is very unfriendly to hardware efficient CPR algorithms which, in turn, greatly limit its application in practical systems.

2.1.1 System Model

A novel probabilistic shaping (PS) scheme based on the quasi-Maxwell–Boltzmann (quasi-MB) distribution model is proposed in order to solve the incompatibility between PS and carrier phase recovery (CPR) algorithms, such as blind phase search (BPS) and principal component-based phase estimation (PCPE). In the proposed quasi-MB model, the same occurrence probability is assigned to each constellation point on the same square-ring, rather than on the same circle. Signals obeying the quasi-MB model have superior CPR friendliness compared to traditional PS signals. Although this approach will slightly deviate the source distribution from the Gaussian model, the drawback in channel capacity is negligible compared to the performance improvement of CPR algorithms.

The flow charts of the two CPR algorithms evaluated in this Letter are shown in Fig. 2.1.

As shown in Fig. 2.2a, the main idea of this scheme is to first generate two PS pulse amplitude modulation (PAM) signals, and then combine them to construct a PSQAM signal. To generate MB-distributed PAM symbols, a constant composition distribution matcher (CCDM) is commonly used to transform uniformly distributed input bits to the positive amplitudes of PS PAM signal (a "half-PAM" constellation). After that, the parity bits generated by a systematic forward error correction (FEC) encoder are used as sign bits to create the PS PAM with desired probability distribution [11]. For the proposed quasi-MB distribution, the shaping and coding scheme needs to be modified. As shown in Fig. 2.2b, one possible solution is

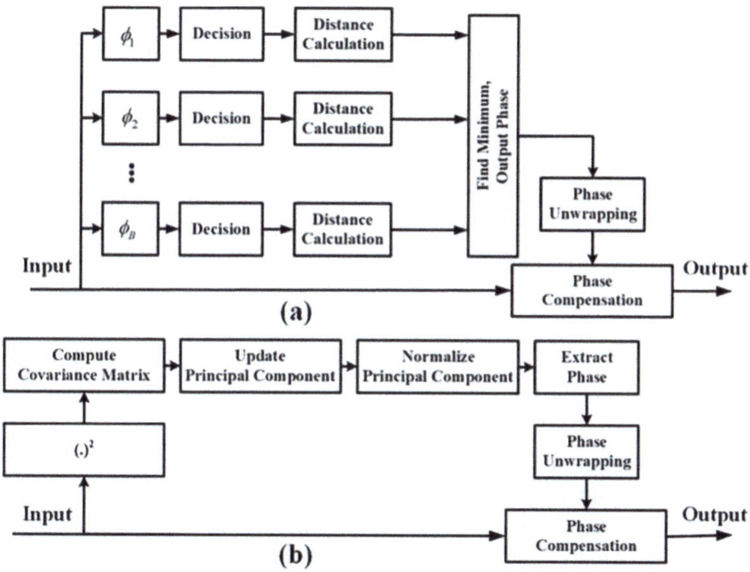

Fig. 2.1 Block diagram of (**a**) BPS and (**b**) PCPE

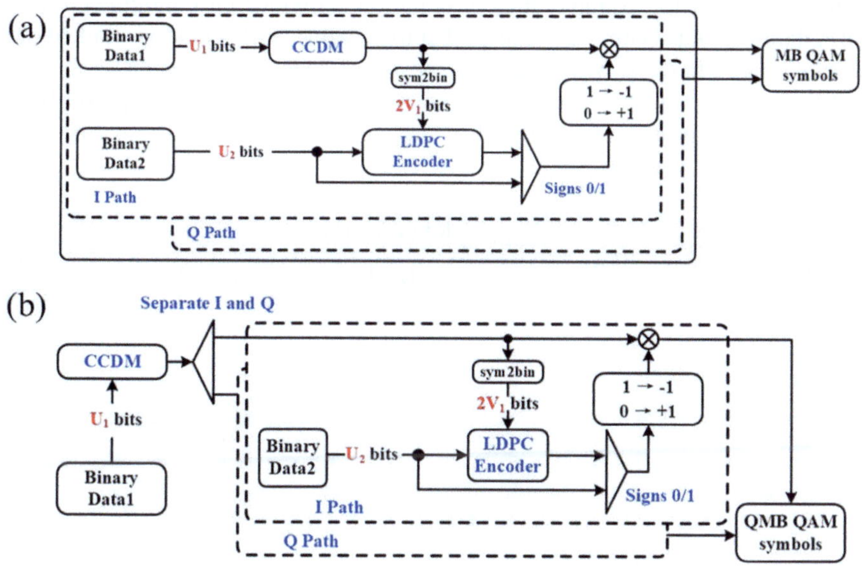

Fig. 2.2 Block diagram of shaping and coding scheme for PS-64QAM following (**a**) MB distribution and (**b**) quasi-MB distribution

to utilize a CCDM to generate one quadrant of the constellation diagram and then apply the parity bits to flip the constellation points around the in-phase and quadrature axes, respectively

2.1.2 Algorithm Principle

The calculation formula of NGMI can be expressed as (2.1)

$$NGMI = 1 - [H - GMI]/m \tag{2.1}$$

where H is the source entropy of M-order quadrature amplitude modulation (QAM), and $m = \log_2 M$. The selected modulation format was PS-64QAM with a shaping parameter of 0.05.

In order to solve the incompatibility between PS and CPR algorithms, we propose a novel PS scheme based on quasi-MB distribution model, where same occurrence probability is assigned to each constellation points on the same square-ring, rather than on the same circle, compared to traditional PS based on MB distribution model. Figure 2.3 illustrates the principals of the conventional PS and the proposed quasi-MB-based PS.

Take 64 QAM for example, the constellation is divided into multiple circles of constellation points with the same probability, while each circle is marked with the same color. In quasi-MB model, the power of constellation points in the same circle is no longer identical, so it is necessary to calculate the average power before determining the probability assigned to the current circle. The average power of each circle A_j is given by

$$P_{A_j} = \frac{\sum_{x_i \in A_j} \mathrm{Re}\,(x_i)^2 + \mathrm{Im}\,(x_i)^2}{|A_j|} \tag{2.2}$$

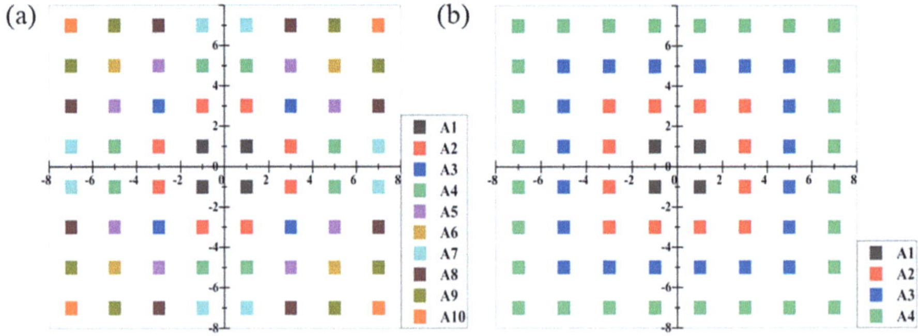

Fig. 2.3 PS-64QAM based on the (**a**) MB model and (**b**) quasi-MB model

where and represent the real and imaginary parts of a complex constellation point x_i belonging to the set A_j. Using $|A_j|$ and M to denote the number of constellation points in $|A_j|$ and the modulation format order, then the subscripts i and j satisfy the following conditions, $1 \leq i \leq |A_j|$ and $1 \leq j \leq \sqrt{M}/2$. The probability of occurrence for each constellation point x_i in A_j is given by

$$\Pr(x_i) = \frac{\exp(-\nu P_{A_j})}{\sum_{k=1}^{\sqrt{M}/2} \exp(-\nu P_{A_k})|A_k|} \quad (2.3)$$

where the shaping parameter ν determines the degree of PS and the corresponding entropy given by

$$H = -\sum_{k=1}^{\sqrt{M}/2} \sum_{x_i \in A_k} \Pr(x_i) \log_2 \Pr(x_i) \quad (2.4)$$

2.1.3 System Setup and Results

In order to clearly express the performance degradation caused by the CPR unfriendliness of MB-based PS signals, a numerical system was carried out. To start with, PS symbols following the MB distribution or the proposed quasi-MB distribution were generated with 1 sample per symbol at a symbol rate of 32 Gbaud. The phase noise consisted of two parts; one was a constant phase rotation of $\pi/6$, and the other was random phase noise introduced by the Wiener process. After multiplying the phase noise on the complex signal, additive white Gaussian noise was also added to guarantee a desired SNR. In the phase estimator, three different algorithms, namely BPS, PCPE, and the hybrid PCPE-BPS [7], were performed to achieve phase recovery. The normalized generalized mutual information (NGMI) performance of the algorithms mentioned above was evaluated considering 2^{19} symbols.

The experimental verification of the proposed quasi-MB scheme is carried out in a 10 GBaud single polarization coherent detection system, as shown in Fig. 2.4.

At the transmitter end, PS-64QAM signals corresponding to MB and quasi-MB models is generated. The shaping factor of the MB distribution is 0.05, while the shaping factor of the quasi-MB distribution is 0.049, which is consistent with the setting. Digital-to analog conversion is realized by an arbitrary waveform generator (AWG) with a sampling rate of 80 GSa/s. The symbol sequence length of is 8192, which is determined by the AWG memory size. The analog signals are then amplified by two electrical amplifiers and fed into the IQ modulator. The continuous light wave is generated by an external cavity laser with a 100 kHz linewidth. An Erbium-doped fiber amplifier (EDFA) and a variable optical attenuator (VOA) are

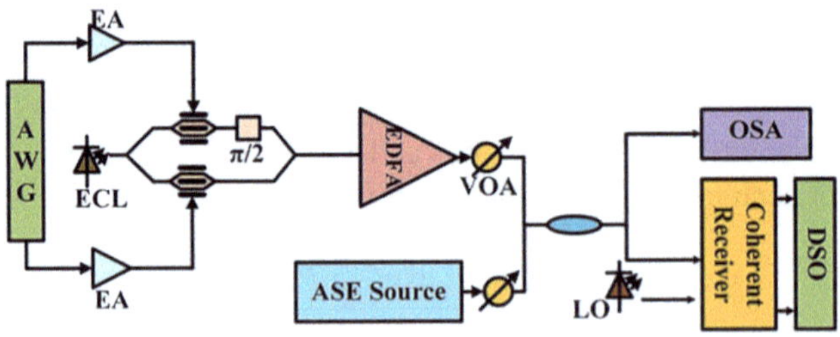

Fig. 2.4 Experiment setup. AWG, arbitrary waveform generator; EA, electrical amplifier; ECL, external cavity laser; EDFA, erbium-doped fiber amplifier; VOA, variable optical attenuator; ASE source, amplified spontaneous emission source; LO, local oscillator; OSA

used to adjust the signal power. An amplified spontaneous emission (ASE) source and another VOA are used to change the optical signal-to-noise ratio (OSNR) of the system for NGMI measurement. At the receiver end, after adding ASE noise, the desired signal is sent into the coherent receiver to beat with the local oscillator (LO) signal. And then, an 80 GSa/s digital storage oscilloscope (DSO) is used to capture the detected signals. Offline digital signal processing is then applied to recover the transmitted signal, including low-pass filtering, the Gram–Schmidt orthogonalizing process, clock recovery, frequency offset estimation, CPR, and NGMI calculation. It should be noted that in the CPR process, the block length of all algorithms is 64.

As shown in Fig. 2.5a–c, for Wiener random phase noise corresponding to a combined linewidth of 200 kHz, PCPE can ensure the shaping gain brought by PS under the condition of an SNR less than 10 dB, regardless of the value of block length N. The NGMI degradation caused by BPS can exceed the capacity gain expected by PS with block length less than 256. Although increasing the block length can alleviate this problem to some extent, it will increase the complexity of the hardware implementation. For an SNR higher than 15 dB, the performance of BPS is almost consistent with the ideal value, while the incompatibility of PCPE and PS becomes serious. Even with the best result of PCPE, that is when N=128, its NGMI cannot reach 0.9 with an SNR high enough. The hybrid algorithm PCPE-BPS can improve the performance of PCPE in the high SNR range, but this improvement is not enough to fully compensate for the NGMI loss caused by the incompatibility between PCPE and PS.

For the case of a 2 MHz combined linewidth, as shown in Fig. 2.5d–f, the fast accumulation of random phase noise requires that the block length must be reduced when performing CPR; otherwise, the corresponding NGMI cannot be approximated to 1 even when the SNR is high enough. In this scenario, PCPE can outperform BPS when the SNR is less than 11 dB; yet, the incompatibility between PCPE and PS will be extremely serious for higher SNRs.

Taking PS-64QAM as an example, the relationship between the shaping parameter and the entropy is shown in Fig. 2.6. It can be seen that using the proposed quasi-MB model can achieve fine-grained rate adaptability similar to conventional PS based on MB model.

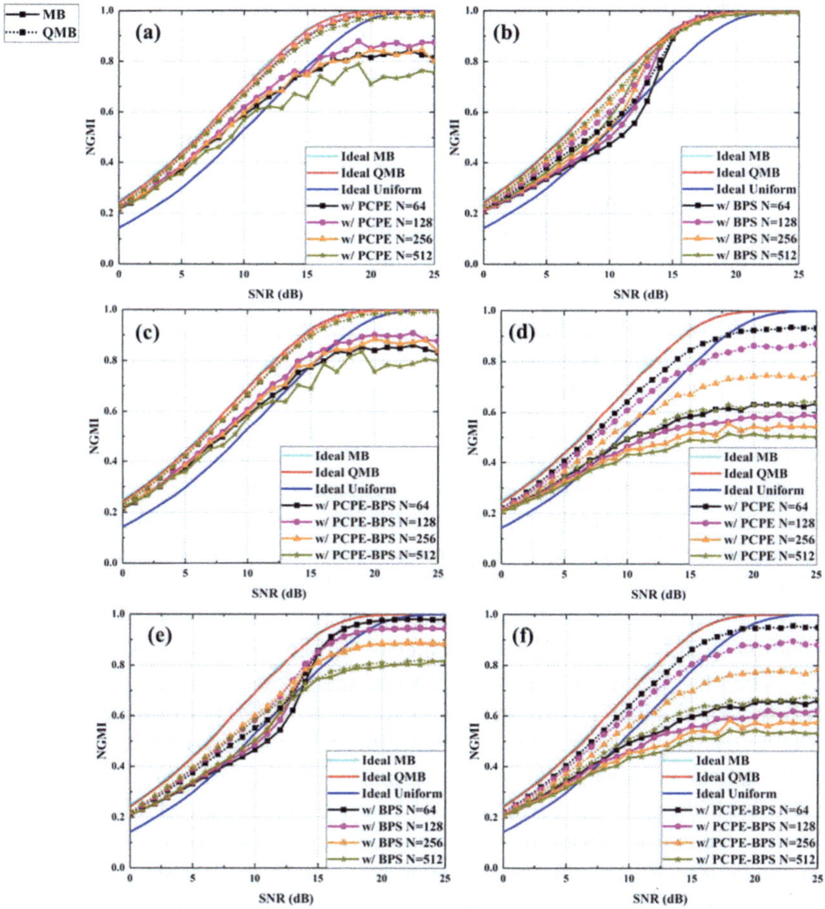

Fig. 2.5 NGMI versus the SNR for PS-64QAM with different CPR algorithms and different linewidths. (**a**)–(**c**) 200 kHz and (**d**)–(**f**) 2 MHz

Fig. 2.6 Entropy versus shaping parameter for PS-64QAM

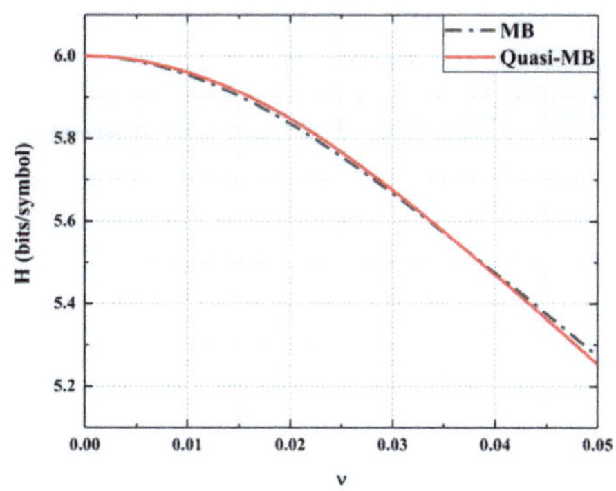

The NGMI performance of BPS, PCPE, and PCPE-BPS for quasi-MB-based PS-64QAM signals is evaluated, as shown in Fig. 2.6 (dot lines). In order to ensure a fair performance comparison, the shaping parameter chosen for quasi-MB-based PS-64QAM was 0.049, so that the entropy values of both distributions are consistent. It can be seen that, by adopting the quasi-MB-based PS technique, the NGMI performance of all three different CPR algorithms has been greatly improved. First, for the case where the combined linewidth was 200 kHz, the phase recovery performance of PCPE and PCPE-BPS is almost consistent with the ideal curve, regardless of the block length and SNR value. While the performance of BPS has also been improved for a low SNR, which can ensure the shaping gain brought by quasi-MB-based PS technique, achieving higher NGMI compared to uniform shaped signals. On the other hand, for a total linewidth up to 2 MHz, performance improvement of all algorithms can also be observed. For PCPE and PCPE-BPS with a block length of 64, the NGMIs at SNRD25 dB have been increased from 0.64 and 0.66 to 0.93 and 0.95. As for BPS, all four block lengths measured can all meet the threshold when the SNR is high enough. In the case where SNR<15 dB, the shaping gain brought by PS can be better preserved thanks to the quasi-MB model. Among all the results, the maximum NGMI enhancement ratio for BPS is 20.70

Figure 2.7 gives the measured NGMI versus OSNR in our experimental demonstration for signals with MB model and quasi-MB model. For both probability distribution models, the NGMI performance of BPS increases as the OSNR becomes larger. When the OSNR is less than 23 dB, using quasi-MB model instead of MB model can help the BPS algorithm to achieve higher NGMI performance. For PCPE and PCPE-BPS algorithms, their incompatibility with the MB model is very obvious when the OSNR is greater than 25 dB. In contrast, the quasi-MB model can be well matched with these two algorithms, and can achieve higher NGMI performance throughout the entire measured OSNR range. In short, the experiment results again verify the better CPR friendliness of the proposed quasi-MB model.

In conclusion, a novel CPR friendly PS scheme based on quasi-MB distribution model is proposed and validated by experiment. The main feature of quasi-MB is to assign the same probability to constellation points on the same square-ring.

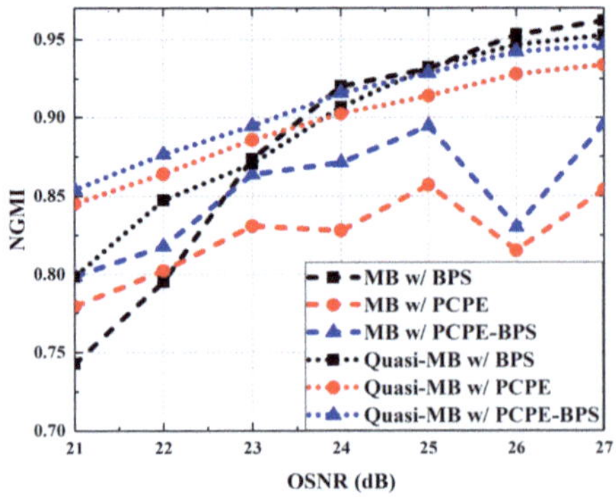

Fig. 2.7 NGMI versus the OSNR for 10 GBaud PS-64QAM

This property can greatly improve the NGMI performance of hardware-efficient CPR algorithms. For practical optical communication systems, the proposed quasi-MB model can be a better alternative to guarantee the shaping gain brought by PS technology.

2.2 Adaptive Reed-Solomon Coding and Envelope Detection of Photonic Vector Signal in V-Band Radio Over Fiber System

With the rapid increasing demand of large capacity and long transmission distance in future broadband wireless access networks, radio-over-fiber (RoF) technology has attracted much attention due to its large bandwidth, high mobility, low propagation loss and simplified base station in the past few years [12–16]. ROF technology is transparent to the transmission rate and protocol of the signal, which is suitable for the future improved technologies [17, 18]. The important application scenarios of RoF technology are where the radio frequency (RF) link is hard to reach, such as the montana, jungle, tunnel, subway, gymnasium. In addition, the high order modulation format signals such as quadrature amplitude modulation have been employed to improve spectral efficiency and capacity [19–23].

For high carrier frequency transmission in ROF systems, signal suffer from high wireless propagation loss and signal-to-signal beating interference (SSBI), which will increase the complexity of base station and severe bandwidth limitation. Besides, a rate-adaptive RS code varies code rates and redundancies which substantially improves the flexibility about transmission capacity and distance. Hence the performance of the system can be adjusted for complicated channel conditions. Recently, a variety of systems using flexible code rates of RS coding scheme has been proposed [24–26].

The down-conversion of RF signal is the essential part in ROF systems. The electrical mixer down-conversion with local oscillator (LO) is widely applied. However, as the carrier frequency increases, the power of LO in mixer-based down-conversion will be limited due to the amplifier, which makes the base station more complex, expensive and power-hungry [27–29]. Therefore, ED-based down-conversion of millimeter-wave signal used in ROF systems has been proposed [30–32], which can only process the amplitude information and shows its insensitivity to phase noise. Nevertheless, it should be noted that the transmission bit rate is limited by the bandwidth of the envelope detector. A higher bit rate can be transmitted if an envelope detector with larger bandwidth is available.

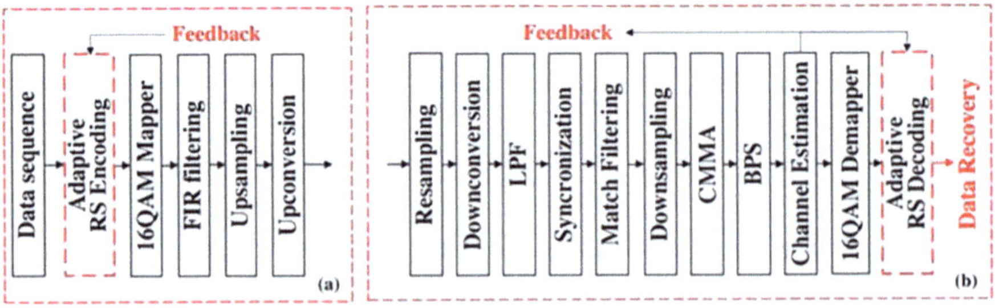

Fig. 2.8 Block diagram of (**a**) TX-, (**b**) RX- DSP

2.2.1 System Model

In this section, a V-band (57–64 GHz) 16QAM-modulated ROF system with adaptive RS coding scheme and ED-based down-conversion is proposed and experimentally demonstrated. Two free-running non-narrow line-width distributed feedback lasers are used to generate the stable V-band RF signals. Envelope detection instead of mixer down conversion is applied in the base station for the sake of simplicity. RS-coded 16-QAM vector signals with the bit rate of 2 and 4-Gb/s at 60-GHz over 20-km standard single mode fiber (SSMF) and 2-m wireless distance are successfully achieved. SSBI are theoretically investigated and effectively mitigated by reasonable set up. Adaptive RS coding module with flexible code rate and redundancies is implemented according to the feedback information of channel estimation. The post-FEC BER less than 1×10^{-8} can be achieved, which shows the use of RS codes will be a beneficial addition to improve the receiver sensitivity.

Figure 2.8 elaborates the offline program in the transmitter and receiver, respectively. In the transmitter, the baseband transmission data is fed to the adaptive RS coding module and then mapped to a sequence of complex 16-QAM symbols. Subsequently, a root raise cosine (RRC) filter with roll-off factor of 0.3 is utilized to realize pulse shaping. After up-sampling and up-conversion by ω_{IF}, the original data has been pre-coded to be a series of electrical vector signal at f_{IF} by programming and uploaded into an arbitrary waveform generator (AWG) with 10-bit vertical resolution operating at the sampling rate of 10-Gs/s, which is performed as a digital-to-analog (DAC). A continuous-wave (CW) light wave output from an external cavity laser (ECL) at 1551.878 nm with 10.39-dBm output is modulated by the generated electrical vector signal at f_{IF} using a MZM biased at its maximum transmission point. The output optical signal is boosted by a polarization-maintaining Erbium-doped fiber amplifier (PM-EDFA) to compensate for modulation loss and insertion loss and then combined with another ECL with center wavelength of 1552.358 nm, output power of 8.94-dBm via a polarization-maintaining optical coupler (PM-OC). ECL2 functions as an optical LO and has 60 GHz frequency offset relative to ECL1. The corresponding optical spectrum (at

0.01-nm resolution) after the PM-OC is presented in Fig. 2.15a. After transmitted through 20-km standard single mode fiber (SSMF) and an optical power attenuator (OPA), which can act as a variable optical attenuator (VOA) and monitor the output power simultaneously, the modulated IF vector signal is detected and up-converted to be a V-band RF signal by a 70-GHz photo-detector (PD). Subsequently, the RF signal is boosted by a low noise amplifier (LNA) and sent to air by a V-band lens horn antenna with 37-dB gain.

After 2-m wireless delivery, the electrical RF signal is captured by the antenna with the same specification and amplified again by a LNA. Then, a V-band envelope detector is utilized to down-convert the received RF signal to the IF-band. Finally, the IF signal is boosted by an broadband electrical amplifier (EA) and captured by an oscilloscope with 50-GHz sampling rate, 20-GHz bandwidth. The spectrum is plotted in Fig. 2.15b. Finally, the original data can be recovered after offline DSP, including resampling, down conversion, low-pass filtering (LPF), synchronization, match filtering, down sampling, cascaded multi-modulus algorithm (CMMA) for 16-QAM modulation, blind phase search algorithm (BPS) for phase recovery, 16-QAM demapping and RS decoding. It is worthy of note that estimated CER feedback information is sent back to the Tx-PC via Internet cable and TCP/IP socket module to realize adaptive coding.

2.2.2 *Algorithm Principle*

Adaptive Reed-Solomon Coding Module

Reed–Solomon codes are non-binary cyclic codes worked in GF(2^m) that is an extension field of GF(2) and made up of m-bits symbols. For RS (n, k) code, k and n represent the number of symbols before and after being encoded, respectively. The symbol-error correcting capacity of the code can be expressed as $t = 1/[2(n-k)]$, which means that the burst symbol errors can be corrected if the number of errors is less than t.

In this section, the adaptive RS codes selection strategy is explained. We consider two RS types, RS(255,239) and RS(255,223). The codeword error rate (CER) is defined as ε/n, where ε is the number of errors per RS code before error correction. It is obvious that RS codes with better performance generally have higher redundancy. However, higher redundancy will reduce the net bit rate with the same symbol rate. Rate-adaptive RS codes will be selected under different channel states and actual demands. For example, RS (255,239) is the first choice to maximum the net bit rate. After optical fiber and wireless transmission, the channel state can be estimated and generate a feedback information. When the performance evaluation is not in anticipation (e.g. the post-FEC BER is not low enough), RS(255,223) with higher correction capability will be utilized in the transmitter.

Figure 2.9 shows the structure of the adaptive RS coding scheme. The signal is encoded in transmission order by multiple RS encoders with symbol

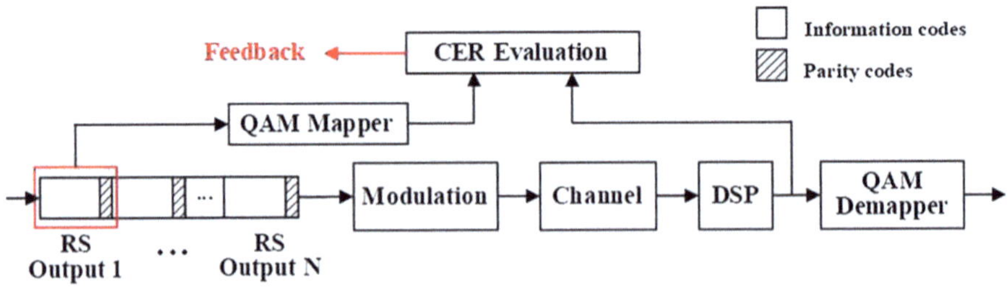

Fig. 2.9 Structure of the adaptive RS coding scheme. (DSP: digital signal processing)

interleaving/de-interleaving applied. Output of the first RS encoder is extracted to be a Ref.d sequence and is used for synchronization, channel estimation and CER evaluation, with no additional bits added. Considering the channel is relatively steady in such a short time with the latency existing between the feedback and transmission, the CER feedback can effectively work in case that the channel state is changed. For adaptive RS(n, k) codes, n is fixed and k is adaptively changed with the CER evaluation feedback. When the channel state is steady, the adaption will be performed on every RS encoders. Thus the most suitable RS codes is adaptively chosen.

Generation of RS-Coded QAM-Modulated RF Vector Signal

Figure 2.10 shows the principle of RS-coded QAM-modulated intermediate frequency (IF) vector signal generation. In the encoding block, the length of input data is firstly detected. A string of zeros will be appended if the length is not an integer multiple of k*m. After that, the new data is transformed into symbols in GF(2^m) by each m-bits. Adaptive RS coding module is blocked outside of the high-order quadrature amplitude modulation (QAM) and changes flexibly according to the feedback information. For RS(n,k) codes, multiple RS encoders are employed parallel by means of (n-k)-stage linear feedback shift register (LFSR) circuits, as shown in Fig. 2.10a. The parity polynomial can be described as the Eq. (2.5).

$$p(x) = (m(x) \cdot x^{n-k}) \bmod g(x) \tag{2.5}$$

where m(x) represents the input symbols and g(x) is the generator polynomial. The RS encoding process is then accomplished by placing the parity symbols behind the input symbols $[mp]$, forming codewords in a systematic format. After parallel-to-serial conversion, 16QAM mapper maps every four bits into one symbols. In this case, the baseband signal of RS-coded 16QAM formats is generated and the code rate will be k/n.

The data stream after RS coding module is divided into two sequences by the odd and even number. Then these two sequences are mapped to −3, −1, 1, 3,

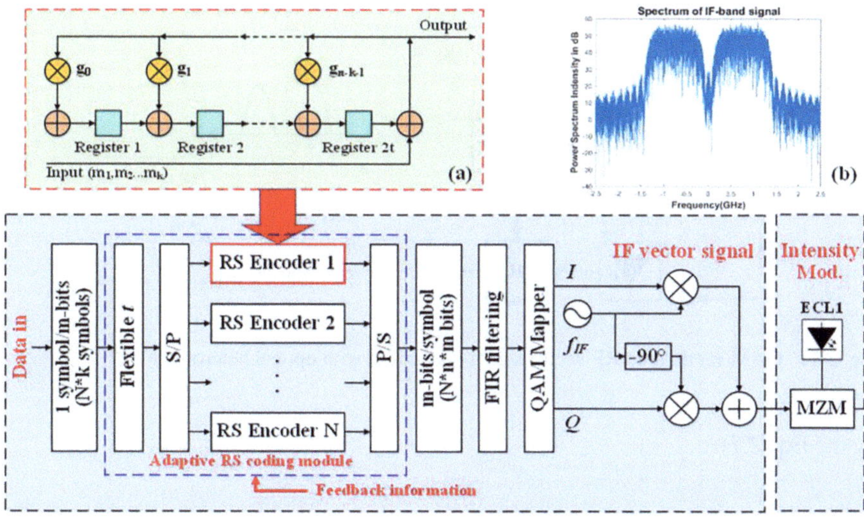

Fig. 2.10 Block diagram of the generation of the RS-coded QAM-modulated IF vector signal. ECL: external cavity laser, MZM: Mach-Zehnder modulator

respectively. After pulse shaping by a root raised cosine filter with the roll-off factor of 0.3, the real part x(t) and imaginary part y(t) of the baseband 16-QAM signal are produced. By multiplying x(t) and y(t) with two sinusoidal signals, which have the same frequency f_{IF} and a phase difference of 90°, and adding them together, the 16-QAM IF vector signals is generated with both amplitude and phase modulated, which can be expressed as,

$$S_{16QAM\_IF}(t) = x(t)cos(2\pi f_{IF}t) - y(t)sin(2\pi f_{IF}t) \tag{2.6}$$

Here, f_{IF} represents the intermediate frequency. A continuous-wave (CW) light at the wavelength of λ_1 passes through a Mach-Zehnder modulator (MZM) and is modulated by the modulated 16QAM IF vector signals, and is combined with another CW light at the wavelength of λ_2 using a polarization-maintaining optical coupler (PM-OC) subsequently, as shown in Fig. 2.11. After transmission, the modulated IF signal is converted to be 16QAM RF vector signals at f_{RF} by using a photo-detector (PD), heterodyning two free-running laser signals with a wavelength difference of the desired carrier frequency. As Fig. 2.11 depicts, the 16QAM vector signal at microwave/mm-wave bands consists of a RF tone and vector signals, that is suitable to be transmitted in the air by horn antenna and is approximately given

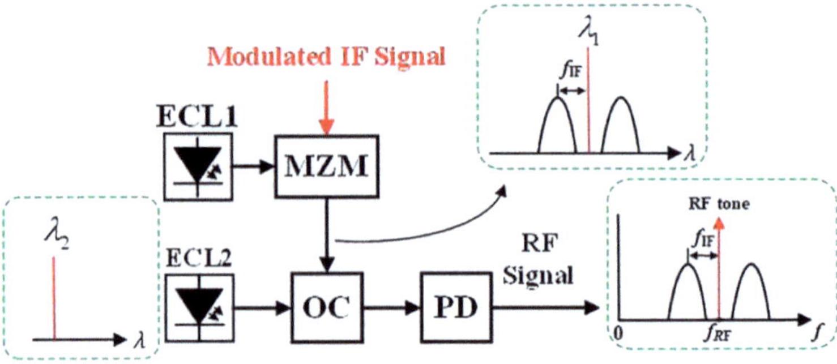

Fig. 2.11 QAM-modulated RF vector signal generation with optical heterodyne

by the Eq.(2.7).

$$\begin{aligned} S_{16QAM\_RF}(t) &\approx A\cos\omega_{RF}t + x(t)\cos(\omega_{RF}+\omega_{IF})t - y(t)\sin(\omega_{RF}+\omega_{IF})t \\ &\quad + x(t)\cos(\omega_{RF}-\omega_{IF})t + y(t)\sin(\omega_{RF}-\omega_{IF})t \\ &= A\cos\omega_{RF}t + 2x(t)\cos\omega_{RF}t\cos\omega_{IF}t \\ &\quad - 2y(t)\cos\omega_{RF}t\sin\omega_{IF}t \\ &= \cos\omega_{RF}t\cdot[A + 2x(t)\cos\omega_{IF}t - 2y(t)\sin\omega_{IF}t] \end{aligned} \tag{2.7}$$

where f_{RF} is equal to $c\cdot\left(\lambda_1^{-1}-\lambda_2^{-1}\right)$, c is the speed of light.

Envelope Detection and Signal Processing

At the receiver, as shown in Fig. 2.12, an envelope detector is adopted to square the RF vector signal received by antenna. Then the signal is divided into several parts, including the 16QAM IF vector signals, DC and high frequency components, which can be written as

$$\begin{aligned} s(t) = \frac{1}{2}A^2 + x^2(t) + y^2(t) + &\quad \rightarrow \text{DC components} \\ A(x(t)\cos\omega_{IF}t - y(t)\sin\omega_{IF}t) + &\quad \rightarrow \text{IF signals} \\ (x^2-y^2)\cos 2\omega_{IF}t - 2xy\sin 2\omega_{IF}t + \cdots &\quad \rightarrow \text{high frequency components} \end{aligned} \tag{2.8}$$

the DC and high frequency components will cause the SSBI, which degrades the system performance. Among those, the energy of the higher harmonics are too small to be considered. f_{IF} is set to an appropriate value in order to reduce the impact of SSBI. A low-pass filter is required to eliminate the high frequency energy and only

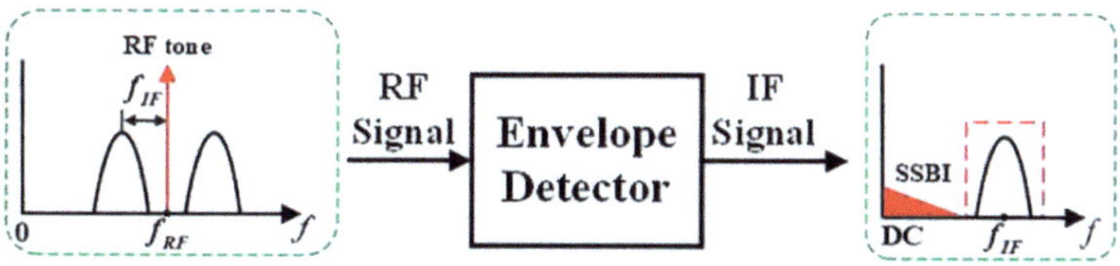

Fig. 2.12 Principle of envelope detection

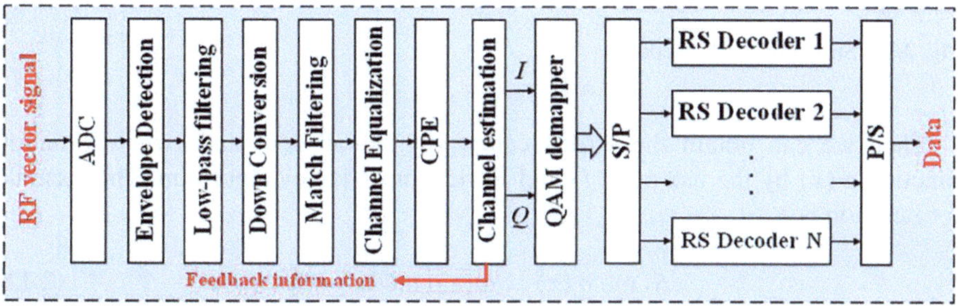

Fig. 2.13 Block diagram of the processing procedure for the received RF vector signal

the signal envelope left. By using an offline program as shown in Fig. 2.13, the RF vector signal is down-converted to obtain the baseband data with little SSBI left. After that, a match filter is necessary to maximize the SNR corresponding to the transmitted signal. In order to eliminate the interference caused by the channel and phase noise generated by devices, channel equalization and carrier phase estimation (CPE) algorithm are employed. After QAM demapping and RS decoding, the demodulation and analysis of the signal are realized. What need to be illustrated is, CER feedback information is calculated after channel equalization and sent back to the personal computer (PC) in the transmitter via TCP/IP socket module. Simultaneously, the parameters of RS decoding module in Rx-DSP change flexibly according to the CER feedback information by a judgment sentence.

The principle of RS decoder is shown in Fig. 2.14. The main purport is to evaluate the error pattern $\hat{e}(x)$ based on the syndrome symbols. The error pattern has two significant components, error location $x(i)$ and error value $Y(i)$, and can be defined as

$$\hat{e}(x) = \sum_{i=1}^{t} Y(i) \cdot x(i) \tag{2.9}$$

Assume that $R(x)$ represents the received corrupted-codewords polynomial. Then the syndrome symbols $S(j)$ can be calculated by Eq. (2.10).

$$S(j) = R(\alpha^j) = \sum_{j=0}^{n-1} \sum_{i=0}^{2t-1} Y(j) \cdot (x(j)\hat{}i), \quad \alpha^j \in GF(2^m) \tag{2.10}$$

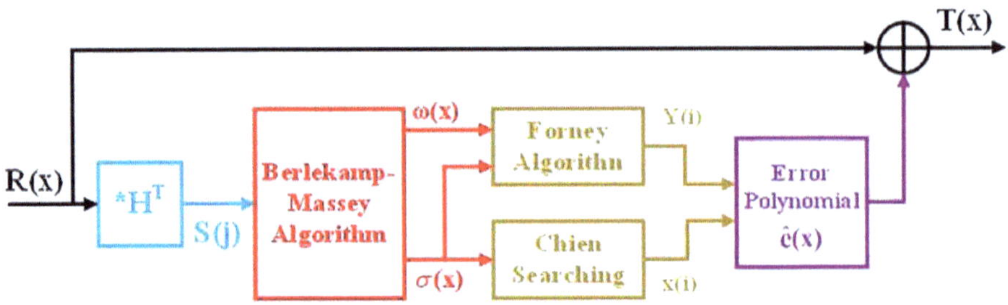

Fig. 2.14 Structure of RS decoder

Then we can obtain the error-locator polynomial $\sigma(x)$ and error estimation function $\omega(x)$ by the use of $S(j)$ and Berlekamp–Massey algorithm. The formula for iteration is

$$S(x) \cdot \sigma(x) \equiv \omega(x)\left(\mathrm{mod} x^{2t+1}\right) \tag{2.11}$$

The roots of $\sigma(x)$ can be solved by Chien Searching and the reciprocal of them represent the error locations $x(i)$. Afterwards, the error values $Y(i)$ can be computed by

$$Y(i) = -\frac{\omega(z)}{\sigma'(z)}\bigg|_{z=x(i)}, \omega(z) = S(z) \cdot \sigma(z) \; (\mathrm{mod} \; x^{2t}) \tag{2.12}$$

Since the error locations and error values are calculated, the error polynomial $\hat{e}(x)$ is then obtained by Eq. (2.9) and the received polynomial $R(x)$ ultimately gets repaired.

$$T(x) = R(x) - \hat{e}(x) \tag{2.13}$$

2.2.3 System Setup and Results

Figure 2.15 shows the experimental setup for the V-band ROF system via single-electrode Mach–Zehnder modulator (SD-MZM) with adaptive RS coding scheme and envelope detection method.

As mentioned in Eq. (2.7), the received mm-wave signals need to be down converted to the valid 16QAM vector signal. However, the signal is also in company with the interference generated by beating, including the DC component, two frequency doubling part and higher harmonics, as Fig. 2.16 shows. The overlapping of these redundant parts with the signal spectrum will cause SSBI, which degrades the performance. When the roll-off factor of the RRC filter is 0.3, the cut-off frequency is equal to 0.65 times the symbol rate. In this case, the SSBI does not

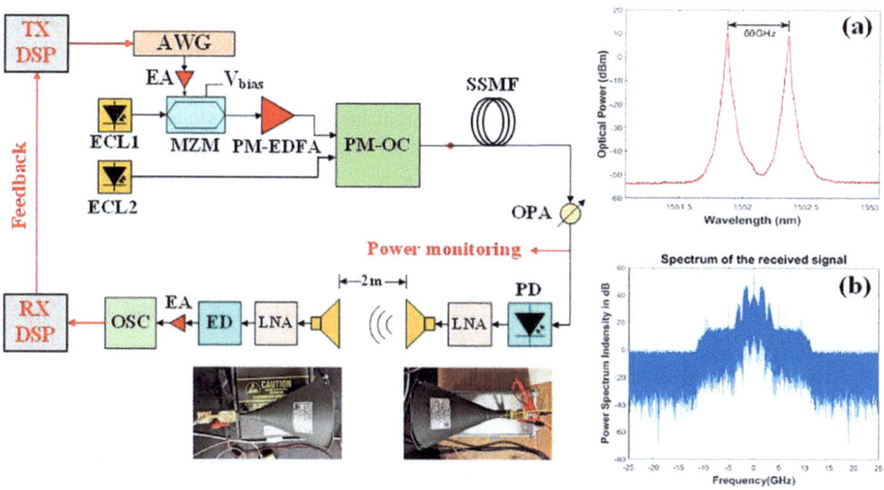

Fig. 2.15 Experimental setup. (**a**) Optical spectrum after PM-OC, (**b**) Spectrum of the received signal. AWG: arbitrary waveform generator, EA: electrical amplifier, PM-EDFA: polarization maintaining erbium-doped fiber amplifier, OPA: optical power attenuator, PD: photo detector

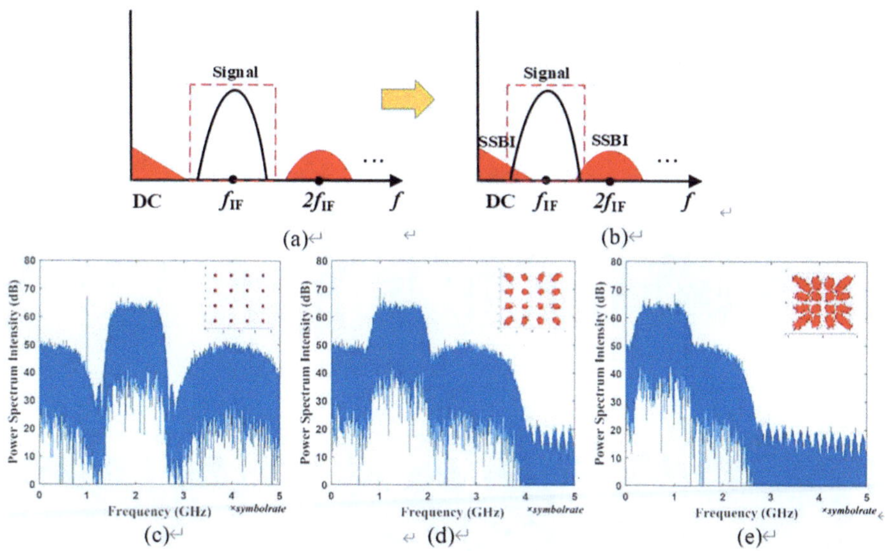

Fig. 2.16 Analysis of the SSBI effect (**a**) $f_{IF} \geq 1.3\times$ symbol rate, (**b**) $f_{IF} < 1.3\times$ symbol rate, The signal spectrum and constellations at the receiver (**c**) $f_{IF} = 2\times$ symbol rate, (**d**) $f_{IF} = 1.3\times$ symbol rate, (**e**) $f_{IF} = 0.75\times$ symbol rate

overlap with the desired signal if the intermediate frequency f_{IF} is greater than 1.3 × symbol rate as shown in Fig. 2.16a. After the low pass filter, the SSBI component is almost filter out. As f_{IF} decreases or the symbol rate promotes, the overlapping parts grow up and much more SSBI will appear and affect the system performance, as shown in Fig. 2.16b.

In the system, the f_{IF} is respectively set to 2 times, 1.3 times and 0.75 times the symbol rate. The signal spectrum and constellations after ED at the receiver are shown in Fig. 2.16c–e. It is obvious that the signal does not overlap with the SSBI when the frequency f_{IF} is 2 times the symbol rate. As f_{IF} decreases, the overlapping parts between the signal and the SSBI grow up and the system performances get worse.

After 20-km standard single mode fiber and 2-m wireless transmission, signals are converted from V-band to IF-band and captured by the oscilloscope, their power spectra are presented in Fig. 2.17a–c. Moreover, Fig. 2.17d–f show the spectrums of the baseband 16-QAM complex signal down-converted from the IF vector signal. It can be clearly seen from these figures that the spectrum intensity reduces and the effect of the SSBI become more and more serious with the increase of the transmission rate, resulting severe signal distortion. When the symbol rate is equal to 8-Gb/s, the signal transmission is seriously affected by the bandwidth constraint, and suffers a lot from the SSBI. The signal distortion makes the recovery impossible.

Figure 2.18 gives the RF spectra and constellations for the 1-Gbaud RS-coded 16-QAM vector signals at the receiver. The input power into PD is −2 dBm. The constellations from left to right in Fig. 2.18 correspond to those before CMMA

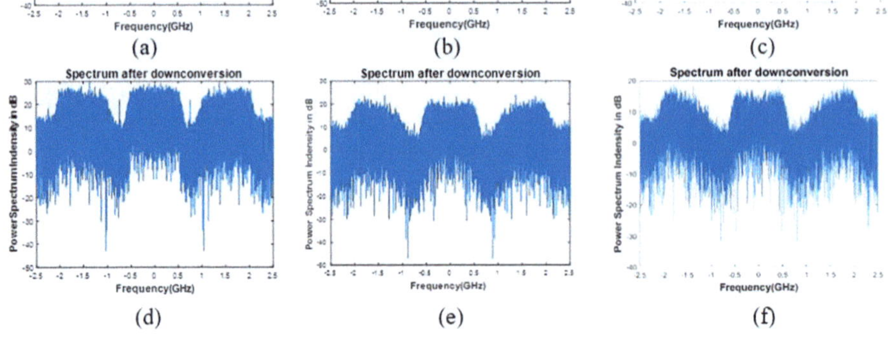

Fig. 2.17 The spectra of the processed signal. After down-sampling (**a**) 2-Gb/s, (**b**) 4-Gb/s, (**c**) 8-Gb/s, After down-conversion (**d**) 2-Gb/s, (**e**) 4-Gb/s, (**f**) 8-Gb/s

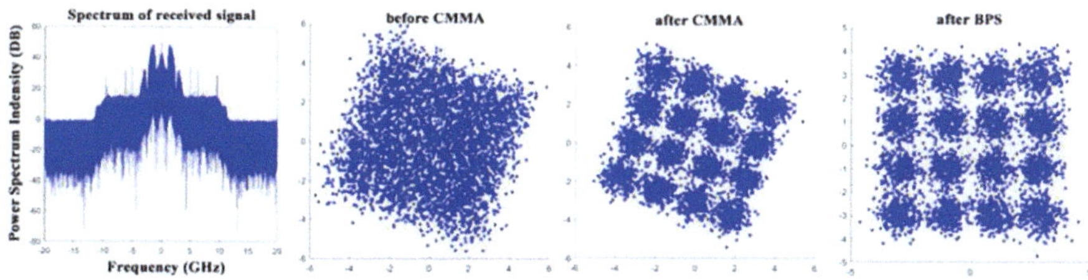

Fig. 2.18 The RF spectra and constellations for RS-coded 16-QAM vector signal at the receiver

Fig. 2.19 BER performances versus received optical power

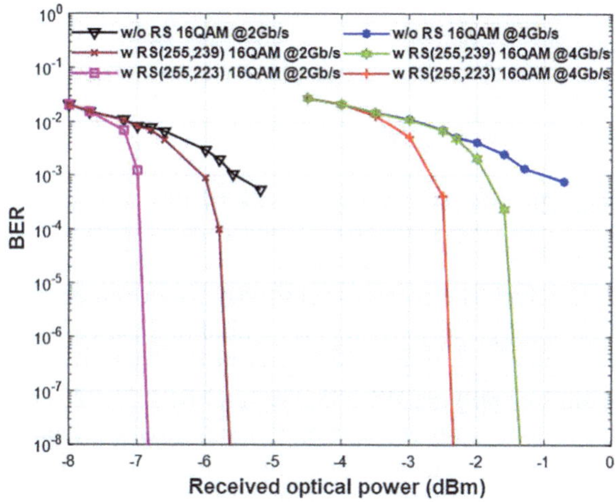

equalization, after CMMA, after BPS for carrier phase estimation, respectively. 20-km SSMF transmission causes very small penalty and the effect of chromatic dispersion (CD) can be neglected without optical dispersion compensation.

Figure 2.19 shows the BER performances versus the received optical power (ROP) for 2-Gb/s and 4-Gb/s 16QAM vector signal transmission across 20-km SSMF and 2-m wireless distance. The RoF System of 20-km SSMF and 2-m wireless distance may be served as the short-reach access network.

The 16QAM signal without RS code is firstly transmitted at the symbol rate of 0.5-Gbaud (i.e. the total bit rate is 2-Gb/s). The BER performance and CER estimation are experimentally measured. The received optical power is −5.6 dBm at the BER of 1×10^{-3}. It can be seen clearly that BER of the system is uniformly high and the curve decreases very slowly. Then the adaptive RS coding module works according to the previous CER to achieve the BER as low as 1×10^{-8}. For RS(255,239), the net bit rate is about 1.87-Gb/s, the post-FEC BER can reach up to 1×10^{-8} at the ROP of −5.6 dBm. When the channel state gets worse (e.g. the received optical power degrades or the channel noise and interference increase), RS(255,223) with higher redundancy will be applied to guarantee the BER performance adaptively when the CER information feeds back. The ROP at the post-FEC BER of 1×10^{-8} is −6.8 dBm with the net bit rate reducing to

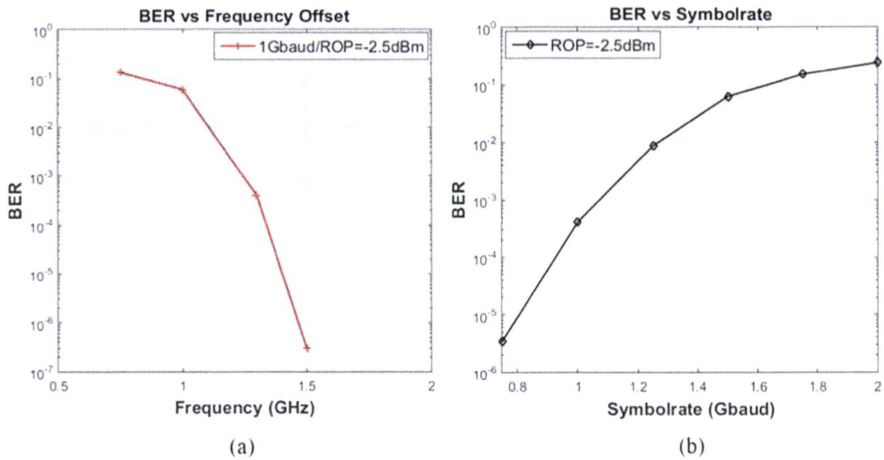

Fig. 2.20 (**a**) BER vs frequency offset, (**b**) BER vs Symbol rate (@ROP= −2.5 dBm)

1.75-Gb/s. The receiver sensitivity improvement is obtained by 1.2 dB compared to RS(255,239) code. As the symbol rate increases to 1-Gbaud (i.e. the total bit rate is 4-Gb/s), the ROPs at the post-FEC of 1×10^{-8} are respectively −1.3 and −2.3 dBm by applying RS(255,239) and RS(255,223) codes as shown in Fig. 2.19. No error floor is observed, which indicates that adaptive RS coding module process a stable and excellent error correction performance with proper CER feedback.

When the ROP is fixed at −2.5 dBm, the system performance under different channel frequency offsets and symbol rates are experimentally investigated. As Fig. 2.20a shows, when setting the symbol rate to 1-Gbaud, the BERs at the frequency of 1.3 and 1.5-GHz are 4.12×10^{-4} and 3.1×10^{-7} respectively, in which SSBI can be neglected just as the analysis in principle. Moreover, the BER performance get worse due to the increasingly channel distortion when the symbol rate promotes and the frequency is fixed, as shown in Fig. 2.20b

2.3 LDPC-Coded Generalized Frequency Division Multiplexing for Intensity-Modulated Direct-Detection Optical Systems

Due to exponential increase in triple play services and high data rate demands by the end users, multi-carrier modulation scheme is a very popular technique for optical access network. Orthogonal frequency division multiplexing (OFDM) plays a constructive role in multicarrier transmission which has great resistance to inter-symbol interference (ISI), high spectral efficiency and one-tap equalization [33–38]. However, several drawbacks of OFDM hinder its further and extensive application [39–42]. In the OFDM system, subcarriers are shaped by rectangular pulse shaping filter in time domain where the side lob-es of Sinc(f) in frequency domain decompose slowly. Thus, OFDM modulation system h-as high spectrum leakage and is very sensitive to carrier frequency offset. OFDM also has high peak-to-average power ratio (PAPR) and the overhead brought by the necessary cyclic prefix (CP) degrade the spectrum efficiency to a certain degree.

Advanced multi-carrier modulation formats have been investigated for the future communication system, such as Filter Bank Multicarrier (FBMC)/Universal Filtered Multicarrier (UFMC)/Generalized Frequency Division Multiplex (GFDM). Generally speaking, the key differences between these waveforms and OFDM is that they utilize filtering techniques, at the subcarrier and/or resource block level, to modify the spectral properties of their signals, leading to lower out-of-band (OOB) emissions [43]. Among those technologies which is brought up by the group of 5GNOW in German, has some advantages in improving the system performance compared with OFDM technique, such as its lower out-of-band radiation and higher robustness against sampling time offset [43, 44]. In GFDM-based system, adjustable pulse shaping filter is required which has lower OOB radiation to the individual subcarriers, such as raised cosine (RC) filter, root-raised cosine (RRC) filter, Dirichlet filter [36–39]. A two-dimensional data structure is introduced to group data symbols across several subcarriers and time slots to blocks. Then tail-biting technique which can preserve circular properties across time and frequency domain is required to process these blocks. It should be noted that, the size of these blocks is a variable parameter and permits to implement long filters or to reduce the total number of subcarriers [45, 46]. Meanwhile, the position of adding CP is used before each frame instead of each symbol as in OFDM, which can effectively enhance the spectrum efficiency. GFDM has been experimentally proved to be more suitable in the passive optical network and LTE cellular system [43, 47]. Ref. [43] demonstrates a universally filtered OFDM (UF-OFDM) and GFDM system with 10Gb/s PAM-4 signal applied in a converged 5G PON system, and Ref. [47] investigates the benefits of the GFDM-based system compared with OFDM in LTE cellular system.

It is well known that forward error correction (FEC) code is widely adopted to recover data which can effectively improve the BER performance. Especially, Low-Density Parity-Check (LDPC) codes have showed its strong ability against burst symbol errors for the past few years. As to be a linear block code in essence, LDPC codes maps the information sequences into specific sequences through a generating matrix G or a sparse parity check matrix H, with redundancy added. Iterative decoding algorithm based on Belief-propagation (BP) are most widely utilized in the receiver [48]. Because of the sparse parity matrix, the information bits far away from each other are checked uniformly in the long code, which makes the continuous burst errors have little impact on the decoding, and the coding itself has the characteristics of anti-burst errors. In the field of coding theory, LDPC is able to achieve large coding gain with low decoding complexity.

LDPC-coded OFDM is proved to be an efficient coded modulation technique in optical transmission and mobile communication which can increase the spectrum efficiency and enhance the resistance to inter-symbol interference (ISI) [49, 50].

Several novel modulation schemes based on N-dimensional signal constellation mapping in coherent optical OFDM system are proposed and applied for ultra-high-speed optical transmission, which has high spectral efficiency and strong resistance against the channel impairment [51, 52]. The intensity-modulation/direct-detection (IM/DD) system with simple transceivers and low cost are widely utilized in optical wireless system and fiber-optic communication. Some devices (e.g., lasers and modulators) with excellent performance have been developed and applied in advanced optical transport experiments [53, 54]. Lots of research have been done about IM/DD system based on OFDM [38, 55, 56]. Advanced multicarrier modulation like UFMC/FBMC also have been investigated with IM/DD transmission system in some radio-over-fiber (RoF) scenarios [57, 58]. However, as far as we know, there are very few works that have studied the performance of the LDPC-coded GFDM for IM/DD optical systems. Considering that LDPC has strong ability for error correction with low decoding complexity in high noise environment and GFDM concentrated high-order modulation can provide high spectral efficiency, the research on combining these technologies is worth to be discovered. As to be an advanced multicarrier modulation technology, the proposed scheme will be applied for the large-capacity, high-speed and cost-effective communication, such as passive optical networks (PONs), Internet of things of 5G, and high-definition video dissemination.

In this section, we propose a LDPC-coded optical GFDM (LCO-GFDM) IM/DD system with high order modulation. The LCO-GFDM signal transmitted over 20-km standard single-mode fiber (SSMF) at 2.5-Gbaud and 5-Gbaud symbol rate is experimental demonstrated and analyzed. 16QAM and TC-32QAM formats are respectively investigated in our proposed system to achieve higher spectral efficiency. Furthermore, the performance improvement by using LDPC-coded GFDM and Turbo-coded GFDM in IM/DD optical system are compared under the same experimental setup. The results successfully demonstrated that the LCO-GFDM system with high-order modulation can achieve significant performance optimization and show its potential for the IM/DD optical systems. For further proving the advantages of our proposed LCO-GFDM modulation scheme, a multiband 16QAM-modulated LCO-GFDM IM/DD system is simulated. Results show that the system performance is mainly affected by fiber length, and the proposed LCO-GFDM scheme has better performance compared to the LCO-OFDM scheme in multi-band transmission.

2.3.1 System Model

The block diagram of the proposed LCO-GFDM with high-level modulation scheme for IM/DD system is depicted in Fig. 2.21. The data sequences are transferred from serial to parallel (S/P) and encoded by multiple LDPC encoders. After parallel to serial (P/S) operation and M-QAM/TCM mapping, the modulated data are sent to the GFDM modulator and then upconverted to be a real-valued signal. A continuous

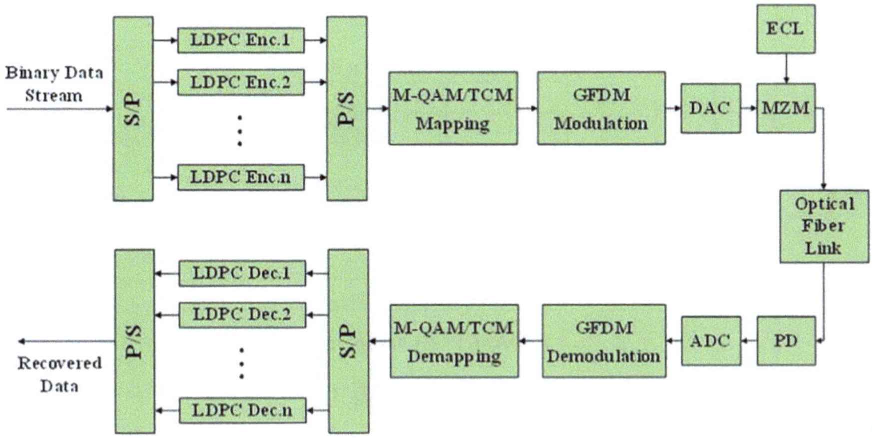

Fig. 2.21 The structure of LCO-GFDM system with M-QAM/TCM (Enc: encoder, Dec: decoder)

wave (CW) output, from an external cavity laser (ECL), is modulated by a RF carrier, which carries the generated LDPC-coded GFDM signals and drives a Mach-Zehnder modulator. After transmitted across the optical fiber link, the LCO-GFDM signals are detected by a photo detector (PD) and captured by an oscilloscope. Data get recovered by the offline DSP, which mainly includes the GFDM demodulation module, M-QAM/TCM demappers and LDPC decoders.

2.3.2 Algorithm Principle

Transmitter of the LCO-GFDM System

Figure 2.22 depicts the generation process of the LDPC-coded GFDM digital signal with high-level modulation.

LDPC-Coded 16QAM-Modulated GFDM System As Fig. 2.22 shows, the random data are concurrently processed by multiple LDPC encoders with the aid of serial-to-parallel and parallel-to-serial transformation. The essential part of LDPC encoders, generating matrix G can be obtained by row and column transformation of parity-check matrix H achieved from IEEE 802.16e standard, where G is a $k \times n$ matrix and the code rate is k/n. Instead of encoding the binary data by H in general, such as lower upper (LU) decomposition coding and partially iterative coding algorithm, data directly multiply with matrix G which can reduce the pretreatment complexity

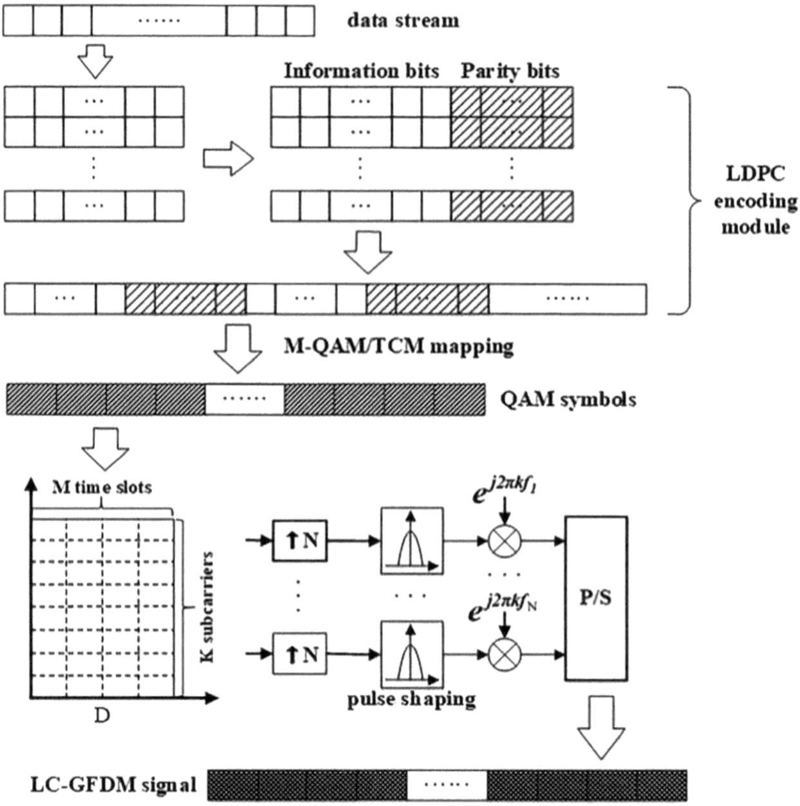

Fig. 2.22 Generation of LDPC-coded GFDM signal with M-QAM/TCM

of coding. The data sequences are encoded by the Eq. (2.14) as follows,

$$y_{l\times n} = \mathrm{mod}\left(\begin{pmatrix} x_{11} & \cdots & x_{1k} \\ \vdots & \ddots & \vdots \\ x_{l1} & \cdots & x_{lk} \end{pmatrix} \cdot G_{k\times n}, 2\right) \cdot I = \frac{\mathrm{length(data)}}{k} \tag{2.14}$$

where *mod* represents the modulo operation and the information bits are reshaped to the matrix $\left[x_{\ddot{y}}\right]_{l\times k}$ by line. Afterwards, parity codes are added and interweaved with information bits by row. 16QAM with Gray mapping are then employed to transfer bits into symbols.

The linear symbol stream after high-level modulation is mapped to a data matrix $D = \{d_k(m)\}_{K\times M}$, where K and M represents the number of subcarriers and time-slots, and $d_k(m)$ represents the complex valued data symbols transmitted on the k_{th} subcarrier and the m_{th} time-slot, $k = 0, \ldots, K-1, m = 0, \ldots, M-1$ are given. After up-sampled by N and digital pulse shaping, the transmitted

LDPC-coded high-level modulated GFDM signal is generated and can be expressed as,

$$x[n] = \sum_{m=0}^{M-1} \sum_{k=0}^{K-1} \Big[d_k[m]\delta\big[n - mN\big] \cdot \tilde{g}_{Tx}\big[n\big]\Big] e^{j2\pi \frac{kn}{N}} \tag{2.15}$$

The subcarrier filter $g_{TX}(n)$ can be achieved by circular shifting of with a period of $(nmodMN)$ which is used to facilitate tail biting at the transmitter. The Dirac function $\delta(n - mN)$ act as a time delay, and $e^{j2\pi kn/N}$ make the signal shifted by k/N in frequency domain. Dirichlet pulse shaping filter is utilized as the subcarrier filter, where a perfect rectangle function is applied in the frequency domain and its passband is centered at the DC bin. Dirichlet pulse shaping can not only get the same BER performance as OFDM system, but also can effectively suppress the OOB radiation. Compared with RC and RRC filter, Dirichlet pulse creates no self-interference. The phase-shifted characteristic matrix G is explained in the Ref. [17], then we can get $g_{TX}(n)$ by IFFT transformation for G.

$$[\bar{G}]_{k,I} = \begin{cases} 1, & 1 \le I < \lceil M/2 \rceil \\ e^{-j2\pi k/K}, & \lceil M/2 \rceil \le I < M \end{cases}, \forall 0 \le k < K \tag{2.16}$$

The GFDM transmitter include the frequency-domain implementation, frequency-domain convolution in time domain as element-wise vector multiplication, and the block circularity of matrices involved in modulation, which involves four steps: M sets of K-point inverse-DFTs (IDFTs), K sets of M-point DFTs, element-wise multiplication with a $K \times M$ matrix, and K sets of M-point IDFTs. The number of complex multiplications (CMs) can be calculated by the formula (2.17).

$$\begin{aligned} C_{GPBM} &= M\frac{K}{2}\log_2 K + K\frac{M}{2}\log_2 M + KM + K\frac{M}{2}\log_2 M \\ &= KM\left(\frac{1}{2}\log_2 KM^2 + 1\right) \end{aligned} \tag{2.17}$$

Supposing the GFDM and conventional OFDM transmitters use the same block size KM, the computational complexity of the conventional OFDM will be $M\frac{K}{2}\log_2 K + K\frac{M}{2}\log_2 M = \frac{1}{2}KM\log_2 KM$.

In the system, the high-level modulated LDPC-coded GFDM signal is up-converted and uploaded into an arbitrary waveform generator (AWG) to drive a MZM and modulate the CW output. Then the LDPC-coded 16QAM-modulated optical GFDM signal is generated, as shown in Fig. 2.23.

The proposed LCO-GFDM scheme is compared with the traditional OFDM technique and demonstrated in Table 2.1 with full details, where M represents the

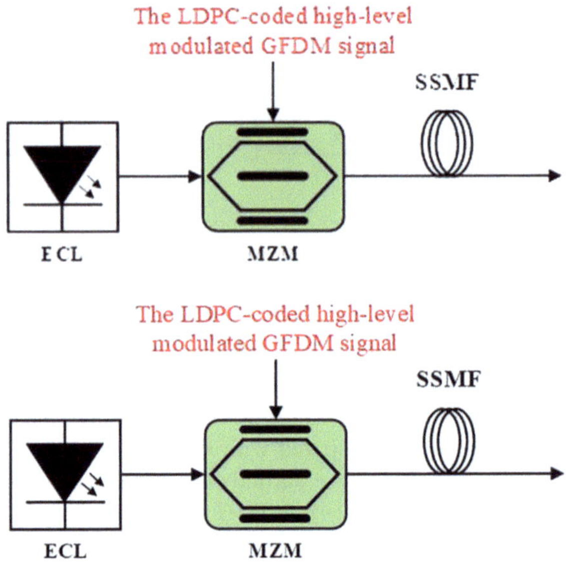

Fig. 2.23 Principle of the LDPC-coded high-level modulated GFDM signal generation

Table 2.1 System properties

	Traditional OFDM	LCO—GFDM
Subcarriers	Orthogonal	Non-orthogonal
Prototype filter	Rectangle	Dirichlet, RC, RRC
Convolution style	linear	circle
Cyclic prefix	LCP	LCP/M
w or w/o FEC	w/o	w

number of time-slots. It indicates that the LCO-GFDM scheme has shorter CP, higher spectrum efficiency and stronger ability to correct burst symbol errors.

LDPC-Coded TC-32QAM-Modulated GFDM System It is common knowledge that trellis-coded modulation (TCM) technology combines modulation with coding and transmits information without bandwidth expansion, which can both have high efficiency, ability of correcting burst errors and is effectively applied in some special conditions, e.g., band-limited channels. So it is worth to be investigated with traditional high-level modulation formats applied in our proposed LCO-GFDM system. In TCM modulator, diversity mapping is a very important part that will not increase the transmission bandwidth and keep the Euclidean distance large enough to make the decision accuracy even if the redundancy is added. Taking the trellis-coded with 32-quadrature amplitude modulation (TC-32QAM) for example, the diversity mapping scheme of TC-32QAM is as Fig. 2.24 shows. Appropriate convolutional encoder should also be designed. Figure 2.25 indicates the optimum structure of 64-state rate 4/5 convolutional encoder that is worth to be compared with traditional 16QAM formats. The system basically remains unchanged except that we use TC-32QAM modules instead of 16QAM blocks.

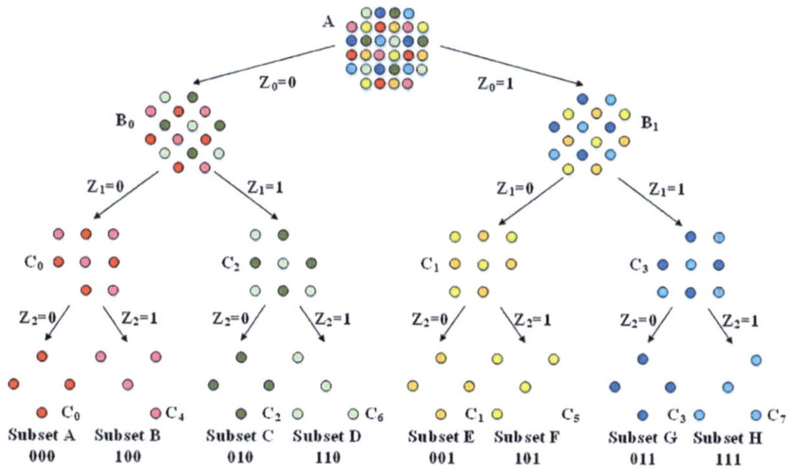

Fig. 2.24 Diversity mapping scheme of TC-32QAM

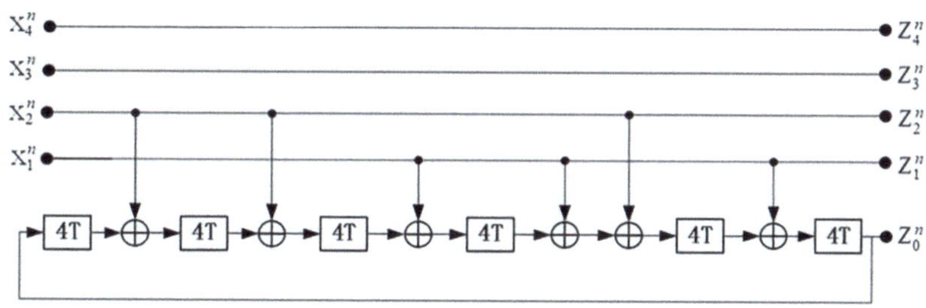

Fig. 2.25 The optimum structure of 64-state rate 4/5 convolutional encoder

Receiver of the LCO-GFDM System

After the fiber link and photodetector (PD) detection, the transmitted signal is firstly processed by GFDM demodulation module performed in frequency domain which is opposite to the modulation module, shown in Fig. 2.26. Suppose $y[n]$ is the signal obtained at the receiver, then the resulting signal at $n = mN$ is given by the expression,

$$d_k^{\%}[m] = \left(y[n] e^{-j2\pi \frac{kn}{N}} \right) \otimes g_{Rx}[n] \Big|_{n=mN} \tag{2.18}$$

where $\otimes g_{Rx}[n]$ denotes circular convolution with matched filter (MF) relative to n which is essential for tail biting at the receiver. In frequency domain, the Eq. (2.17) will be calculated by FFT and IFFT conversion flexibly. Except for the similar procedure at the transmitter, frequency-domain one-tap equalization which has $KM\left(\log_2 KM + 1\right)$ CMs is introduced at the receiver. So the total number of

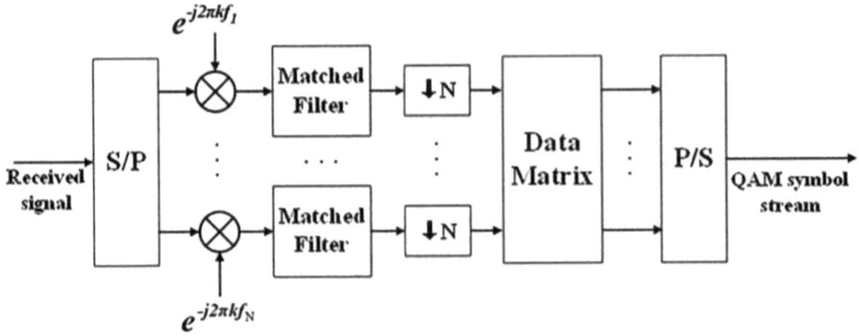

Fig. 2.26 Principle of GFDM demodulation system

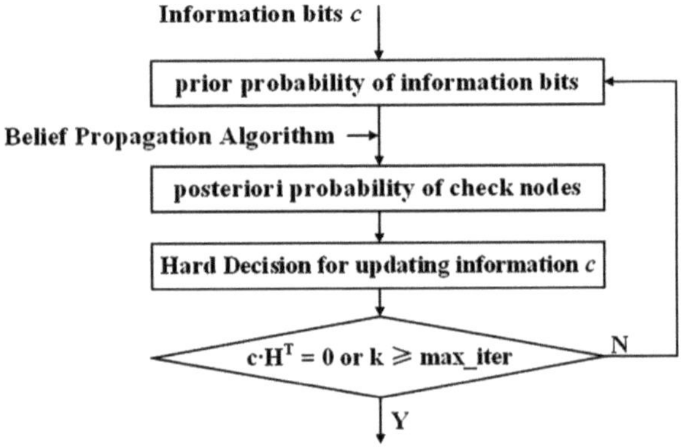

Fig. 2.27 Structure of LDPC decoder based on BP decoding algorithm

CMs at the GFDM receiver will be $KM\left(\frac{1}{2}\log_2 KM^2+1\right)+KM\left(\log_2 KM+1\right)$. The conventional OFDM receiver implementation will take $\frac{1}{2}KM\log_2 KM+KM$ CMs.

After digital Rx-filtering, the signal is down-sampled turning sample index into QAM symbol index. The symbol index is then sent to the high-level modulation demappers. What needs to be illustrated is, in the TCM decoders, 'fast subset decoding' algorithm get the distance and distance index from the receiving point to the eight subsets firstly, which can reduce the complexity and storage space of subsequent Viterbi decoding algorithm. And then soft-decision Viterbi decoding module is employed by comparing the Euclidean distance with the receiving point and the various Ref. points. The closest point is identified and chosen.

The demapping outputs are fed to LDPC iterative decoders based on the belief propagation algorithm. The structure of iterative decoder is shown in Fig. 2.27, where the 'max_iter' represents the maximum number of iterations. Since the average of the random information is 0.5, different from the traditional BPSK signal, the prior probability of information bits should be modified as,

$$q_{ij}(1)=P_i=\Pr\left(x_i=1|y\right)=\frac{1}{1+e^{-2(y_i-0.5)/\sigma^2}} \tag{2.19}$$

$$q_{ij}(0)=1-P_i = \Pr\left(x_i=0\middle|y\right) \tag{2.20}$$

where $q_{ij}(b)$ indicates the probability that the i_{th} information node $c_i = b$ and other check nodes provide external information except the j_{th} check node. After calculating the posteriori probability of check nodes and information nodes successively, a hard decision is required to recover the sequence c of 0, 1. The iterative process of the BP algorithm is completely parallel. Moreover, the decoding process will directly break up if the condition is satisfied or the designed maximum number is reached.

2.3.3 Experimental Setup and Results

A. Experimental Setup

We experimentally demonstrated the proposed LCO-GFDM IM/DD systems with the setup shown in Fig. 2.28. The aforementioned LDPC-coded GFDM signal with high-order modulation is generated by Tx-DSP as Fig. 2.29a depicts. In detail, the random binary bits are coded by LDPC encoding module and then mapped to M-

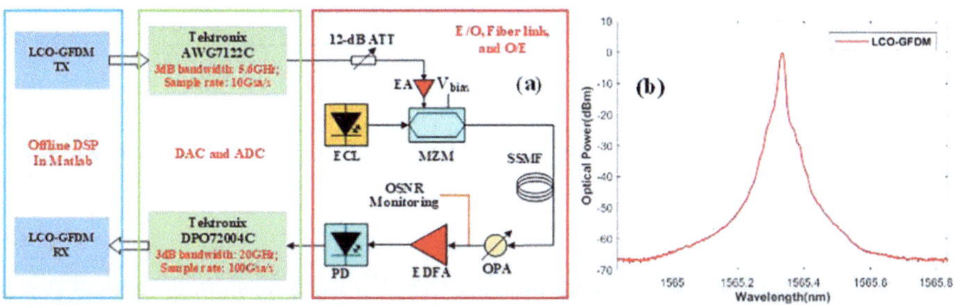

Fig. 2.28 (**a**) Experimental setup of the LCO-GFDM IM/DD system, (**b**) Optical spectrum of the LCO-GFDM signal (EA: electrical amplifier, OPA: optical power attenuator, EDFA: erbium doped fiber amplifier)

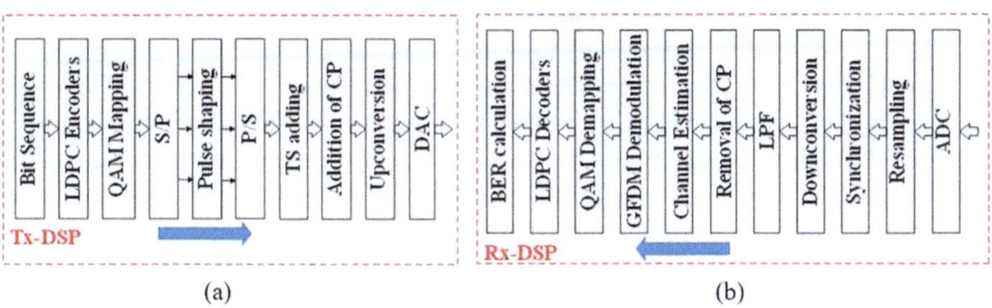

Fig. 2.29 The procedure of offline DSP. (**a**) Transmitter (TX), (**b**) Receiver (RX)

QAM/TCM symbols with Gray mapping. After serial-to-parallel (S/P) converting into a $K \times M$ data matrix ($K = 1024$, $M = 20$), a 1024-point IFFT is used to convert the signal from frequency domain to time domain, then digital pulse shaping is performed subcarrier-wise which is crucial to yield low out of band radiation. Simultaneously, the paralleled data are converted to serial stream. Afterwards, two training sequence (TS) symbols is added to the data stream before symbol synchronization and zero-forcing equalization which can eliminate the impact of the channel. A single cyclic prefix (CP) for an entire GFDM block that contains multiple sub-symbols is required. After 2-times up-sampling and up-conversion by a carrier frequency f_c , the data is uploaded into an arbitrary waveform generator (AWG) operating at 10 GSa/s to realize digital-to-analog conversion (DAC). An external cavity laser (ECL) is adopted to generate the optical carrier with a frequency of 191.52 THz. A Mach–Zehnder modulator (MZM) with a 3-dB bandwidth of 10 GHz which is biased at quadrature point and driven by electrical amplifier is used to modulate the optical carrier with the generated LCO-GFDM signal with high-order modulation. At the output of AWG, 12-dB electrical attenuator is required to limit the peak to peak voltage that can ensure the modulator working on its linear workspace after the signal amplified by EA. The net bit rate can be calculated by the Eq. (2.21),

$$\begin{aligned} &Net\ bit\ rate = 4\ bits/symbol \times 10GS/s\ (sampling\ rate) \\ &\times \frac{512\ (payload\ subcarriers)}{1024\ (total\ subcarriers)} \times \frac{1}{1 + 20\%\ (FEC\ overhead)} \\ &\times \frac{20\ (payload\ symbols)}{20\ (payload\ symbols) + 2\ (training\ symbols) + 1\ (cyclic\ prefix)} \\ &\approx 14.49Gb/s \end{aligned} \tag{2.21}$$

After 20-km standard single mode fiber (SSMF) transmission, an optical power attenuator is adopted, which can act as a variable optical attenuator (VOA) monitoring the received of optical power (ROP). An Erbium doped fiber amplifier (EDFA), is used to amplify the generated LCO-GFDM signal and add amplified spontaneous emission (ASE) noise. Then, the received optical signal is converted into the electrical signal by a photodetector (PD) and captured by a real-time oscilloscope operating at 50 Gsa/s to realize analog-to-digital conversion (ADC). Following the ADC, the offline Rx-DSP are performed. The received data are down-sampled from 50 Gsa/s to 10 Gsa/s to match the ADC. Synchronization is employed using training sequence (TS) to align the up-converted data symbols. After being down-converted by f_c , the signal is filtered out with the high-frequency component following by the FIR low-pass filter (LPF). The second synchronization is required when the added CP is removed. After that, TS is used to do zero-forcing equalization for sake of channel estimation and phase recovery. After S/P converting and channel equalization, matched filter (MF) receiver is utilized to realize GFDM

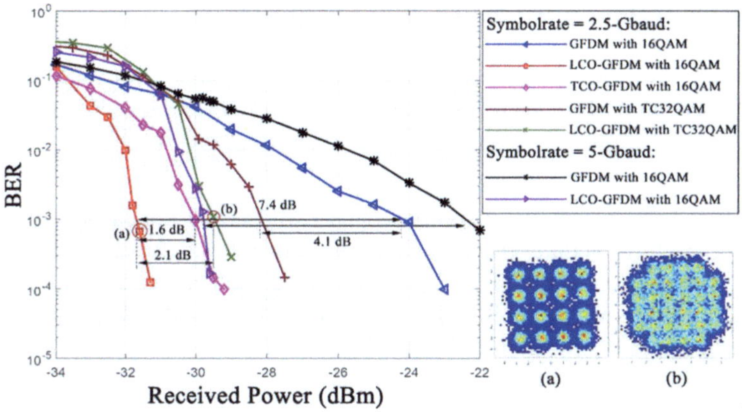

Fig. 2.30 The BER performance of GFDM IM/DD system with different modulation formats

demodulation. Finally, data get recovered by M-QAM/TCM demappers and LDPC decoders.

B. Results and Discussion

The experimental results are measured from oscilloscope by varying the ROP with the help of optical power attenuator, and calculated by offline DSP algorithm. Figure 2.30 shows the BER performances versus the received optical power (ROP) for un-coded/coded GFDM IM/DD system with high-level modulation transmitted through 20-km SSMF at 2.5-Gbaud and 5-Gbaud symbol rate.

LDPC-Coded 16QAM-Modulated GFDM IM/DD System

As shown in Fig. 2.30, when the measured BER is fixed at 1×10^{-3}, the ROPs of the LDPC-coded 16QAM-modulated GFDM system and the uncoded 16QAM-modulated GFDM system at 2.5-Gbaud are respectively −31.6 and −24.2 dBm, which shows LDPC-coded scheme can get 7.4 dB improvement in this experiment. As the symbol rate increases to 5-Gbaud, the ROP is measured to be about −29.8 dBm for LDPC-coded 16QAM-modulated GFDM IM/DD system, which shows the applicability of the proposed model. The constellation diagrams of received signal are shown in Fig. 2.30a.

In our experiment, there exists phase difference caused by up-conversion operation which trigger the phase rotation. By using TS to do the zero-forcing equalization, phase rotation can be eliminated as is shown in Fig. 2.30. The system performance and the symbol rate are restricted due to bandwidth limitation. Bandwidth limitation of our experiment is mainly because the restriction of essential components, such as the intensity modulator, electrical amplifier (EA), PD, and is need to be improved by increasing the spectral efficiency.

LCO/TCO-GFDM IM/DD System with 16-QAM/TC-32QAM

When baud rate is fixed at 2.5-Gbaud and the experimental setup is all the same, the transmission performances of the optical GFDM IM/DD system after 20 km standard single mode fiber transmission are investigated in the following three cases:

(1) As Fig. 2.30 shows, in our proposed GFDM IM/DD optical system without LDPC coding module, the ROPs are respectively −24.2 and −28.3 dBm for 16QAM-modulated and TC-32QAMmodulated scheme at BER of 1×10^{-3}, which can prove the ability for correcting errors and 4.1 dB performance improvement of TC-32QAM technology. What needs to be illustrated is, TC-32QAM encoders with 64-state and the trellis code of rate 4/5 is adopted considering the coding gain, redundancy and time delay.
(2) We also experimentally demonstrate the LDPC-coded TC-32QAM-modulated GFDM system with the same experimental setup. The constellation diagrams of received signal are shown in Fig. 2.30b. From Fig. 2.30 we can get that the ROP of the TC-32QAM-modulated scheme is −29.5 dBm which shows worse receiver sensitivity by 2.1 dB than the LDPC-coded 16QAMmodulated GFDM system at BER of 1×10^{-3}. It's because the constellation in TC-32QAM formats are enlarged which increase the difficulty of the hard-decision part. BER performance get worse in lower OSNR condition and will cause interference in LDPC decoding. Take the high efficiency and considerable power penalty into account, TC-32QAM-modulated scheme can be applied in some band-limited channels.
(3) By using Recursive system convolutional coder (RSC), random interleaver and soft output iterative decoding algorithm, Turbo coding module is realized and applied in our proposed system as an extension study. Turbo codes is also widely used in coding fields because it is almost as close to the Shannon limit as that of LDPC codes. However, it is susceptible to be interfered by low code weight codes and convolvers, which increases the complexity and reduces the BER performance. Turbo-coded 16QAM-modulated GFDM for IM/DD optical system is successfully achieved. As shown in Fig. 2.30, the ROP at the BER of 1×10^{-3} was measured to be about −30.0 dBm. It shows that Turbo-coded scheme has about 1.6 dB power penalty compared to LDPC-coded scheme. Therefore, LDPC codes is more suitable in our proposed system (Fig. 2.31).

Multi-Band Transmission Based on the LCO-GFDM Scheme

GFDM systems have better bandwidth roll-off performance than OFDM, which can decrease the inter-carrier interference (ICI) and improve the performance in multiband transmission. For further proving the advantages of our proposed LCO-GFDM modulation scheme, a multiband 16QAM modulated LCO-GFDM IM/DD system is demonstrated. As shown in Fig. 2.32, four strings of data are represented

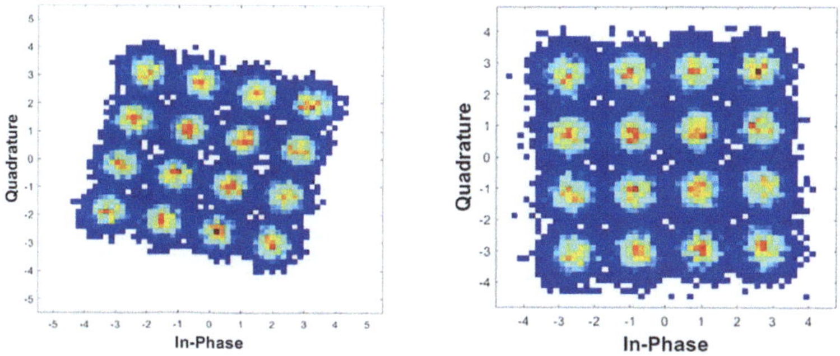

Fig. 2.31 Constellation after GFDM demodulation (**a**) before equalization. (**b**) after equalization

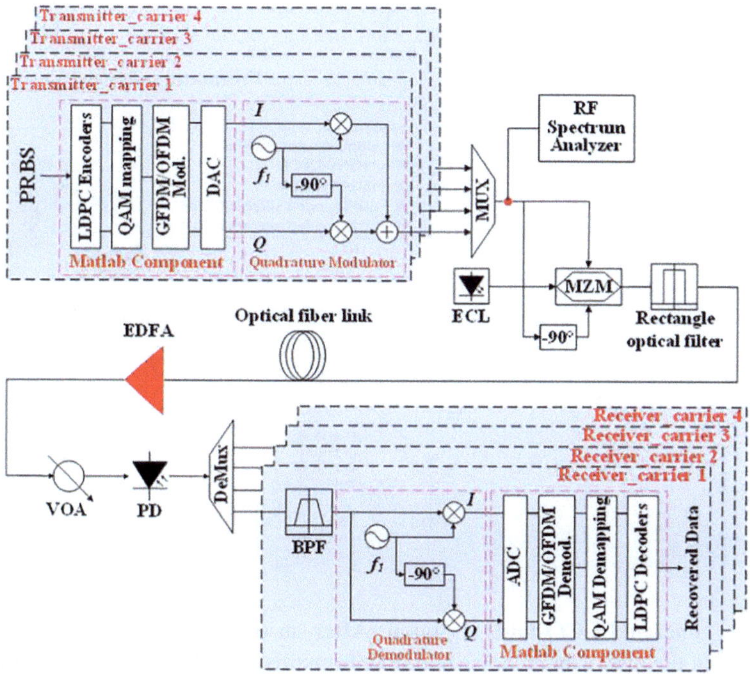

Fig. 2.32 Block diagram of multi-band 16QAM-modulated LCO-GFDM IM/DD system

by the 10 Gb/s pseudorandom bit sequence generated from the PRBS. The electrical LDPC-coded 16QAM-GFDM/OFDM signal is generated in the transmitter and is converted from baseband to RF band by a quadrature modulator. Four quadrature modulators at different center frequency of 10/13/16/19 GHz are used to realize the multi-band transmission by a 4×1 multiplexer. The output electrical spectrum for the LDPC-coded GFDM signal is displayed by a RF spectrum analyzer, as shown in Fig. 2.33. Results show that GFDM-based system has about 20 dB OOB suppression compared to OFDM-based system. A LiNb MZM is used to modulate a CW laser

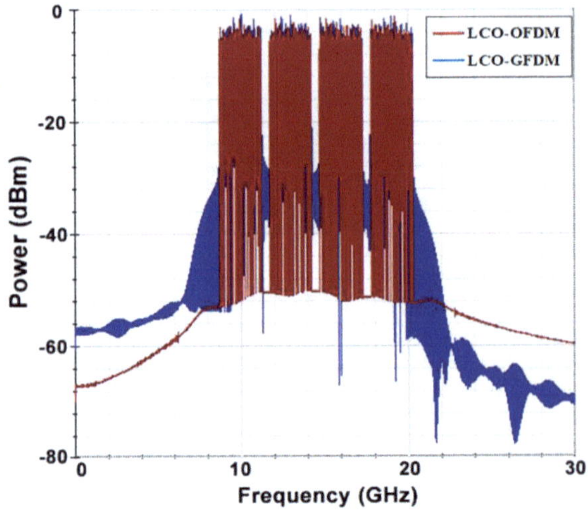

Fig. 2.33 Electrical spectrum for the multi-band 16QAM-modulated LCO-GFDM/OFDM signals

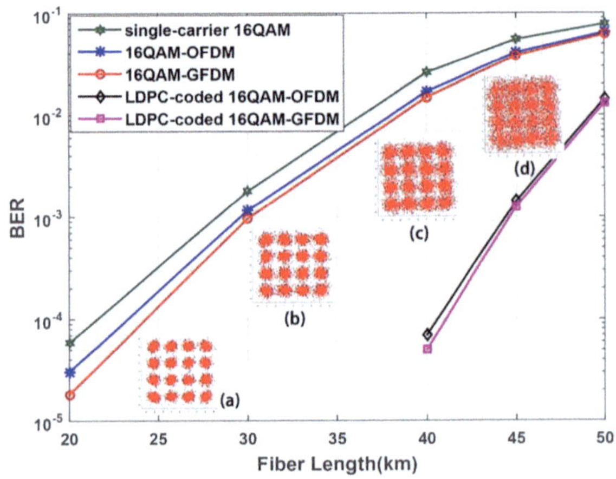

Fig. 2.34 BER performance versus fiber length at 2.5-Gbaud (for carrier2). (**a**) 20 km-, (**b**) 30 km-

with the generated LCO-GFDM signal. After an optical band pass filter (OBPF) filtering out the out-of-band noise and fiber transmission, the signal will be amplified and detected. And then, the signal is split into four parts and each part will pass a BPF with specified center frequency at $f_1 - f_4$ respectively. Then the RF signal is down-converted to the baseband by a quadrature demodulator and data get recovered including ADC, GFDM demodulation, 16QAM demapping and LDPC decoding.

In the system, the fiber parameters are set according to SSMF, the attenuation is 0.2 dB/km, the dispersion coefficient is 16 ps/nm/km. By sweeping the fiber length from 20 to 50 km, the BER performance and constellations are calculated and illustrated in Fig. 2.34, and it shows that the multi-band 16QAM-modulated LCO-GFDM/OFDM IM/DD system was successfully realized at BER of 4.6×10^{-5} when the signal transmit over 40 km fiber. It can be observed that system performance gets worse with the fiber length increasing, which is mainly affected by the fiber

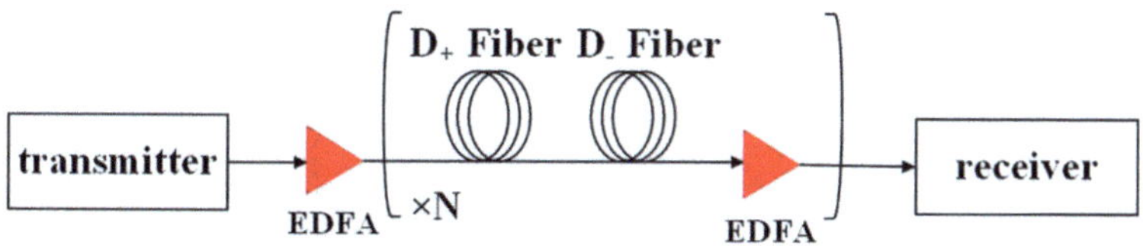

Fig. 2.35 System setup with N spans fiber link

Table 2.2 Fiber parameters

	D_+ Fiber	D_- Fiber
Fiber length	30	6
Dispersion [ps/(nm · km)]	16	−80
Dispersion slope [ps/(nm^2 · km)]	0.075	0.075
Effective cross—sectional area [μm^2]	80	80
Nonlinear refractive index [m^2/W]	2.6×10^{-20}	2.6×10^{-20}
Attenuation coefficient [dB/km]	0.2	0.2

dispersion. The performance of single-carrier 16QAM system is slightly worse than that of 16QAM-OFDM/GFDM system because it cannot effectively resist to the inter-symbol interference (ISI). Besides, multi-band transmission of the LCO-GFDM signal have lower BER than that of LCO-OFDM, which indicates the benefits of our proposed scheme.

To verify the relevance of the proposed scheme, longer SSMF distance is performed as Fig. 2.35 shows. The new fiber link is composed of N spans of length L = 36 km, consisting of 30 km D+ fibers followed by 6 km D- fibers and EDFA. The fiber Parameters are listed in Table 2.2. The length of D+ fibers is five times of D- fibers so that the effect of chromatic dispersion can be ignored without corresponding algorithm added in the receiver. In this case, channel distortion is mainly introduced by some nonlinear effect such as self-phase modulation (SPM) and cross-phase modulation (XPM). When the optical power into PD is fixed at 0-dBm, the BER curve versus number of spans for LCO-GFDM optical IM/DD system is shown in Fig. 2.36.

2.4 Summary

In this chapter, we introduce and discuss several advanced modulation technologies, which can effectively improve the performance and bandwidth of fiber optic communication systems. First, we propose a novel symbol division multiplexing technology by multiplexing/de-multiplexing of multiple QAM symbols onto one complex constellation point which can achieve an additional ...

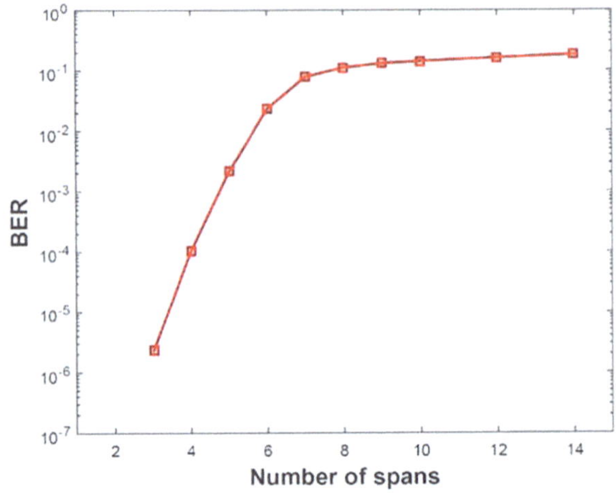

Fig. 2.36 BER curve versus number of spans for LCO-GFDM optical IM/DD system

References

1. X. Wang, Q. Zhang, J. Yu, et al., Carrier phase recovery friendly probabilistic shaping scheme based on a quasi-Maxwell-Boltzmann distribution model. Opt. Lett. **45**(17), 4883–4886 (2020)
2. D. Guo, Q. Zhang, X.J. Xin, et al., Adaptive Reed-Solomon coding and envelope detection of photonic vector signal in V-band radio over fiber system. Opt. Commun. **439**, 210–217 (2019)
3. D. Guo, W. Zhang, F. Tian, et al., LDPC-coded generalized frequency division multiplexing for intensity-modulated direct-detection optical systems. IEEE Photon. J. **11**(2), 1–15 (2019)
4. D. Semrau, T. Xu, N.A. Shevchenko, et al., Achievable information rates estimates in optically amplified transmission systems using nonlinearity compensation and probabilistic shaping. Opt. Lett. **42**(1), 121–124 (2017)
5. X. Chen, Y. Chen, M. Tang, et al., Hybrid constellation entropy loading for adaptively partitioned SSB-DMT systems. Opt. Express **27**(19), 26295–26306 (2019)
6. B. Liu, X. Li, Y. Zhang, et al., Probabilistic shaping for ROF system with heterodyne coherent detection. APL Photon. **2**(5), 1–1 (2017)
7. J. Cho, P.J. Winzer, Probabilistic constellation shaping for optical fiber communications. J. Lightw. Technol. **37**(6), 1590–1607 (2019)
8. D.A.A. Mello, F.A. Barbosa, J.D. Reis, Interplay of probabilistic shaping and the blind phase search algorithm. J. Lightw. Technol. **36**(22), 5096–5105 (2018)
9. V.N. Rozental, D. Kong, B. Foo, et al., Cycle-slip-less low-complexity phase recovery algorithm for coherent optical receivers. Opt. Lett. **42**(18), 3554–3557 (2017)
10. J.C.M. Diniz, Q. Fan, S.M. Ranzini, et al., Low-complexity carrier phase recovery based on principal component analysis for square-QAM modulation formats. Opt. Express **27**(11), 15617–15626 (2019)
11. M. Kong, X. Li, J. Zhang, et al., High spectral efficiency 400 Gb/s transmission by different modulation formats and advanced DSP. J. Lightw. Technol. **37**(20), 5317–5325 (2019)
12. X. Li, et al., Performance improvement by pre-equalization in W-band (75–110 GHz) RoF system, in *Optical Fiber Communication Conference and Exhibition (OFC)* (Optical Society of America, Washington, 2013), p. OW1D.3
13. J. Zhang, et al., Fiber-wireless integrated mobile backhaul network based on a hybrid millimeter-wave and free-space-optics architecture with an adaptive diversity combining technique. Opt. Lett. **41**(20), 1909–1912 (2016)
14. X. Li, J. Xiao, J. Yu, Long-distance wireless mm-Wave signal delivery at W-band, in *International Conference on Wireless Communications & Signal Processing IEEE* (2015)

15. X. Li, et al., W-band 8QAM vector signal generation by MZM-based photonic frequency octupling. IEEE Photon. Technol. Lett. **27**(12), 1257–1260 (2016)
16. Xiao Jiangnan, J. Yu, X. Li, W-band vector millimeter-wave signal generation based on phase modulator with photonic frequency quadrupling and precoding. J. Lightwave Technol. **35**(13), 2548–2558 (2017)
17. A. Kanno, et al., 40 Gb/s W-band (75–110 GHz) 16-QAM radio-over-fiber signal generation and its wireless transmission. Opt. Express **19**(26), B56–B63 (2011)
18. Y. Shao, et al., Generation of 16-QAM-OFDM signals using selected mapping method and its application in optical millimeter-wave access system. IEEE Photon. Technol. Lett. **24**(15), 1301–1303 (2012)
19. M. Chen, et al., Demonstration of software-reconfigurable real-time FEC-enabled 4/16/64-QAM-OFDM signal transmission in an X-band RoF system. IEEE Photon. J. **8**(2), 1–8 (2016)
20. J. Zhang, et al., WDM transmission of single-carrier 120-GBd ETDM PDM-16QAM signals over 1200-km terrestrial fiber links. J. Lightwave Technol. **35**(20), 1033–1040 (2017)
21. M. Kong, W. Zhou, Delivery of 12QAM single carrier signal in a MIMO radio-over-fiber system at 60 GHz. IEEE Photon. J. **9**(3), 7202707 (2017)
22. A.L. Yi, et al., Frequency offset compensation and carrier phase recovery for differentially encoded 16-QAM vector signal in a 60-GHz RoF system. IEEE Photon. J. **5**(4), 7200807–7200807 (2013)
23. H.T. Huang, et al., High spectral efficient W-band OFDM-RoF system with direct detection by two cascaded single-drive MZMs. Opt. Express **21**(14), 16617–16620 (2013)
24. Y. Wei, et al., An approach enabling adaptive FEC for OFDM in fiber-VLLC system. Opt. Commun. **405**, 329–333 (2017)
25. S. Kenya, et al., A practicable rate-adaptive FEC scheme flexible about capacity and distance in optical transport networks, in *Optical Fiber Communications Conference & Exhibition IEEE* (2016)
26. Y. Li, et al., Adaptive FEC-based lightpath routing and wavelength assignment in WDM optical networks. Opt. Switch. Network. **14**(8), 241–249 (2016)
27. M. Chen, et al., Real-time generation and reception of OFDM signals for X-band RoF uplink with heterodyne detection. IEEE Photon. Technol. Lett. **29**(1), 51–54 (2017)
28. X. Li, et al., Photonic vector signal generation at W-band employing an optical frequency octupling scheme enabled by a single MZM. Opt. Commun. **349**, 6–10 (2015)
29. T. Zhang, et al., High-spectral-efficiency photonic frequency down-conversion using optical frequency comb and SSB modulation. IEEE Photon. J. **5**(2), 7200307 (2013)
30. C. Li, et al., W-band OFDM RoF system with simple envelope detector down conversion, in *Optical Fiber Communications Conference & Exhibition IEEE* (2015)
31. X. Li, J. Yu, W-band RoF transmission based on optical multi-carrier generation by cascading one directly-modulated DFB laser and one phase modulator. Opt. Commun. **345**, 80–85 (2015)
32. M. Chen, J. Yu, X. Xiao, Real-time Q-band OFDM-RoF systems with optical heterodyning and envelope detection for downlink transmission. IEEE Photon. J. **9**, 7902007 (2017)
33. N. Cvijetic, OFDM for next-generation optical access networks. J. Lightw. Technol. **30**(4), 384–398 (2012). https://doi.org/10.1109/JLT.2011.2166375
34. J. Zhou et al., Interleaved single-carrier frequency-division multiplexing for optical interconnects. Opt. Exp. **25**(9), 10586–10596 (2017). https://doi.org/10.1364/OE.25.010586
35. F. Li, X. Li, J. Zhang, J. Yu, Transmission of 100-Gbps VSB DFT-spread DMT signal in short-reach optical communication systems. IEEE Photon. J. **7**(5), 7904307 (2015)
36. F. Li, J. Zhang, J. Yu, X. Li, Transmission and reception of PDM dual-subcarrier coherent 16QAM-OFDM signals. Opt. Fiber Technol. **26**, 201–205 (2015). https://doi.org/10.1016/j.yofte.2015.09.016
37. F. Li, X. Li, J. Yu, Performance comparison of DFT-spread and pre-equalization for 8 × 244.2-Gbps PDM-16QAMOFDM. J. Lightw. Technol. **33**(1), 227–233 (2015)
38. J. Zhou, Y. Qiao, Low-PAPR asymmetrically clipped optical OFDM for intensity-modulation direct-detection systems. IEEE Photon. J. **7**(3), 7101608 (2015)

39. J. Zhou et al., 256-QAM interleaved single carrier FDM for short-reach optical interconnects. IEEE Photon. Technol. Lett. **29**(21), 1796–1799 (2017). https://doi.org/10.1109/LPT.2017.2752215
40. X. Li, J. Xiao, F. Li, Y. Xu, L. Chen, J. Yu, Large capacity optical wireless signal delivery at W-band: OFDM or single carrier? in *Proceedings of Optical Fiber Communication Conference Exhibtion, Anaheim, CA, USA* (2016). Section TU2B.6
41. X. Li et al., Real-time demonstration of over 20Gbps V- andW-band wireless transmission capacity in one OFDM-RoF system, in *Proceedings of Optical Fiber Communication Conference Exhibtion, Los Angeles, CA, USA* (2017). Section M3E.3
42. J. Shi, Y. Zhou, Y. Xu, J. Macbeth, J. Yu, N. Chi, 200-Gbps DFT-S OFDM using DD-MZM-based twin-SSB with a MIMO-volterra equalizer. IEEE Photon. Technol. Lett. **29**(14), 1183–1186 (2017)
43. C. Browning et al., 5G wireless and wired convergence in a passive optical network using UF-OFDM and GFDM, in *Proceedings of IEEE International Conference on Communications Workshops, Paris, France* (2017), 386–392
44. N. Michailow et al., Generalized frequency division multiplexing for 5th generation cellular networks. IEEE Trans. Commun. **62**(9), 3045–3061 (2014)
45. N. Michailow, I. Gaspar, S. Krone, M. Lentmaier, G. Fettweis, Generalized frequency division multiplexing analysis of an alternative multi-carrier technique for next generation cellular systems, in *Proceedings of the International Symposium on Wireless Communication Systems, Paris, France* (2012), pp. 171–175
46. G. Fettweis, M. Krondorf, S. Bittner, GFDM-generalized frequency division multiplexing, in *Proceedings of the IEEE 69th Vehicular Technology Conference, Barcelona, Spain* (2009)
47. M. Danneberg, R. Datta, A. Festag, G. Fettweis, Experimental testbed for 5G cognitive radio access in 4G LTE cellular systems, in *Proceedings of IEEE 8th Sensor Array Multichannel Signal Processing Workshop, A Coruna, Spain* (2014), pp. 321–324
48. J. Chen, MPC Fossorier, Near optimum universal belief propagation based decoding of low-density parity check codes. IEEE Trans. Commun. **50**(3), 406–414 (2002). https://doi.org/10.1109/26.990903
49. B. Djordjevic, B. Vasic, LDPC-coded OFDM in fiber-optics communication systems. J. Opt. Netw. **7**(3), 226–226 (2008)
50. B. Djordjevic, H.G. Batshon, LDPC-coded OFDM for heterogeneous access optical networks. IEEE Photon. J. **2**(4), 611–619 (2010)
51. B. Djordjevic, M. Arabaci, L. Xu, T. Wang, Generalized OFDM (GOFDM) for ultra-high-speed optical transmission. Opt. Exp. **19**(7), 6969–6979 (2011)
52. B. Djordjevic, H.G. Batshon, L. Xa, T. Wang, Four-dimensional optical multiband-OFDM for beyond 1.4 Tb/s serial optical transmission. Opt. Exp. **2**(2), 876–876 (2011)
53. S. Chen et al., Electrically pumped continuous-wave iii-v quantum dot lasers on silicon. Nat. Photon. **10**(5), 307–322 (2016)
54. K. Xu, Monolithically integrated Si gate-controlled light-emitting device science and properties. J. Opt. **20**(2), 024014 (2018)
55. L. Zhang, B. Liu, X. Xin, Secure optical generalized filter bank multi-carrier system based on cubic constellation masked method. Opt. Lett. **40**(12), 2711–2714 (2015)
56. P.M. Anandarajah, R. Zhou, C. Browning, L.P. Barry, Performance investigation of IMDD compatible SSB-OFDM systems based on optical multicarrier sources. IEEE Photon. J. **6**(5), 7903110 (2014)
57. L. Zhang, S. Xiao, M. Bi, L. Liu, X. Chen, FFT-based universal filtered multicarrier technology for low overhead and agile datacenter interconnect, in *Proceedings of the International Conference on Transparent Optical Networks* (2016), pp. 1–4
58. J. Wang et al., Delta-sigma modulation for digital mobile fronthaul enabling carrier aggregation of 32 4G-LTE / 30 5G-FBMC signals in a One λ 10-Gb/s IM-DD Channel, in *roceedings of the Optical Fiber Communication Conference and Exhibition, Anaheim, CA, USA* (2016), Section W1H.2

Chapter 3
Advanced Modulation Technology

Advanced modulation technology is an effective way to improve the performance and bandwidth of fiber optic communication systems. Key technologies include high-order Quadrature Amplitude Modulation (QAM) modulation, multi-dimensional modulation, and other advanced modulation formats. First, we propose a novel symbol division multiplexing technology (SDM) by multiplexing/de-multiplexing of multiple quadrature amplitude modulation (QAM) symbols onto one complex constellation point which can achieve an additional 40% gain for effective data rate and system spectral efficiency [1]. Otherwise, we investigate the introduction of the delta-sigma modulation (DSM) technique into a wireless mm-wave transmission system and propose a photonic-aided dual-vector radio-frequency (RF) signal generation and detection scheme enabled by bandpass DSM and heterodyne detection, which can overcome the insufficient power and receiver sensitivity problems and thus to realize high-level QAM signal transmission [2]. Finally, to further improve transmission capacity and performance, we propose a scheme of probabilistic shaping 4D ultrahigh order modulation format with pilot aided carrier phase recovery [3].

3.1 Symbol Division Multiplexing in Optical Fiber Communication Systems

Recent advances in optical fiber communication systems have facilitated the explosive growth in transmission capacity required by modern information services, such as cloud computing, virtual reality applications, and high-definition video services [4–6]. In turn, the dramatic increases in data services and applications have stimulated higher bandwidth requirements for optical channels. In order to satisfy such an unprecedented bandwidth need, researchers have been exploring efficient and innovative ways to broaden the channel capacity by optimizing multiplexing

X. Xin, *Advanced Optical Fiber Transmission Systems*, Wireless Networks,
https://doi.org/10.1007/978-3-032-09516-9_3

technologies, which primarily include (1) Effective phase multiplexing employing high-order quadrature amplitude modulation (QAM) (beyond 16QAM) and high symbol rate (>100Gbaud) cooperated with advanced digital signal processing (DSP) algorithms for impairments compensation [7–10]. (2) Time division multiplexing (TDM) [11, 12] to enable one single bit stream formed by multiple bit streams by means of assigning a recurrent time slot. (3) Frequency division multiplexing (FDM), wavelength division multiplexing (WDM) [13–15], and the derivatives like orthogonal frequency-division multiplexing (OFDM), Nyquist-WDM, and all-optical OFDM. (4) Optical polarization division multiplexing (PDM) [16–18] to double transmission capacity by transmitting two independent signals carried by orthogonally-polarized beams in coherent optical transmission systems. (5) Spatial division multiplexing [19–25] referring to establish multiple spatially distinguishable data paths carrying multi-data streams in parallel over multi-mode and multicore fibers. All types of multiplexing technologies are still under extensive research for the ever-increasing channel bandwidth demand. To further enhance the spectral efficiency (SE) and transmission capacity in different systems, it is of great importance to explore another innovative multiplexing technology which can be employed separately or in combination with the existing multiplexing techniques based on the applicability and other criteria such as transmission distance, access cost, network latency, and energy efficiency.

Herein, we propose a novel symbol division multiplexing (SDM) technology in symbol dimension to further improve the bandwidth efficiency of optical fiber communication systems. The SDM performs the optical modulation by multiplexing/de-multiplexing of four symbols per complex valued constellation point based on many-to-one (MTO) mapping. The MTO mapping was originally outlined by Gallager as an assumption to increase the channel capacity, and had recently been considered as a promising candidate for the probabilistic shaping (PS) scheme to approach Shannon's limit [26–31]. However, all efforts for MTO mapping are focused on the probability mass function (PMF) optimization of QAM signals at the cost of excessive forward error correction (FEC) redundancy. Hence, a valuable goal enabled by MTO mapping comes to carrying information multiplexed in phase-coded symbols, resulting in yet another significant improvement of channel capacity and system SE. As a typical instance of our scheme, the SDM-32QAM is multiplexed from a 7-bit 128QAM symbol, which achieves an extra 40% gain for effective data rate and SE compared to the typical 5-bit 32QAM signals with the same symbol rate and physical link. Moreover, since SDM evidently reduces the order of QAM, the transmission reliability can be effectively improved by alleviating the impairments caused by fiber chromatic dispersion and Kerr nonlinearity. For SDM de-multiplexing (De-SDM), an optimized joint iteration between SDM-QAM demapping and FEC decoding is explored free from any additional FEC redundancy.

3.1.1 Algorithm Principle

Basic Principle of SDM

Figure 3.1 shows the optical fiber communication system adopting SDM amplitude/phase modulation (SDM-QAM) format. The processing of SDM mainly consists of FEC coding, digital interleaving, and SDM-QAM modulation based on many-to-one (MTO) mapping.

The data source $u = (u_1, u_2 \cdots u_i)$, $u_i \in \{0, 1\}$, firstly goes through the LDPC based FEC encoder to provide the mathematical mechanism for the whole bit codewords.

$$c = u \cdot G \tag{3.1}$$

where G indicates the generation matrix of LDPC following the 5G-NR standard. After digital interleaving, the data stream $v = (v_1, v_2 \cdots v_i)$ is divided into u parallel data streams to form the original 128QAM symbols.

$$s_u(t) = \gamma(v_t^{7u}, v_t^{7u+1}, \mathrm{L} v_t^{7u+6}) \quad (u = 0, 1, \mathrm{L}\frac{n}{7} - 1) \tag{3.2}$$

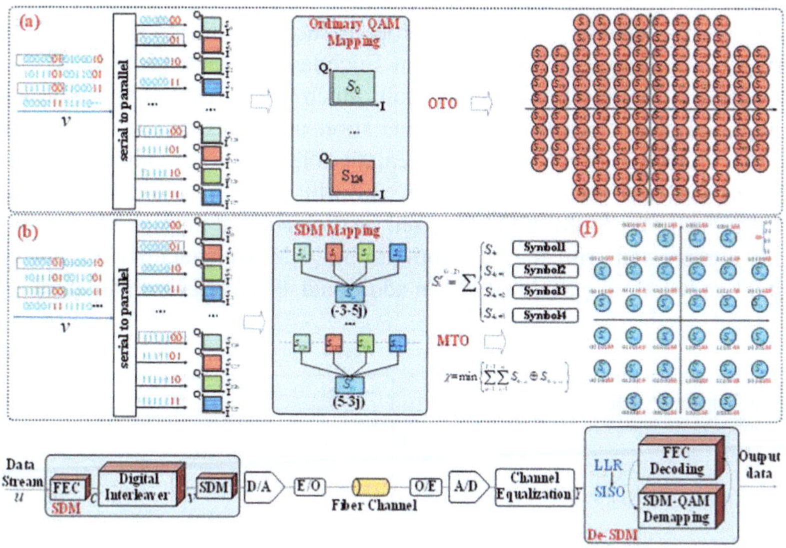

Fig. 3.1 Optical fiber communication system. $2^N (N = 7)$ parallel data streams are interpreted and coded as (**a**) Classical 128QAM, (**b**) SDM-32QAM with four QAM symbols multiplexing per one complex constellation point. *OTO* one-to-one mapping, *MTO* many-to-one mapping, *SDM/De-SDM* symbol division multiplexing/de-multiplexing, *DAC* digital-to-analog conversion, *ADC* analog-to-digital conversion, *E/O* electrical-to-optical conversion, *O/E* optical-to-electrical conversiSon

where $\gamma(\cdot)$ represents the Gary mapping rule, n denotes the length of codewords. Exemplified by the $2^N(N = 7)$ bit streams shown as Fig. 3.1a, there are 128 complex valued symbols necessary to fully carry the 2^7 bit information. Then, every four $s_u(t)$ are multiplexed and modulated per single complex valued symbol $S_i'(t)$ sequentially.

$$S_i'(t) \overset{(c,\chi)}{=} \sum_{u=4i}^{4i+3} s_u(t) = A_i(t)e^{j\theta_i(t)} \quad (i = 0, 1\text{L}\frac{1}{4} \times \frac{n}{7} - 1) \tag{3.3}$$

where $A_i(t)$ and $\theta_i(t)$ represent the amplitude and phase components of SDM-32QAM, respectively. And (c, χ) represents the joint source channel coding/ QAM modulation. As shown as the inset (I) in Fig. 3.1a, the x is the mapping rule based on minimum Hamming distance (HD) in both intra- and inter- SDM symbols

$$\begin{aligned} \chi &= \min\left\{\sum_{u=1}^{T-1}\sum_{i=1}^{u} s_{4-u} \oplus s_{4-u+i}\right\} \\ &= \min\{s_3 \oplus s_4 + s_2 \oplus s_3 + s_2 \oplus s_4 + s_1 \oplus s_2 + s_1 \oplus s_3 + s_1 \oplus s_4\} \end{aligned} \tag{3.4}$$

T = 4indicates that the number of the multiplexed symbols, and i and u are used for counting exclusively. By doing so, the 128QAM is restructured as a SDM-32QAM profile, enabling the QAM order reduction from 2^7 to 2^5. Due to the quadruple enlargement of the minimum Euclidean distance (ED) between inter-symbols, the SDM-32QAM benefits from much lower resolution requirement of optical/electrical devices for signal generation and transmission compared with 128QAM, as well as much greater systematic tolerance to the impairment of fiber dispersion and Kerr nonlinearity. Figure 3.2 illustrates the theoretical analysis of channel capacity for SDM-32QAM based on the maximal mutual information (MI) [28]. In comparison with the conventional 32QAM with identical signal symbol rate and bandwidth, the SDM offers an additional 40% (N is with an extra 2 out of

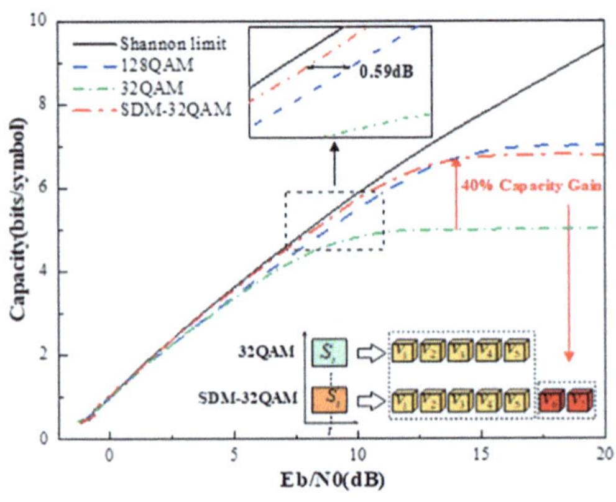

Fig. 3.2 Theoretical analysis of channel capacity with the same symbol rate

5) gain for symbol capacity. In the meantime, there is an additional 0.59-dB gain obtained by SDM in the lower SNR/power region compared with the 128QAM.

For clarity, the applied LDPC for FEC follows the 5th Generation (5G) standard [29] with the code rate of 2/3 and check metric of (1248, 2288). The moderate short coding length introduce an acceptable processing delay and corresponding low latency. It is interesting to note that the FEC plays a key role in De-SDM in two aspects, (1) providing higher signal transmission reliability in optical fiber channels [33]; (2) establishing a strong mechanism between the bit-segment in both intra- and inter- SDM-QAM symbols, so as to theoretically support the iterative decoding of De-SDM.

Basic Principle of De-SDM

After a digital-to-analog conversion (DAC), the QAM signals are subsequently modulated onto an optical carrier for electrical-to-optical (E/O) conversion. The optical signal is transmitted over a fiber link before the optical-to-electrical conversion (O/E) and digital channel equalization processes at the receiver. The received symbols can be depicted as:

$$S_{SDM}^{Rx}(t) = h(t)S_i^{'}(t) + n(t) \tag{3.5}$$

where h(t) denotes the channel response of the optical fiber channel, comprised of chromatic dispersion and Kerr nonlinearity, and n(t) represents the noise components. After the digital channel equalization, the equalized signal $S_{SDM}^{Rx}(t)$ is processed by jointly iterative De-SDM and FEC decoding based on bit-interleaved coded modulation and iterative decoding (BICM-ID). As shown in Fig. 3.3, the overall process of De-SDM can be represented by

$$S_{De-SDM}(t) = S_{SDM}^{Rx} {}^{-Eq.}(t) \cdot \psi^{-1}(\lambda, LLR) \tag{3.6}$$

where the function ψ^{-1} represents the opposite process of SDM, LLR denotes the priori information of symbol decision, and λ represents the soft decision based on minimal Euclidean distance (ED) with the ideal SDM-32QAM symbols $S_i^{'}$

$$\lambda = \min \left\| S_{SDM}^{Rx} {}^{-Eq.} - \varsigma l_t s \right\|^2, s \in S_i' \quad i = 0, 1, \mathrm{K}\ 31 \tag{3.7}$$

Where $\varsigma l_t = 1$ in the intensity modulation direct detection (IM/DD) system. Multi-form LLR information exchanges between the FEC decoder and SDM-QAM demapper to realize the De-SDM. Based on the maximum posterior probability (MAP) algorithm, the output LLR for each bit of symbol $S_{De-SDM} =$

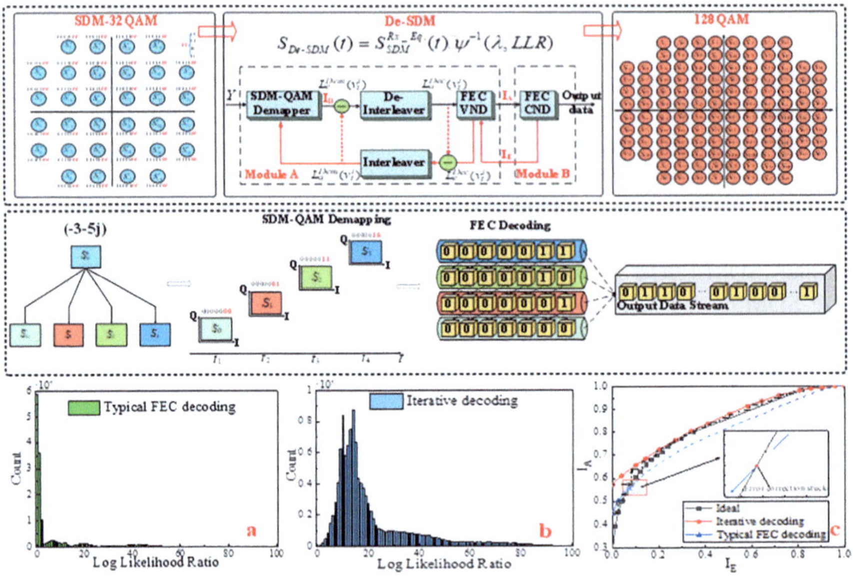

Fig. 3.3 Block diagram of the De-SDM.FEC decoding reliability performance for the inspected SDM-32QAM by (**a**) typical FEC decoding and (**b**) iterative decoding, (**c**) mutual information by various decoding methods. *Dem* SDM-QAM demapping, *Dec* FEC decoding

$\gamma(v_t^0, v_t^1, Lv_t^6)$ is calculated as the original standard 128QAM by iterative decoding

$$LLR(v_i^t) = \log \frac{p(v_i^t = 0 \mid S_{SDM}^{Rx} -^{Eq.} (t))}{p(v_i^t = 1 \mid S_{SDM}^{Rx} -^{Eq.} (t))}$$

$$= \log \frac{\sum_{s\in S_t^{v_t^i=0}} \exp\left(\frac{-\left(s_{SDM}^{Rx} -^{Eq.} - s\right)^2}{2\sigma^2}\right) p(S_{SDM}^{Rx} -^{Eq.})}{\sum_{s\in S_t^{v_t^i=1}} \exp\left(\frac{-\left(s_{SDM}^{Rx} -^{Eq.} - s\right)^2}{2\sigma^2}\right) p(S_{SDM}^{Rx} -^{Eq.})} \quad (3.8)$$

where $s_t^{v_t^i=0}$ and $s_t^{v_t^i=1}$ refer to the coded bits at time slot t with the label value $v_t^i = 0$ and $v_t^i = 1$ respectively. The $p(S_{De-SDM})$ denotes the probability of De-SDM symbol, which is refreshed along with the updated prior information $L_a^{Dem}(v_t^i)$ for SDM-QAM demapping

$$p(v_t^i = 1) = 1 - p(v_t^i = 0) = \frac{e^{L_a^{Dem}(v_t^i)}}{1 + e^{L_a^{Dem}(v_t^i)}} \quad i = 0, 1, \text{K}\ 6 \quad (3.9)$$

$$p(s_t^u) = \prod_{i=0}^{6} p(v_t^i) \quad i = 0, 1, \text{K}\ 6 \quad (3.10)$$

The extrinsic information $L_e^{Dem}(v_t^i)$ obtained from SDM-QAM demapper can be obtained by

$$L_e^{Dem}(v_t^i) = \log \frac{p(v_i^t = 0 \mid S_{SDM}^{Rx} - ^{Eq.}(t))}{p(v_i^t = 1 \mid S_{SDM}^{Rx} - ^{Eq.}(t))} - L_a^{Dem}(v_t^i)$$
$$\times (t = 1, \mathrm{L}T; i = 0, 1, \mathrm{L}6) \tag{3.11}$$

The initial value of $L_a^{Dem}(v_t^i)$ is set as zero in the first external iteration. As the feedback to the FEC VND, a prior information $L_a^{Dec}(v_t^i)$ is equal to the external information $L_a^{Dem}(v_t^i)$ after de-interleaving. The SISO information iteratively transferring between FEC variable node decoder (VND) and FEC check node decoder (CND) is named as "inner iteration". By re-operating the soft information outputted from FEC VND, the updated prior information $L_a^{Dec}(v_t^i)$ can be obtained for the next "external iteration", defined for the transferring SISO information in Module A.

The joint iteration processing between the FEC decoder and the SDM-QAM demapper to obtain more reliable LLRs will not be terminated until the output comes to certain convergence. The feedback LLR is obtained by typical sum-product algorithm (SPA) and exchanged via independent bit interleavers and de-interleavers to effectively improve the transfer reliability. The property of the LLR value enables the key parameter to illustrate the De-SDM performance. Figure 3.3a and b shows the number and value of LLR counting for SDM-32QAM signals over 20-km SSMF transmission. We can see that there are an enormous amount of low-valued LLR in conventional FEC decoding, which means the failed decoding for SDM symbols. The joint iteration contributes to optimal/larger-valued LLR, which guarantee the reliability of the simultaneous FEC decoder and De-SDM.

We introduce the mutual information (MI) to illustrate the validity and the convergence performance for De-SDM. Initially, the MI for SDM-QAM demapper is defined as

$$I_D = I(v, D) = \frac{1}{2} \sum_{v_t^i} \int_{-\infty}^{+\infty} p_D(\xi \mid v_t^i) \cdot \log_2 \frac{2 p_D(\xi \mid v_t^i)}{p_D(\xi \mid 0) + p_D(\xi \mid 1)} d\xi \tag{3.12}$$

The MI for module A, consisting of VND and SDM-QAM demapper, is defined as

$$I_A = \sum_i \mu_i J(\sqrt{(i-1)(J^{-1}(I_E)^2) + J^{-1}(I_D)^2}) \tag{3.13}$$

Similarly, the MI for module B (FEC CND) is denoted as

$$I_E = 1 - \sum_i \varepsilon_i J(\sqrt{(i-1)} J^{-1}(1 - I_A)) \tag{3.14}$$

where the function J(ψ) is defined as

$$J(\psi) = \int_{-\infty}^{\infty} \frac{1}{\sqrt{2\pi\psi}} e^{\frac{(x-\frac{\psi^2}{2})^2}{2\psi^2}} \log \frac{2}{1+e^{-x}} dx \tag{3.15}$$

and $\varepsilon(\cdots)$, $\mu(\cdot)$ are the LDPC degree distributions in FEC VND and CND, respectively.

As shown in Fig. 3.3c, the MI by jointly iterative decoding is approaching the theoretical (ideal) FEC decoding convergence due to the successful De-SDM. For the error correction stuck, it means the failure of decoding by using the typical FEC scheme at the beginning of the De-SDM procedure due to the ambiguous symbols.

3.1.2 Experiments Setup and Results

The experimental setup for optical fiber communication system employing SDM-32QAM discrete multi-tone (DMT) signal over 20-km SSMF transmission is illustrated in Fig. 3.4. An arbitrary waveform generator (AWG) with 50-GSa/s sampling rate is used to generate the electrical SDM-32QAM DMT signal The DMT modulation process comprises N-point FFT for DFT-spread processing, M-point IFFT, and cyclic prefix (CP) adding [32, 33]. The SDM is performed in time domain by means of transmitting-side digital signal processing (DSP). A 40 GHz intensity Mach-Zehnder modulator (MZM), driven by the power amplified DMT signal, is used to modulate the continuous light-wave generated by an external cavity laser (ECL). After a 20-km SSMF transmission link with a nonlinear Kerr coefficient of $1.27\,\mathrm{W}^{-1}\,\mathrm{km}^{-1}$ and a dispersion coefficient of $17\,\mathrm{ps}\cdot\mathrm{nm}^{-1}\,\mathrm{km}^{-1}$, the optical-to-

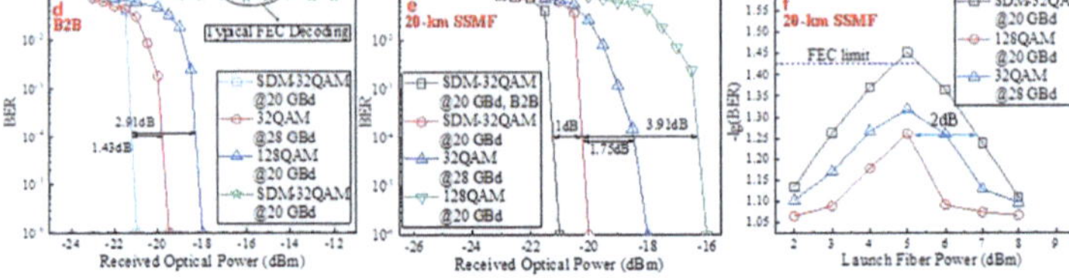

Fig. 3.4 Experimental setup. (**a**), (**b**), (**c**): Constellation diagrams for the inspected signals with the same effective data rate, (**d**), (**e**), (**f**): BER measurements for B2B, 20-km SSMF transmission and fiber nonlinear effect cases

electrical (O/E) conversion is implemented by a 20-GHz photodetector (PD) with integrated pre-amplifier (TIA). The detected signals are captured by a digital storage oscilloscope (OSC) with 50-GSa/s sampling rate, and subsequently processed via offline DSP comprised of synchronization, resampling, DMT demodulation, De-SDM, FEC decoding, and final BER calculation.

Linear BER Performance

We firstly focus on signal impairment by optical chromatic dispersion and noise (relative intensity noise and amplified spontaneous emission noise). The BER performance of the optical SDM-32QAM DMT is shown in Fig. 3.4d and e. For a fair comparison, i.e. equivalent effective data rate, the BER performance of optical DMT systems employing 20Gbaud 128QAM and 28Gbaud 32QAM are also measured under the same physical link and the identical FEC code rate of 2/3. Figure 3.4d refers to the back-to-back (B2B) BER performance. Relative to SDM-32QAM at the post-FEC BER of $1\times10\text{-}5$, the 2.91 dB power penalty for 128QAM mainly comes from the impairment by the resolution limitation of the EA and the pre-amplifier (TIA) commonly integrated in PD. External pre-amplifier receivers with high enough resolution or low-noise avalanche photodiode (APD) may be utilized to decrease the power penalty. While the power penalty for 32QAM is 1.43 dB due to the high-frequency fading effect caused by DAC/ADC with limited bandwidth.

As is predicted for the SDM-32QAM DMT by typical FEC decoding, there exists a BER floor due to the absence of the De-SDM. Figure 3.4e shows the BER performance over 20-km SSMF link at the launch fiber power of 3 dBm. The improvement by SDM for power penalties caused by chromatic dispersion and Kerr nonlinear effect are 1.75 and 3.91 dB compared with 32QAM and 128QAM DMT signals, respectively.

Fiber Kerr Nonlinear Transmission Performance

It is well known that the fiber-optic communication channel might not be simply emulated with AWGN mode. For both the power-constrained optical access networks and the long-distance multi-span transmission systems, the non-negligible portion of signal distortion caused by fiber nonlinearity is an essential factor, which is addressed partially by the super-Gaussian distribution scheme but leaves huge space for improvements. We investigate the limitations imposed by Kerr nonlinearity in the optical SDM-32QAM DMT system. As shown in Fig. 3.4f, the BER performance of various optical channels as a function of launch fiber power is measured with a fixed received optical power of -19 dBm. All the cases are involved in a 20-km SSMF transmission with a nonlinearity coefficient of $1.27\,\text{km}^{-1}\text{W}^{-1}$. The curves represent three kinds of modulation formats with the equivalent effective data rate. As seen from the figure, the BER improves with the launched fiber power until it reaches the optimal value at the launch fiber power of 5 dBm. After that, it begins to deteriorate with the growing nonlinear distortion caused by the Kerr effect. Similar BER variation tendency can be observed in SDM-32QAM and 32QAM DMT as well, however, the BER for 128QAM shows a sharp deterioration. While the SDM scheme shows a 2-dB launch fiber power optimization in terms of a superior systematic tolerance to the fiber Kerr nonlinearity due to the QAM-order reduction.

3.1.3 Application Complexity

For SDM, the essential process of the MTO mapping is similar to the traditional Gray mapping, which can be easily implemented via the functions of AND, OR, and exclusive OR circuits of look-up tables (LUTs) with minor complexity penalty. The key challenge for the application of De-SDM is the implementation complexity and power consumption by joint feedback iteration for FEC decoding and SDM-QAM demapping, which is primarily based on traditional belief propagation (BP) algorithm [34]. While the major system complexity load in BP comes from the hyperbolic tangent function of iterative SPA in commercial FEC [35]. However, as an extra feedback section in BICM-ID system, the complexity for the separated and softly updated extrinsic information Le based on the MAP algorithm can be logically neglected compared with that of SPA. Furthermore, the iterative Le by SDM-QAM demapping can help to provide much more accurate LLRs compared with that in conventional BICM system, thus resulting in the optimized and reduced iteration number for FEC decoding.

3.2 Photonic-aided W-band Dual-vector RF Signal Generation and Detection Enabled by Bandpass Delta-sigma Modulation and Heterodyne Detection

The mobile communication is expanding from the lower-frequency microwave band to the higher-frequency millimeter-wave (mm-wave) band and even Terahertz band to carry the ever-increasing data traffic [36–40]. The higher-frequency mm-wave band not only has inherent large bandwidth, but also provides the advantage of easily realizing the seamless integration of fiber-optic communication and wireless communication. However, the wireless communication link at the mm-wave band is typically power-limited due to large path loss. Therefore, it can only cope with the relatively simple and low-spectral-efficiency data modulation, such as on-off-keying (OOK) and quadrature-phase-shift-keying (QPSK), but it is very difficult to directly deliver high-spectral-efficiency high-level quadrature-amplitude-modulation (QAM) signals, which have high signal-noise-ratio (SNR) and receiver sensitivity requirements. Recently, the technique of the delta-sigma modulation (DSM) has been intensively studied for its application in the 5G/6G mobile fronthaul [41–44]. Based on oversampling and noise shaping, the DSM technique has the capability of digitizing the relatively complicated waveform into a simple waveform with only 2 or 4 levels, which is more compatible with the power-limited and receiver-sensitivity-limited transmission systems; Moreover, the original waveform can be easily retrieved at the receiver side by a filter [41–44]. Therefore, it is interesting to investigate the introduction of the DSM technique into a wireless mm-wave transmission system to overcome the insufficient power and receiver sensitivity problems and thus to realize high-level QAM signal transmission.

On the other hand, the existing photonic-aided dual radio-frequency (RF) signal generation, wireless transmission, and detection typically need multiple wireless transmitters as well as the wireless multiple-input multiple-output (MIMO) link, which significantly increases the system complexity [45]. Moreover, some of the existing photonic-aided dual RF signal generation and detection schemes based on dual single-sideband (dual SSB) modulation require optical bandpass filtering for the separation of two sideband signals and two parallel photodiodes (PDs) for optical-to-electrical conversion [45–47]. Other schemes based on dual SSB modulation removes the requirement of optical bandpass filtering with the aid of digital signal processing (DSP) algorithms [47–49], but they can only realize dual RF signal generation and detection with relatively low-level modulation, such as QPSK. Furthermore, these schemes [45–50] can only generate dual RF signals with relatively low carrier frequencies at microwave band or lower mm-wave band no more than 40 GHz, due to the limited optical in-phase/quadrature (I/Q) modulator bandwidth.

3.2.1 Proposed Scheme and Principle

In the Letter, we propose a photonic-aided dual vector RF signal generation and detection scheme enabled by band-pass delta sigma modulation and heterodyne detection. With the aid of the band-pass delta sigma modulation, our proposed scheme is transparent to the modulation format of the dual vector RF signals and can support the generation, wireless transmission, and detection of both single-carrier (SC) and orthogonal-frequency-division-multiplexing (OFDM) vector RF signals with high-level QAM modulation. With the aid of the heterodyne detection, our proposed scheme can support up to W-band dual vector RF signal generation and detection. Based on our proposed scheme, we experimentally demonstrate the simultaneous generation of a SC-64QAM signal at 94.5 GHz and a SC-128QAM signal at 93.5 GHz and their error-free transmission over 20-km single-mode fiber-28 (SMF-28) and 1-m single-input single-output (SISO) wireless link at W-band.

Figure 3.5 shows the principle of our proposed photonic-aided dual-vector RF signal generation scheme enabled by optical QPSK modulation and bandpass delta-

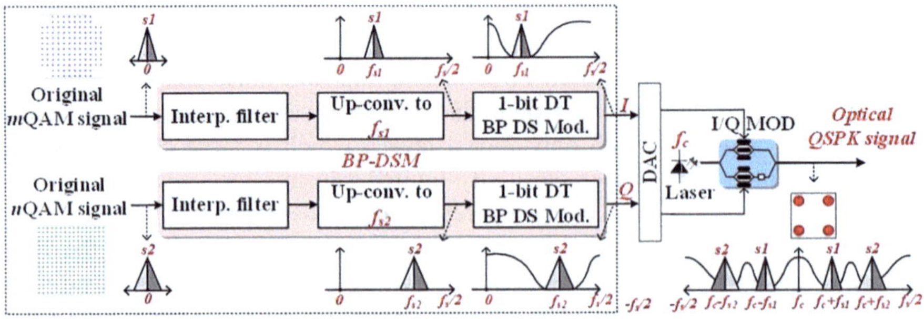

Fig. 3.5 Principle of photonic-aided dual-vector RF signal generation enabled by optical QPSK modulation and bandpass delta-sigma modulation

sigma modulation. At the transmitter side, two independent baseband complex signals with identical or different QAM modulation formats (denoted the m QAM signal and the n QAM signal, respectively) are generated with the aid of DSP. Here, both m and n are integers, and their values are equal to 2k (k = 2, 3, 4, 5, ...). Each baseband complex signal can employ SC modulation or OFDM modulation. That is, both baseband complex signals can be the SC ones or the OFDM ones, or one baseband complex signal is the SC one while the other is the OFDM one. Then, the two baseband complex signals are simultaneously sent into two bandpass delta-sigma modulators (BP DSMs) in parallel. Each BP DSM is composed of an interpolation filter, a digital upconverter, and a 1-bit discrete-time BP DSM. Here, the interpolation filter essentially implements an up-sampling operation with an oversampling ratio (OSR) factor that matches the OSR setting of the following discrete-time BP DSM, as well as a subsequent anti-image filtering operation to suppress the image induced by up-sampling. For the mQAM signal, the digital upconverter implements the up conversion from the baseband to the carrier frequency of f_{s1}. For the nQAM signal, the digital upconverter implements the up conversion from the baseband to the carrier frequency of f_{s2}. f_{s1} and f_{s2} can be identical or different. The discrete-time BP DSM employs 1-bit quantization and finally converts the mQAM signal at a carrier frequency of f_{s1} or the nQAM signal at a carrier frequency of f_{s2} into a digital sequence including levels of +1 and -1 only, i.e., an OOK signal. In terms of its physical interpretation, the BP DSM pushes the in-band noise out of band, and out-of-band shaped noise is therefore induced. The two generated OOK signals are uploaded into a digital-to-analog converter (DAC) and then used to drive an I/Q modulator to implement optical QPSK modulation. An optical continuous-wavelength (CW) light wave from a laser with an operating frequency of f_c is fed into the optical input of the I/Q modulator. Finally, the I/Q modulator outputs an optical QPSK signal including both the m QAM signal at a carrier frequency of $f_c \pm f_{s1}$ and the nQAM signal at a carrier frequency of $f_c \pm f_{s2}$.

Figure 3.6 shows the principle of our proposed photonic aided dual-vector RF signal transmission and detection scheme enabled by heterodyne detection and

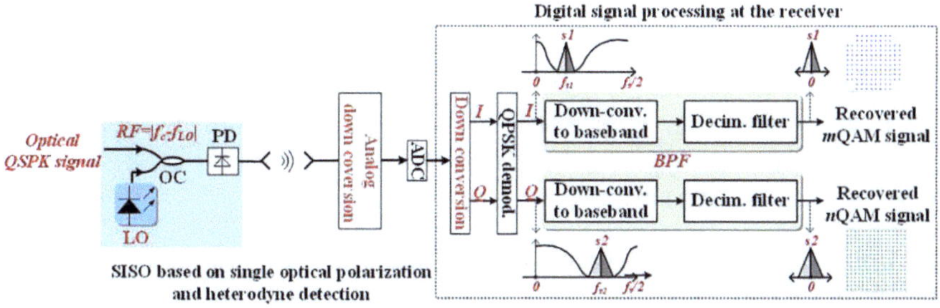

Fig. 3.6 Principle of photonic-aided dual-vector RF signal transmission and detection enabled by heterodyne detection and digital bandpass filtering

digital bandpass filtering. First, the optical QPSK signal generated by the system shown in Fig. 3.5 is converted into an electrical QPSK-modulated mm-wave signal by optical heterodyne detection. That is, the optical QPSK signal is coupled with an optical local oscillator (LO) signal by an optical coupler (OC) and then sent into a high-speed PD for optical-to-electrical conversion. The frequency spacing between the optical QPSK signal and the optical LO signal is the carrier frequency of the converted electrical mm-wave signal. At the receiver side, the received mm-wave signal is first down converted into an electrical intermediate-frequency (IF) signal with a lower carrier frequency by analog down conversion. Then, the electrical IF signal is captured by an analog-to-digital converter (ADC) before being processed by the receiver DSP. During the receiver DSP, after digital down conversion from IF to the baseband and QPSK demodulation, the two transmitted OOK signals are recovered from the received IF signal and are then simultaneously sent into two bandpass filters (BPFs) in parallel. Each BPF includes a digital downconverter and a decimation filter. For the I-component OOK signal, the digital downconverter implements the down conversion from the carrier frequency of f_{s1} to the baseband. For the Q-component OOK signal, the digital downconverter implements the down conversion from the carrier frequency of f_{s2} to the baseband. The interpolation filter essentially implements an anti-aliasing operation as well as a subsequent down-sampling operation with a factor of OSR. In terms of its physical interpretation, the interpolation filter, which performs low-pass filtering (LPF), effectively suppresses the out-of-band shaped noise while reserving the original mQAM or nQAM signal. That is, after the two parallel BPFs, the original m QAM signal and the original m QAM signal are recovered at the same time. The following experimental demonstration verifies the simultaneous generation, fiber-wireless-integration transmission, and detection of the SC-64QAM signal at 94.5 GHz and the SC-128QAM signal at 93.5 GHz. The OSR setting for both the SC-64QAM and the SC-128QAM signals is 32. It is worth noting that the generation and detection of a dual-vector RF signal employing some other higher-level QAM modulation or OFDM modulation is also applicable in our experimental system.

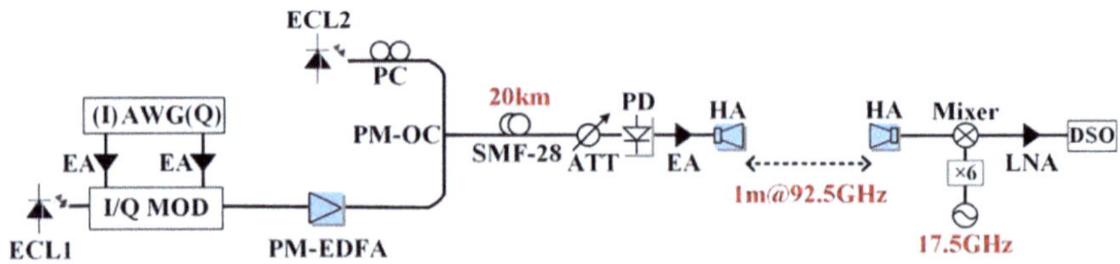

Fig. 3.7 Experimental setup of photonic-aided W-band dual vector signal generation, wireless transmission, and detection

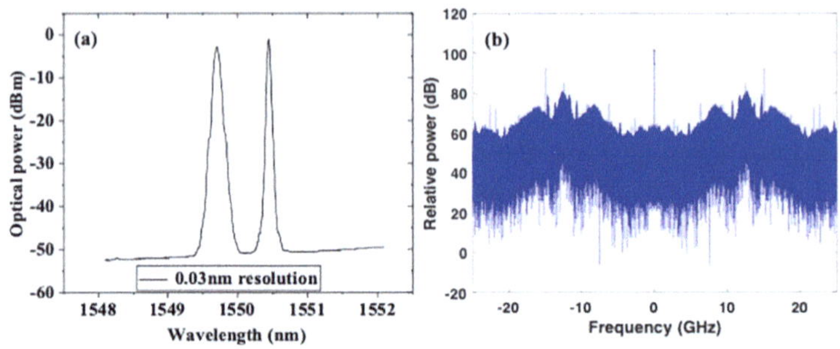

Fig. 3.8 (**a**) Measured output optical spectra of the I/Q modulator. (**b**) Measured electrical spectra captured by the DSO

3.2.2 Experiments Setup and Results

Figure 3.7 shows the experimental setup of the photonic-aided W-band dual vector signal generation, wireless transmission, and detection based on our proposed scheme. At the transmitter side, with the aid of the DSP, a SC-64QAM signal and a SC-128QAM signal, each with a baud rate of 156.2Mbaud, are separately generated and then digitized into two OOK signals, each with a baud rate of 10Gbaud, by the two BP DSMs in parallel. The detailed DSP procedures at the transmitter side will be introduced later. Then, the two 10-Gbaud OOK signals are simultaneously uploaded into a 10-GSa/s arbitrary waveform generator (AWG) for digital-to-analog conversion, then boosted by two parallel electrical amplifiers (EAs), and finally used to drive a I/Q modulator. The I/Q modulator has an optical bandwidth of about 30 GHz and an insertion loss of about 7 dB. The optical input of the I/Q modulator is a CW lightwave from an external cavity laser (denote by ECL 1 in Fig. 3.7). The ECL 1 has a linewidth of <100kHz and an output optical power of 12 dBm. The I/Q modulator output is therefore an optical QPSK signal, including both SC-64QAM and SC-128QAM data information.

Figure 3.8a shows the measured optical spectrum at 0.03-nm resolution of the generated optical QPSK signal. After that, the generated optical QPSK signal is boosted by a polarization-maintaining Erbium-doped fiber amplifier (PM-EDFA), and then coupled by a polarization-maintaining OC (PM-OC) with another CW lightwave from ECL2. The ECL2, with a linewidth of <100kHz and an output optical power of 12 dBm, functions as an optical LO source here. A polarization controller (PC) is used after the ECL2 to manually align the polarization state of the optical LO signal with that of the optical QPSK signal. The frequency spacing between ECL 1 and ECL 2 is 92.5 GHz, and therefore the coupled optical QPSK signal and optical LO signal after the PM-OC can be considered as an optical mm-wave signal at 92.5 GHz. The optical mm-wave signal is delivered over 20-km SMF-28 with a chromatic dispersion (CD) coefficient of 17 ps/km/nm at 1550 nm. After passing through an optical tunable attenuator (ATT), the optical mm-wave signal is sent into a high-speed PD with 100-GHz 3-dB optical bandwidth and 0.6-A/W responsivity. Here, the ATT is used to assist the bit-error ratio (BER) measurement by adjusting the input optical power into the PD. The 100-GHz PD converts the optical mm-wave signal into the 92.5-GHz electrical mm-wave signal, which is then boosted by a W-band EA with a gain of 35 dB and radiated by a W-band dual-polarized lens-corrected antenna with a gain of 25 dBi into a 1-m SISO wireless link at W-band. It is worth noting that only one polarization of the dual-polarized antenna is on in this experiment.

After 1-m SISO wireless transmission, the 92.5-GHz electrical mm-wave signal is received by another identical dual-polarized lens-corrected antenna with a gain of 25 dBi. Similarly, only one polarization of the receiver dual-polarized antenna is on in this experiment. After analog down conversion enabled by a 17.5-GHz LO sinusoidal source, a×6 frequency multiplier, and a mixer, the received 92.5-GHz electrical mm-wave is down-converted into a (17.5×6−92.5) =12.5-GHz IF signal. The down-converted 12.5-GHz IF signal is boosted by a low-noise amplifier with a gain of 25 dB, and then captured by a 50-GSa/s digital storage oscilloscope (DSO) with 23-GHz bandwidth for further offline DSP. Figure 3.4b gives the captured 12.5-GHz IF signal spectra by the DSO. With the aid of the offline DSP, the original SC-64QAM signal and SC-128QAM signal can be recovered from the 12.5-GHz IF signal. The detailed DSP procedures at the receiver side will also be introduced later.

Figure 3.9a and b show the detailed DSP procedures at the transmitter and receiver sides, respectively. During the transmitter-based DSP in Fig. 3.7a, a SC-64QAM signal and a SC-128QAM signal, each with a baud rate of 156.2 Mbaud, are simultaneously generated after the procedures of pseudo-random binary sequence (PRBS) generation, 64QAM or 128QAM mapping, and 2× up-sampling. Then, with the aid of two parallel BP-DSMs, the SC-64QAM and SC-128QAM signals are simultaneously converted into two 10-Gbaud OOK signals. The two 10-Gbaud OOK signals can be considered as the I and Q components of a 10-Gbaud QPSK signal, respectively. The OSR and sampling rate for each BP-DSM are 32 and 10 GSa/s, respectively. Moreover, each BP-DSM employs the fourth-order one-bit-quantization structure. The BP-DSM for the 64QAM signal has a center frequency of 2 GHz, while that for the 128QAM signal has a center frequency of 1 GHz. For the receiver-based DSP in Fig. 3.7b, the detailed procedures include QPSK demodulation, BPF, and 64/128QAM demodulation. The QPSK demodulation includes down conversion from 12.5 GHz to the baseband, resampling,

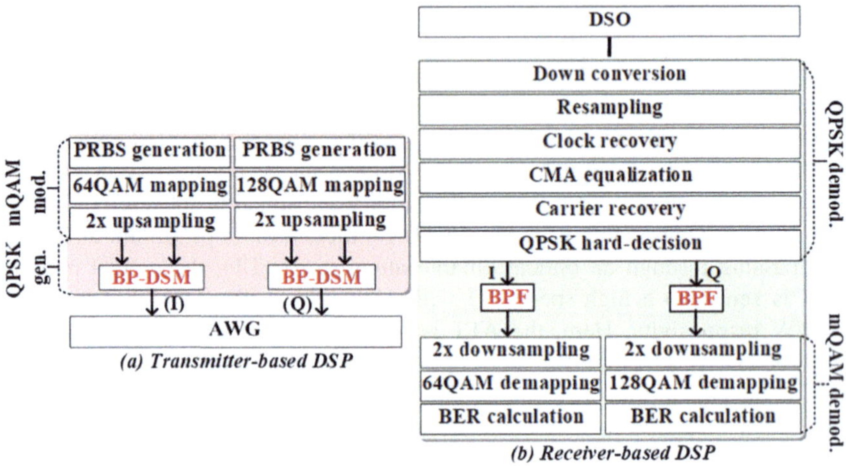

Fig. 3.9 (**a**) Transmitter-based DSP. (**b**) Receiver-based DSP

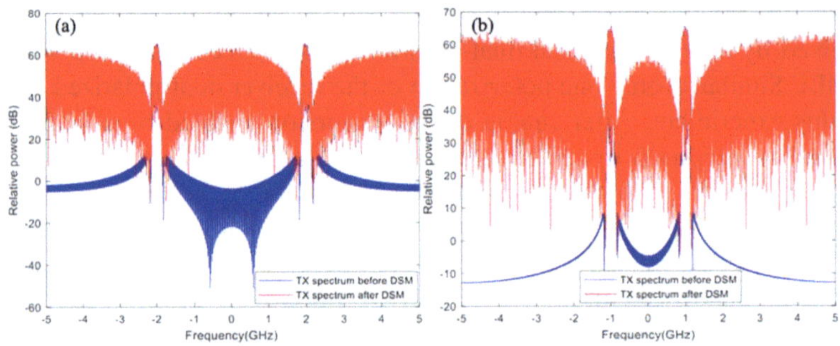

Fig. 3.10 (**a**) SC-64QAM signal spectrum before and after BP-DSM at 2 GHz. (**b**) SC-128QAM signal spectrum before and after BPDSM at 1 GHz

clock recovery, constant-modulus-algorithm (CMA) equalization, carrier recovery (including frequency offset estimation and carrier phase estimation), and QPSK hard decision. After the QPSK hard decision, the original two 10-Gbaud OOK signals are recovered. After two BPFs in parallel, the original SC-64QAM and SC-128QAM signals, each with a baud rate of 156.2 Mbaud, are simultaneously recovered. The 64/128QAM demodulation includes 2× down-sampling, 64/128QAM demapping, and BER calculation.

Figure 3.10a and b show the SC-64QAM and SC-128QAM signal spectra before and after the BP-DSM centered at 2 and 1 GHz, respectively. After the BP-DSM, the noise is pushed out of the signal band while the original QAM signal is well reserved. Figure 3.11a–d give the recovered constellations after carrier recovery, hard decision, I- and Q-branch BPF, respectively. Both recovered 64QAM and 128QAM constellations are error free when the input power into the PD is −5 dBm after 20-km SMF-28 and 1-m wireless transmission. The hard decision, which

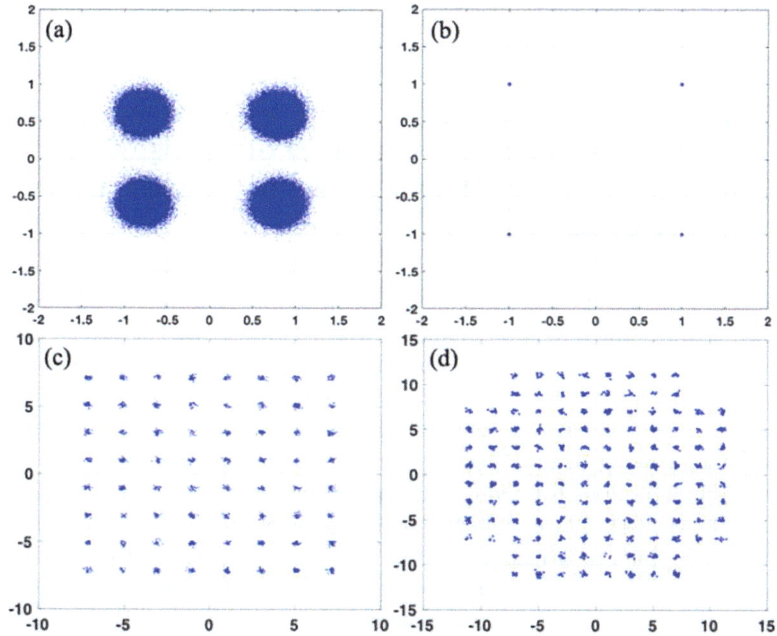

Fig. 3.11 Recovered constellations after (**a**) carrier recovery, (**b**) hard decision, (**c**) I-branch BPF, and (**d**) Q-branch BPF

Fig. 3.12 Measured BER versus input power into the PD

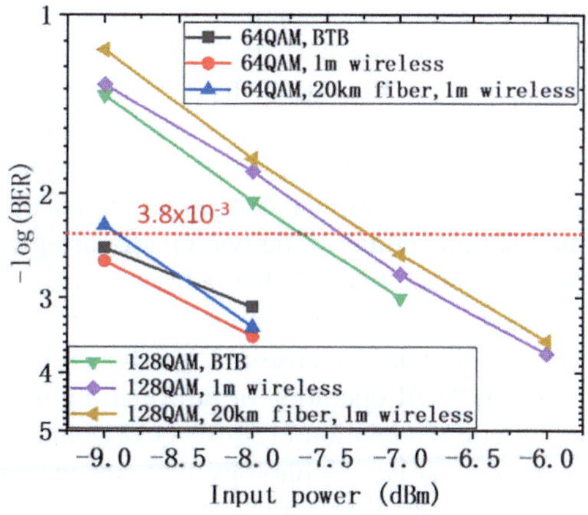

eliminates the phase noise and converges the constellation points to the greatest extent, significantly improves the receiver sensitivity. Figure 3.12 shows the BER versus the input power into the PD in the scenarios of BTB, 1 m of wireless transmission, 20 km SMF-28 and 1 m of wireless transmission, respectively. The CMA tap number is 21, 27, and 33 for the three scenarios, respectively. The

BER can reach the hard-decision forward-error-correction (HD-FEC) threshold of 3.8×10^{-3} for all three scenarios.

3.3 Design of Probabilistic Shaping 4D Ultra High Order Modulation Format With 8APSK Pilot Aided Carrier Phase Recovery

With the rapid development of mobile communication networks, live video broadcast, telemedicine, and other communication services, the internet traffic is growing rapidly for center interconnection. Consequently, the development of high-capacity optical communication is necessary. The technology of ultra-high order modulation (UHM) is used to increase the spectral efficiency (SE) toward the Shannon limit and meet the growing demand for high transmission capacity. Many experiments for ultra-high order signals on single-carrier have been demonstrated [51–60]. The first 1024 QAM single carrier coherent optical transmission is demonstrated in [60], in which a 60 Gbit/s polarization-multiplexed signal at a SE of 13.8 bit/s/Hz was transmitted over 150 km [51]. Similarly, probabilistically shaped 4096-QAM has been transmitted over 50.9 km at a potential SE of 15.9 bits/s/Hz and a net bit rate of 484.4 Gb/s [59]. So far, the highest-order QAM constellations that have been demonstrated in optical systems are 10-GBd probabilistically shaped square 16384-QAM [60] at a net bit rate of 223.8 Gb/s. However, as the modulation format order increases, the minimum Euclidean distance decreases, and thus the ability to resist noise declines. Therefore, it is critical to propose a novel modulation format to expand the Euclidean distance of constellation points and optimize the DSP algorithm to improve the system performance.

Therefore, the four-dimensional (4D) modulation format is an advanced optical modulation format that is generated by drawing into the check bit through modulation coding in the two quadrature-phase components and two polarizations of the optical field. It has attracted a lot of attention in recent years because it can increase the minimum Euclidean distance between symbols [61–64]. The 4D modulation format uses Ungerboeck's set partitioning strategy to generate the parity by XOR operation of information bits [61], such as 128-level SP quadrature amplitude modulation (128-SP-QAM) and 2048-SP-QAM.Moreover, probabilistic shaping (PS) is a technology that can increase the mutual information of signals toward the Shannon limit and provide a flexible platform for rate adaptation. PS-based multidimensional signaling schemes have recently been reported [65, 66] to provide higher fiber nonlinear tolerance. The PS scheme, however, is incompatible with the SP-QAM modulation format due to the presence of parity bits.

Besides, the scheme of UHM is more sensitive for channel noise, especially for phase noise. Moreover, the reported phase recovery algorithms are not suitable for the 4D-PS-MQAM signal. It's necessary to develop the advanced carrier phase recovery (CPR) algorithm to improve transmission performance. Many CPR algorithms have been reported, such as blind search algorithm (BPS), quadrature phase-shift keying (QPSK) partitioning, maximum likelihood estimation (BPS/ML), quadratic approximation (BPS/QA) algorithm and Viterbi-Viterbi(V-V) algorithm [67–71]. The BPS algorithm have the mutual information punishment with the PS signal at low signal-to-noise ratio (SNR) [72]. The QPSK partitioning algorithm based on constellation-dependent is inappropriate to phase recovery for high-order modulation format signals. The two-stage BPS algorithm has large computational complexity after introducing a two-stage configuration. The pilot-aided carrier phase recovery (PA-CPR) algorithm has high noise tolerance and low computational complexity. However, the main drawback is the inherent overhead (OH) which will reduce the SE of the system. Therefore, it is necessary to optimize the PA-CPR scheme for PS-4D-SP-1024QAM.

3.3.1 System Model

Principle of PS-4D-SP-1024QAM

Probabilistic amplitude shaping (PAS) is a mainstream standard probabilistic shaping scheme [73]. As shown in Fig. 3.13, it divides the in-phase and quadrature components of the square M-QAM into two independent $\sqrt{m}$ -level pulse amplitude modulation ($\sqrt{m}$-PAM) signals. The PAS structure of each component includes a constant component distribution matcher (CCDM) and a forward error correction (FEC) encoder [74]. The structure of PS-4D-SP-1024QAM is shown in Fig. 3.14, which can maintain probability distribution characteristics after SP encoding. The n bits enter the processing unit of DP-PAS and pass through the SP to perform the amplitude translation in the first stage which generate m symbols and $2m$ binary label bits respectively. In the second stage, the mentioned $2m$ binary label bits are used to generate 2k symbols and 4k label bits in the same process. Finally, the bits coming from the second stage are mapped into 2k/3 PDM-8APSK symbols, which is used as the pilots for CPR in optical fiber communication systems.

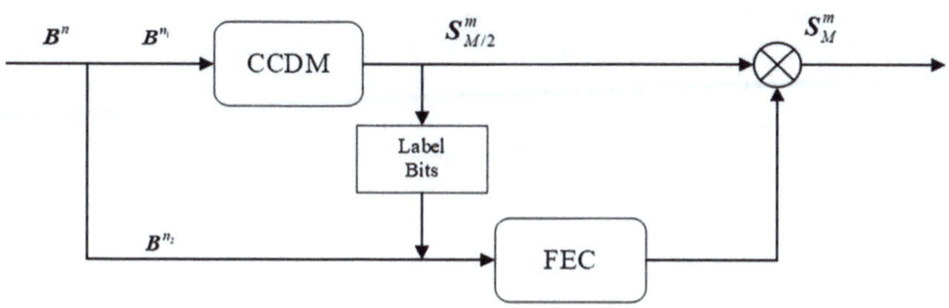

Fig. 3.13 The diagram of PS M-PAM signal generation based on PAS architecture

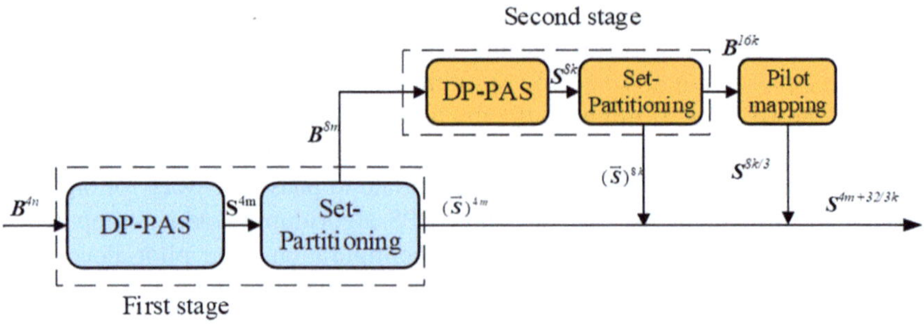

Fig. 3.14 The diagram of PS-4D-SP-1024QAM overall mapping process

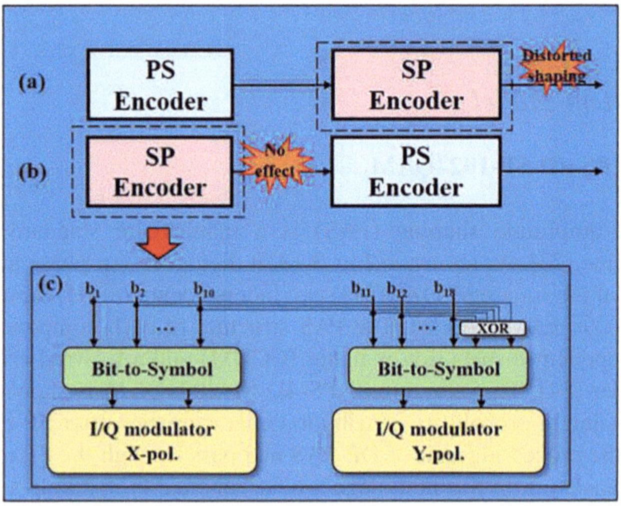

Fig. 3.15 The structure of 4D-SP-1024QAM

Figure 3.15 shows the generation of SP-QAM modulation which is performed by XOR coding on the information bits, cf. Fig. 3.15c. In PS systems, the coding after shaping at the transmitter causes the shaped symbol distribution to be distorted, shown in Fig. 3.15a. On the other hand, XOR coding before shaping at the transmitter, has no practical effect on decoding at the receiver, Fig. 3.15b. In this section, the amplitude translation method is proposed to overcome the incompatibility between SP and PS for ultra-high order constellations.

PS-4D-SP-1024QAM Phase Recovery Scheme

A. Phase Noise

The DSP algorithm in coherent optical communication includes dispersion compensation, clock recovery, polarization demultiplexing, frequency offset estimation, and phase recovery. Except for phase noise, all the channel impairments are compensated in this paper. The received signal at discrete time k can be written as:

$$y_k = x_k e^{j\theta_k} \tag{3.16}$$

where x_k is the transmitted symbol, θ_k is laser phase noise, n_k is AWGN additive white Gaussian noise. The phase noise can be modeled as a Wiener process, and is given by (3.17):

$$\theta_t = \sum_{i=-\infty}^{t} \omega_i \tag{3.17}$$

where ω_i are independent identically distributed Gaussian random variables with zero means and the variance [75]:

$$\sigma_\omega^2 = 2\pi(\Delta f \cdot T_s) \tag{3.18}$$

T_s is the symbol period and Δf is the sum of the linewidths of the Tx laser and the local oscillator.

B. The Principle of Pilot Aided Algorithm

The digital signal processing of the pilot-aided algorithm mainly depends on the known symbols. The frame structure of the signal is shown in Fig. 3.16. It consists of an initial pilot sequence and a periodic payload. The inserted pilot symbols are used for continuous phase tracking and phase estimate. There are p pilot symbols with 8APSK inserted into the N symbols payload sequence at the receiver, and the insertion ratio is p/N. The length of the CPR block can be dynamically adjusted according to the shaping parameters in the PS systems to estimate the optimal phase.

Figure 3.17 shows the block diagram of the PA-CPR scheme. Generally, the pilot will not carry any meaningful information. The labels are carried by 8APSK pilots, which is a kind of meaningful information, because we need to extract these labels at the receiver to demodulate the signals. Therefore, the pilot should be preliminary recover at the receiver, which is the purpose of introducing discrete circle (DC) Viterbi-Viterbi.

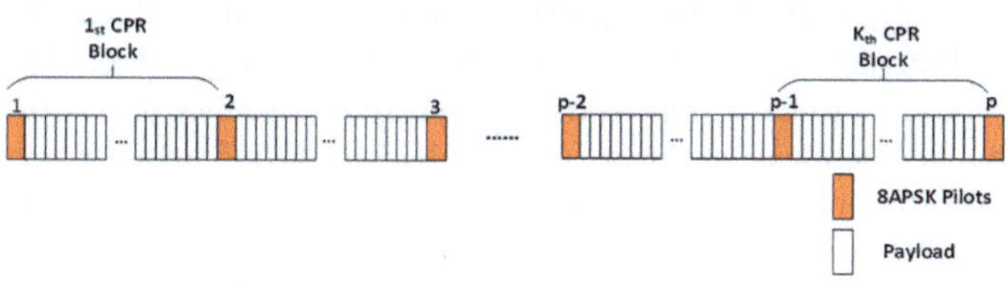

Fig. 3.16 The frame structure of implementing the pilot-aided CPR

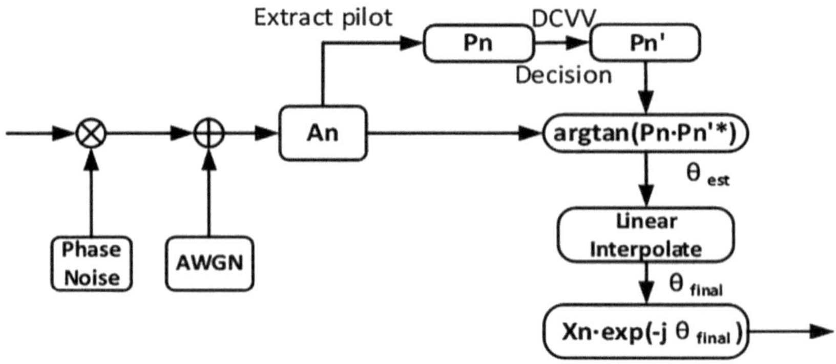

Fig. 3.17 The implementation of pilot aided CPR scheme

3.3.2 Algorithm Principle

Principle of PS-4D-SP-1024QAM

A. Probabilistic Amplitude Shaping
As shown in Fig. 3.13, the n binary bits are divided into n_1 and n_2 parts. Firstly, n_1 bits are entered into CCDM to generate m non-uniform positive amplitude $\sqrt{m}/2$-PAM signals, and the information rate of the obtained positive amplitude $\sqrt{m}/2$-PAM signal is $\beta = n_1/m$ bit/symbol. Then, the obtained symbols from CCDM are reflected into binary labels, which are combined with n_2 bits together to enter the FEC encoder for generating parity bits. These parity bits and n_2 are used as the sign bits to form a $\sqrt{m}$-PAM signal following Maxwell-Boltzmann (MB) distribution [76]. The MB distribution can be expressed as:

$$P_X(x) = e^{-\lambda x^2}/Z\left(x'\right), Z\left(x'\right) = \sum_{x' \in X} e^{-\lambda x'^2} \tag{3.19}$$

where x represents the $\sqrt{m}$-PAM constellation points and λ denotes the shaping factor.

B. Generation of PS-4D-SP-1024QAM
Figure 3.18 shows that the 1024QAM constellation is divided into four subsets (A, B, C, and D) named subset constraint (SSC).The corresponding 4D-1024QAM signals are generated according to the subfamily constraints (SFC) in Table 3.1. The minimum Euclidean distance of the 4D-1024QAM can be expanded by twice. Firstly, the non-uniform amplitude set of one dimensional (1D) signal generated by PS is divided into a subset C11D and C21D shown in Fig. 3.18.Then, the two 1D subsets are pairwise combined into two-dimensional (2D) subset

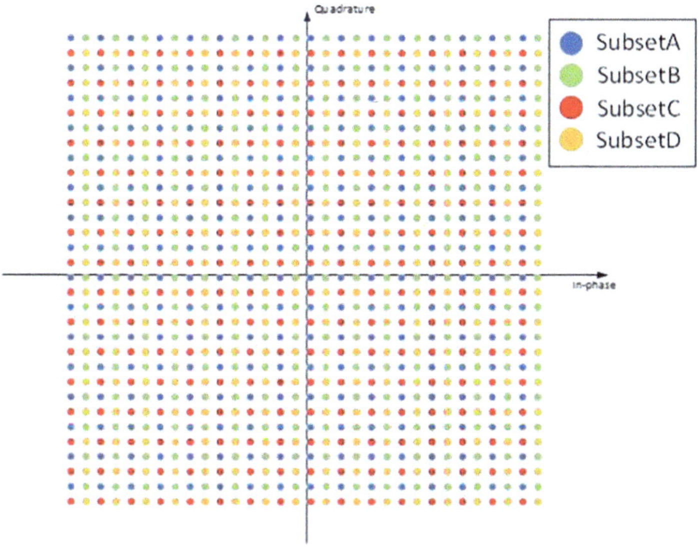

Fig. 3.18 The diagram of 1024QAM constellation Set-partitioning

A, B, C, and D. Finally, 2Dsubsets are combined into four 4D subfamilies (A_D, B_C, C_B, D_A) as shown in Fig. 3.19 corresponding to Table 3.1.

Amplitude translation is used to make the PS signal satisfy the above SFC. After the shaping, the 2D signals on two polarizations corresponding to a time interval are:

$$S_{pol} = S_{pol,I} + iS_{pol,Q} \tag{3.20}$$

where $Spol, l$ and $Spol, Q$ represents the I-component and Q-component signal on *pol* polarization, respectively, and *pol* is the X or Y polarization.

The 2D signals on X polarization is firstly selected and the subset of corresponding Y-polarization 2D signals are judged according to Table 3.1. Then 2D signals are decomposed into 1D signals for analysis. There are two cases:

1. If the 1D signal on the Y polarization mismatches the current SFC, the 1D signal is shifted to the adjacent signal points (forward translation is adopted in this paper) as the red arrow shows in Fig. 3.20, i.e., the complementary 1D subset, and then the binary label bit is set to 1. In the shifting process, there are two special points at both ends of the constellation during the translation. If the signal is beyond the limit of the constellation on translation, the points at the end are translated to the head to form a ring structure, as the blue arrow in Fig. 3.20. Take

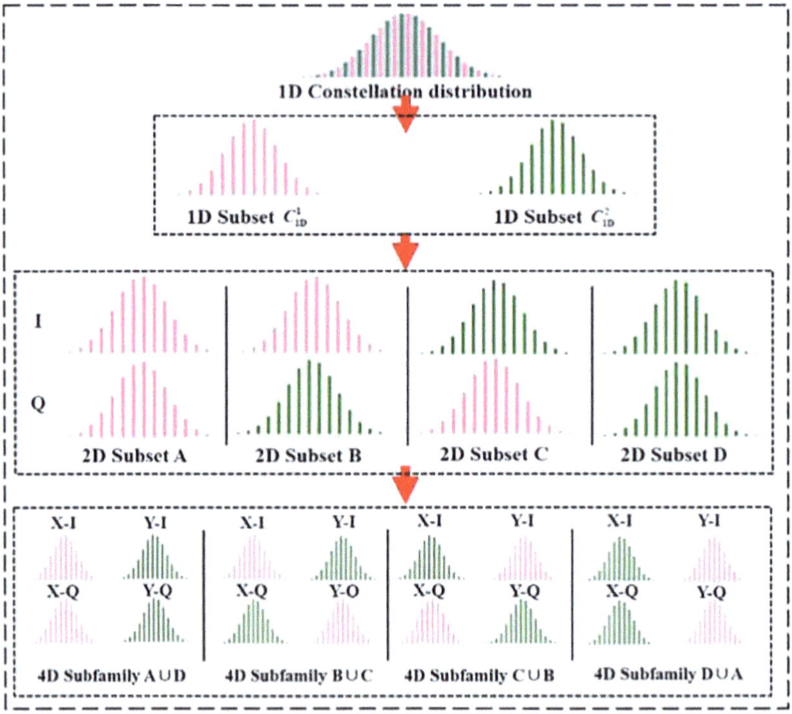

Fig. 3.19 Schematic diagram of non-uniform constellation set partitioning

the in-phase component as an example:

$$label \to 1 : \begin{cases} \underset{s_{y,I}}{\to} = s_{y,I} + d, s_{y,I} < \max\left(C_{1D}\right) \\ \underset{s_{y,I}}{\to} = \min\left(C_{1D}\right), s_{y,I} = \max\left(C_{1D}\right) \end{cases} \tag{3.21}$$

Where $\underset{s_{y,I}}{\to}$ is the translated signal for Y polarization, d represents the Euclidean distance between adjacent constellation points, and $min\,(\cdot)$ and $max\,(\cdot)$ represent the minimum and maximum values of the set respectively.

2. If the 1D signal matches the SFC, the translation is unnecessary and the label bit is 0:

$$\underset{s_{y,I}}{\to} = s_{y,I} : label \to 0 \tag{3.22}$$

Table 3.1 Subfamily constraints

X-Pol	Y-Pol
A	D
B	C
C	B
D	A

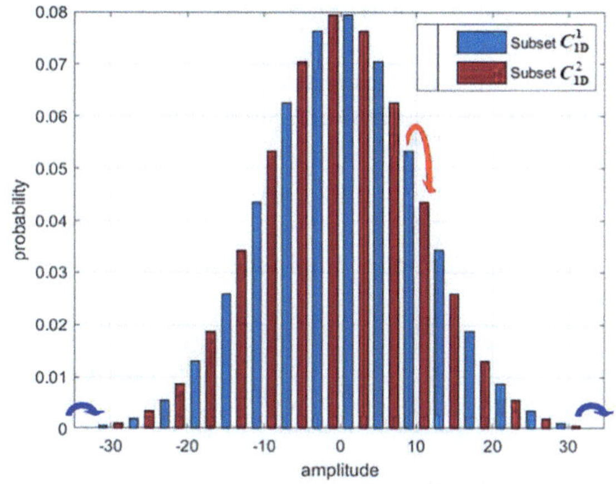

Fig. 3.20 Schematic of amplitude translation

In the process of signal generation, a 4D symbol will generate two binary label bits.

C. Redundancy Processing and Decoding

In this section, the proposed scheme can reduce the loss of transmission efficiency by saving overhead. It is assumed that the overhead of 4D is $OH4D$ and the overhead generated by pilot is OHP, so the total overhead is $OH_{Total}=OHP+OH4D$ in general. When using pilot and multi-dimensional signals, the generated overhead decreases compared to using pilot and multi-dimensional signals separately. The saved overhead can be expressed as $OHC=OHP \cap OH4D$. The total cost is $OH * \text{Total} = OHP + OH4D - OHC$. Figure 3.16 shows that $4n$ bits obtain $4m+8k$ 4D symbols and $16k$ bits after two stages of PAS and amplitude translation. In this paper, $16k$ bits continue to be mapped into $8k/3$ symbols as a part of pilot symbols for phase recovery, which improves the utilization of overhead. These pilots also can be used for demodulation at the receiver.

The 4D signal demodulation is divided into two parts: one is SP demodulation and the other is PS/FEC decoding. Firstly, the pilot signal is demodulated into label bits, a hard decision (HD) for a signal is operated. Then, the signal on the Y polarization is translated into the backward direction according to the label bit. If the label bit is 1, the amplitude of the corresponding signal component is translated in the backward direction. If the binary label bit is 0, there is no need for translation. When the translation of all symbols is completed, the bit metric decoding (BMD) [77] or HD of PS is performed to obtain the label bits in the first stage, and the signal is demodulated.

D. Influence on Constellation Probability Distribution

The probability distribution of the signal on the Y polarization will inevitably produce a certain offset when the amplitude of the signal point is translated, resulting in a constellation following a quasi-MB distribution. The derivation of probability distribution of the Y polarization signal is as follows.

It is assumed that the one-dimensional constellation set $Y = \{\pm 1, \pm 3, \pm 5, \ldots, \pm \sqrt{m}\}$ of Y polarization signal after PAS follows the MB distribution with a parameter of λ. Label bit set $L = \{0, 1\}$ generated by 4D signal follows the (0-1) distribution with parameter 1/2. Y and L are independent of each other. When the label is 0, the distribution of constellation set Y' remains unchanged, that is:

$$P_{Y'}(y_i) = P\{Y = y_i\} = e^{-\lambda y_i^2}/Z(y') \tag{3.23}$$

The signal point is amplitude translated when the label is 1, and the probability is:

$$P_{Y'}(y_i) = \begin{cases} P_Y(y\sqrt{m}) & i = 1 \\ P_Y(y_{i-1}) & 2 < i \le \sqrt{m} \end{cases} \tag{3.24}$$

The probability mass function (PMF) of Y_{-} is:

$$\begin{aligned} P_{Y'}(y_i) &= P\left\{Y' = y_i\right\} \\ &= P\left\{Y' = y_i, L = 0\right\} + P\left\{Y' = y_i, L = 1\right\} \\ &= \begin{cases} e^{-\lambda y_i^2/Z(y')} & i = 1 \\ e^{-\lambda y_i^2} + e^{-\lambda y_{i-1}^2}/2Z(y') & 1 < i \le \sqrt{m} \end{cases} \end{aligned} \tag{3.25}$$

Then the normalized generalized mutual information (NGMI) performance of PS-1024QAM signals following MB distribution and quasi MB distribution with different is simulated and compared.

In Fig. 3.21 For λ =0.2 the 1024QAM constellation is truncated obviously, and the information entropy is greatly affected by the change of distribution. There is a about 5 dB gap between quasi-MB distribution and MB distribution; For $\lambda = 0.02$ the influence of distribution change on information entropy weakens, and the gap decreases to $\sim$ 2 dB; For $\lambda = 0.002$, the degree of shaping is very small, and the change of NGMI can be negligible. Therefore, the constellation scale is large when the shaping degree is small. The NGMI degradation caused by quasi MB distribution can be ignored, and the application in higher shaping degree or lower order modulation format needs to be reconsidered.

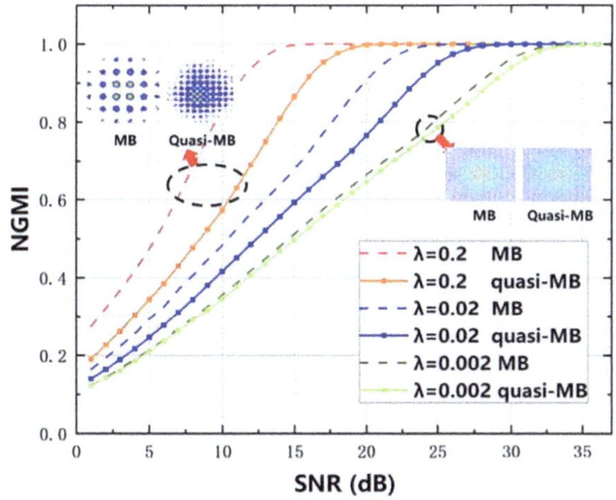

Fig. 3.21 NGMI comparison of signals following MB distribution and quasi-MB distribution with different λ

3.3.3 PS-4D-SP-1024QAM Phase Recovery Scheme

As Fig. 3.21 shows, The signal A_n added white Gaussian noise and phase noise which includes payload signal X_n and pilot signal P_n. The 8APSK pilot signal P_n is extracted from A_n and it's recovered by the DC-VV algorithm and be decided to obtain the original signal P_n'. The phase noise of the pilot signal θ_{est} is estimated by multiplying P_n with the conjugate of the original 8APSK pilot P_n'. The payload signal phase noise θ_{final} is obtained by linear interpolation of θ_{est} in adjacent CPR blocks, as shown in (3.26):

$$\varphi_{a,b} = \varphi_a + b \cdot (\varphi_{a+1} - \varphi_a)/N \tag{3.26}$$

where φ_a is the pilot phase noise of the a_{th} CPR block, $\varphi_{a,b}$ is the b_{th} payload signal phase noise in the ath CPR block, N is the length of the PS-4D-SP-1024QAM payload in each CPR block [78]. Finally, the original payload signal is obtained by subtracting the estimated phase θ_{final} from the received payload signal X_n. The algorithm shows that the PA algorithm will not be affected by the modulation format type.

The amplitude of the pilot signal has different effects on the PA-CPR performance. The pilot signal with a small amplitude is more sensitive to AWGN. Moreover, the power of the payload signal will decrease with the increasing of pilot amplitude. We design a novel scheme with 8APSK-PA to solve the problems. Figure 3.22 shows the pilot structure with the eight different constellation points in the inner circle (IC) and the outer circle (OC). The amplitude levels in the 1024-QAM constellation are $A = \{\pm 1, \pm 3, \pm 5, \ldots, \pm 31\}$. The absolute value of the amplitude level is selected as the amplitude value of the IC and OC (AoIC and AoOC). For QPSK pilot recovery algorithm, the accuracy of single pilot amplitude for phase recovery is reduced. To improve the estimation accuracy, the optimal

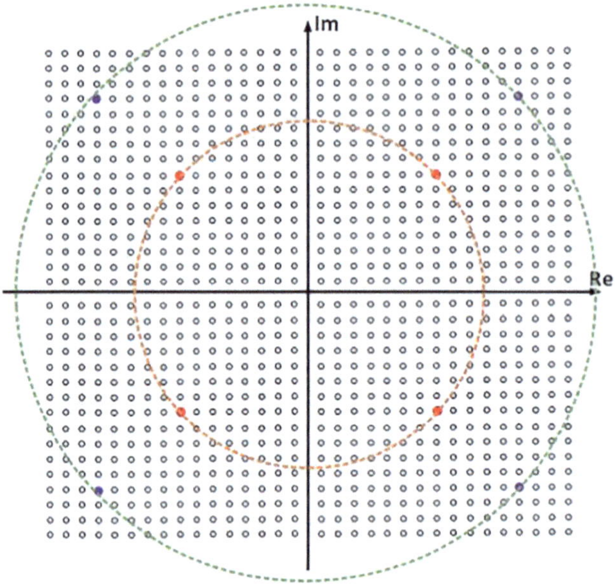

Fig. 3.22 8APSK-pilot constellation point

8APSK pilot amplitude in IC and OC is chosen in phase recovery scheme. In this paper, the 8APSK is selected as of the pilot for phase recovery to be better matched with PS-4D-SP-1024QAM signals. The PS-4D-SP-1024QAM is designed to meet the modulation format redundancy requirements with the number of pilot symbols and control the pilot insertion rate. In addition, the 8APSK pilot scheme can improve the recovery performance of the algorithm without changing the average transmission power of the system compared with QPSK. Moreover, the 8APSK-PA-CPR has lower complexity than blind phase recovery algorithm.

3.3.4 System Setup and Results

Figure 3.23 shows the simulation setup for 20 GBaud PDMPS- 1024QAM and PS-4D-SP-1024QAM transmission systems. The 1250 pilot symbols generated by random generation or amplitude translation are evenly inserted into 33,750 signal symbols to form a block of 35,000 symbols, resulting in a pilot insertion ratio of 1/27. The light is emitted at 1550 nm by a laser, separated by a polarization beam splitter, and modulated by two IQ modulators with a PDM-PS-1024QAM signal or a PS-4D-SP-1024QAM signal at the transmitter. The signal is sent to a variable optical attenuator (VOA) and an EDFA, which provides the ASE noise to control the SNR. Afterward, the signal is transmitted across 25 km G652D standard single-mode fiber. Other impairments in coherent optical communication systems

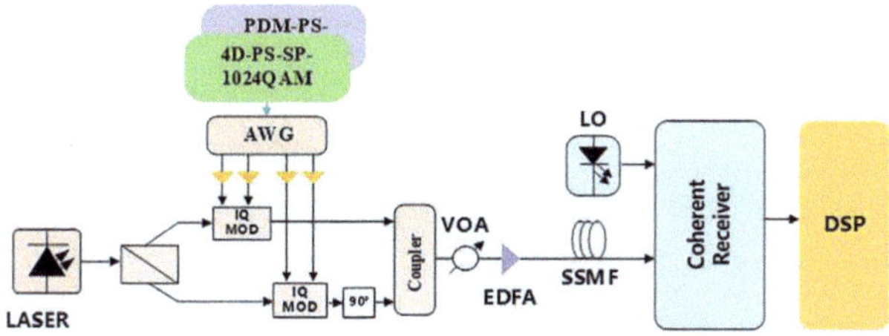

Fig. 3.23 Simulation setup for the 20Gbaud PDM-PS-1024QAM and PS-4DSP-1024QAM systems

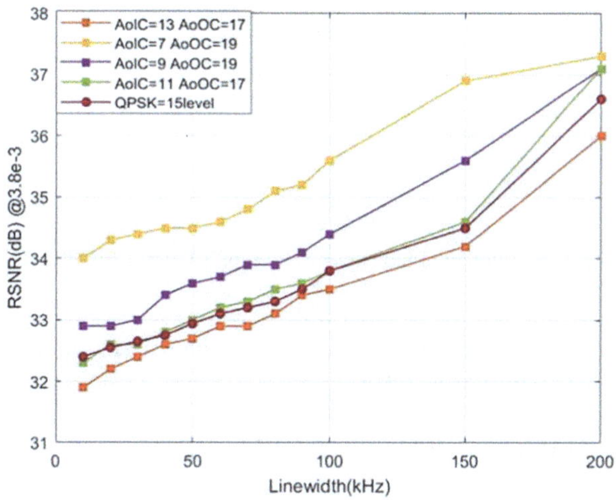

Fig. 3.24 The RSNR at the BER of $3.8 \times 10-3$ versus linewidth for PDM-PS-1024QAM system using different PA-CPR amplitude

are assumed to be completely compensated, leaving only phase noise and AWGN to be considered. The signals are received by an integrated coherent receiver and then processed offline by a DSP algorithm.

A. Pilot Amplitude Optimization

On the PDM-PS-1024QAM, the 8APSK-PA-CPR scheme is compared to the QPSK-PA-CPR scheme at approximately average power. The amplitude of the pilot in two circles is swept and optimized. Figure 3.24 shows the SNR variation with different laser linewidths for QPSK-PA-CPR and 8APSK-PA-CPR with different AoIC and AoOC at approximate average power. The performance of the above algorithms is analyzed by calculating the required SNR at the BER of $3.8 \times 10-3$, which corresponds to a 7% FEC overhead. It is demonstrated that the maximum gain of the AoIC = 13 and AoOC = 17 8APSK-PA-CPR obtained by amplitude optimization is ~0.4 dB compared with 15 level QPSK-PA-CPR at the equal average power. The performance of phase recovery algorithm is affected by the AoIC and AoOC(see Fig. 3.24). This is because the pilot points with small amplitude are more affected by AWGN than those with large amplitude. To find the optimal AoIC and AoOC, we numerically simulate the GMI performance of

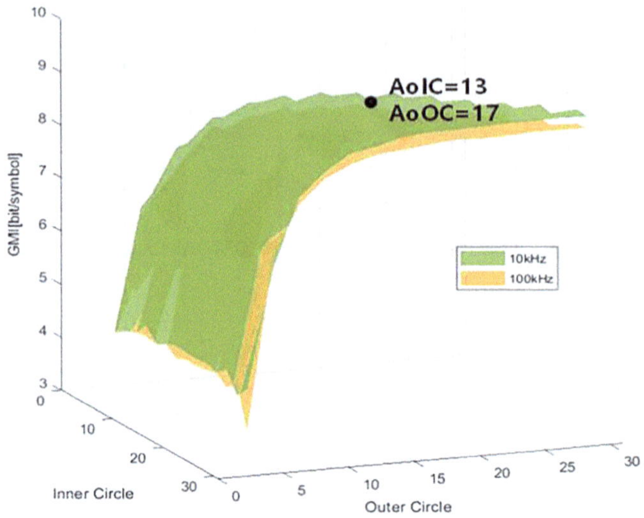

Fig. 3.25 Pilot amplitude optimization of 8APSK-PA-CPR

different IC and OC pilot schemes with laser linewidth of 10 and 100 kHz. HD-GMI is derived as [79]:

$$GMI_{HD} = H\left[1 - (-BER\log_2(BER) - (1 - BER) \times \log_2(1 - BER))\right] \tag{3.27}$$

where H represents constellation entropy. As shown in Fig. 3.25, the PS factor λ is set to 0.01 and the SNR to 31.5 dB in the process of radius optimization. GMI is improved when the AoIC and AoOC increase gradually. However, there will be a GMI ceiling because the effect of AWGN for the pilot format has been minimized when it increases to a certain extent. In, a 15 level is adopted in QPSK-PA-CPR to maintain the average transmitted power in the system. Due to the increase in the average energy of the transmitted signal, the system requires higher transmitted power to keep the same operating SNR. The selection of CPR pilots at the average energy is more appropriate. In Fig. 3.25, the AoIC = 13 and AoOC = 17 is chosen as close as possible to the average power to carry out the amplitude of the 8APSK-PA-CPR scheme to avoid the change of the overall average power.

B. Compatibility Evaluation of 8APSK-PA-CPR in Probabilistic Shaping System

In PS systems, the GMI will suddenly decline at the optimal value of λ when using the BPS algorithm for phase recovery at low SNR. The BPS algorithm has a strong dependence on PS, even for long noise rejection windows. The decisions made inside the BPS algorithm are affected by shaping. The impact of PS on BPS

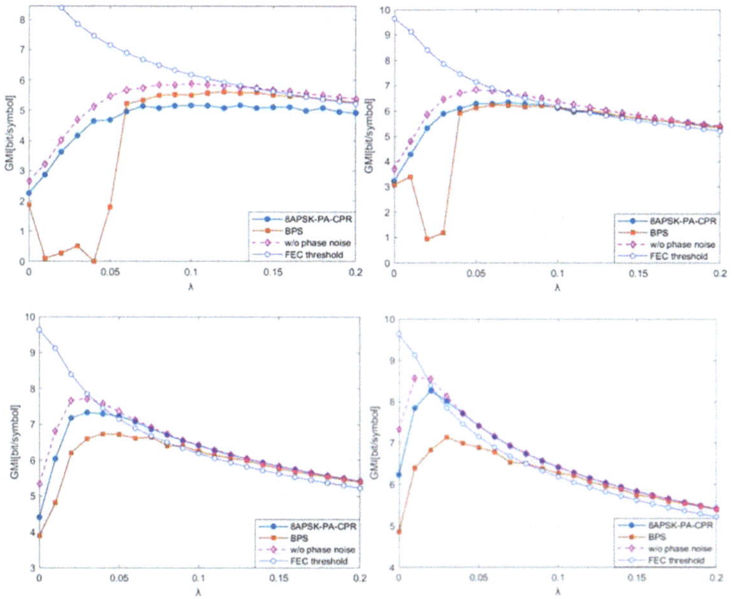

Fig. 3.26 GMI results versus λ with different phase recovery algorithms in PS-1024QAM systems when the SNRs are (**a**) 20 dB. (**b**) 23 dB. (**c**) 26 dB. (**d**) 29 dB

can affect the overall system performance, specially at low SNRs. It reflects the incompatibility of the BPS algorithm with PS, even though increasing the window size reduces the sudden drop in GMI. The impact of different PS factors on GMI in the 8PASK-PA-CPR scheme is evaluated. Figure 3.26 shows the GMI performance of 8PASK-PA-CPR and BPS (N = 100, B = 128) in different PS factors λ. The SNRs are 20, 23, 26 and 29 dB respectively. For SNR = 20 dB (Fig. 3.26a) the sudden drop of GMI under BPS is obvious from 0 to 0.06. 8APSK-PA-CPRdoes not cause a sudden drop in the whole measurement interval of PS factor λ. In the condition of low SNR, the additive white noise is dominant and the 8APSK-PA-CPR has accurate recovery on phase noise. The BPS algorithm is a blind estimation of AWGN and phase noise. The performance of 8APSK-PA-CPR is inferior to that of blind phase recovery algorithm when the AWGN is too large to be accurately estimated. In this situation of SNR, the performance of 8APSK is greatly affected by AWGN compared with BPS, resulting in a gap of 0.5 bit/symbol. Moreover, the FEC threshold of GMI corresponding to $3.8 \times 10-3$ cannot be reached. For SNR = 23 dB (Fig. 3.26b), the sudden drop caused by the BPS algorithm is alleviated, and the GMI of 8APSK-PA-CPR increases by ~0.1 bit/symbol compared with that of BPS. The GMI performance of 8APSK-PA-CPR is better than that of BPS as the value of SNR increases. For example, the maximum GMI increases by 0.7 bit/symbol when SNR = 26 dB, and the maximum GMI increases by 1.3 bit/symbol when SNR = 29 dB, approaching the limit (see Fig. 3.26c and d). The GMI with 8APSK-PA-CPR can

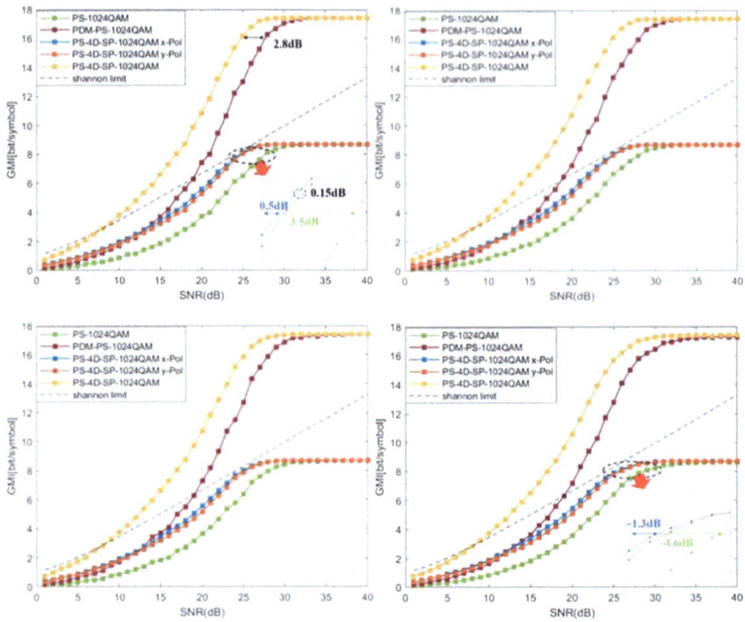

Fig. 3.27 GMI results versus SNR with different polarizations in PDM-PS-1024QAM and PS-4D- SP-1024QAM systems when the value of shaping parameter is 0.02, the linewidths are (**a**) 10 kHz. (**b**) 50 kHz. (**c**) 100 kHz. (**d**) 200 kHz

reach the FEC threshold at this SNR, when the shaping parameter is greater than 0.02. It is acceptable to obtain significant gain at higher SNR (greater than 26 dB).

C. Evaluation of PS-4D-SP-1024QAM and 8APSK-PA-CPR Performance

In the 8APSK-PA-CPR scheme, the overhead generated by the second stage is put into the pilot symbol to obtain additional mutual information and increase the transmission efficiency of the system. Figure 3.27. shows the SNR and GMI curves of PDM-PS-1024QAM and PS-4D-SP-1024QAM applied with the 8APSKPA- CPR scheme. The PS factor is 0.02, and the linewidths are set to 10, 50, 100, and 200 kHz respectively. In the simulation, the HD decoder is used at the receiver and there is a larger gap between PS-1024QAM and Shannon limit The Euclidean distance is expanded by multidimensional set-partition, so the noise tolerance of PS-4D-SP-1024QAM is higher than that of PS-1024QAM. As shown in Fig. 3.27, the SNR gain of PS-4D-SP-1024QAM is ~2.8 dB compared with PDM-PS 1024QAM.The gain of 2.8 dB is the theoretical gain of modulation format. For linewidth=10 kHz (Fig. 3.27a) the gaps between Shannon limit with the GMI of PS-4D-SP-1024QAM and that of PDM-PS-1024QAM is 0.5 and 3.5 dB respectively when the GMI is 8 bit/symbol. For linewidth = 200 kHz (Fig. 3.27d) the gaps rise to 1.3 and 4.6 dB, respectively The sensitivity of PS-4D-SP-1024QAM decreases by only 0.7 dB from the linewidth 10–200 kHz, while the sensitivity of PDM-PS-1024QAM decreases

Table 3.2 Computational complexity of different algorithms

Algorithm	Real multiplier	Real adder	Comparison	Decision
DP-BPS	12NB+2B	12NB+2B	2B	2NB
8APSK pilot-aided	24	16	2	2
QPSK pilot-aided	24	16	0	2

by 1.1 dB, indicating that the proposed scheme has a high phase noise tolerance and better pilot phase recovery performance behavior. The penalty between X polarization and Y polarization of PS-4D-SP-1024QAM format is ~0.15 dB due to the amplitude translation taken in the 4D format. The maximum GMI of PS-4D-SP-1024QAM can be achieved at 17.7971 bit/symbol with the increasing of SNR for the linewidth at the FEC threshold of $3.8 \times 10 - 3$. The final net bit rate is 293.7 Gbit/s considering a 7% FEC overhead, a pilot insertion ratio of 1/27 and redundant symbols from the first stage.

D. Computational Complexity Analysis

The complexity of the DSP algorithm is important for practical application. The low complexity algorithm can effectively reduce the system power consumption and increase the speed of hardware processing. As shown in Table 3.2, the complexity includes four aspects: real multipliers, real adders, comparison, and decision operation. The parameters are selected when the three algorithms are based on optimum implementations. Taking the linewidth of 100 kHz as an example (N = 100, B = 128), the DP-BPS algorithm requires 12NB+2B real multiplier, 12NB+2B real adders, 2B comparison, and 2NB decision operation. In the 8APSK-PA-CPR scheme the DC-VV algorithm is used to estimate the phase of the 8APSK pilot, which increases a small amount of complexity (2 Comparison) compared with QPSK-PA-CPR. The complexity of the 8APSK-PA-CPR algorithm is also reduced by 99.99% compared with the DP-BPS algorithm.

3.4 Summary

In this chapter, we introduce and discuss several advanced modulation technologies, which can effectively improve the performance and bandwidth of fiber optic communication systems. First, we propose a novel symbol division multiplexing technology by multiplexing/de-multiplexing of multiple QAM symbols onto one complex constellation point which can achieve an additional 40% gain for effective data rate and system spectral efficiency. Otherwise, in order to transmit higher-order modulation signal in radio-over-fiber system, we investigate the introduction of the delta-sigma modulation (DSM) technique into a wireless mm-wave transmission system and propose a photonic-aided dual-vector radio-frequency (RF) signal generation and detection scheme enabled by bandpass DSM and heterodyne detection, which can overcome the insufficient power and receiver sensitivity problems and thus to realize high-level QAM signal transmission. Finally, to further improve transmission capacity and performance, we propose a scheme of probabilistic shaping 4D ultrahigh order modulation format with pilot aided carrier phase recovery.

References

1. Z. Dong, J. Yu, Y. Chen, F. Li, X. Xin, Symbol division multiplexing in optical fiber communication systems. Opt. Express **30**(9), 14998–15007 (2022)
2. X. Pan, D. Wu, X. Li, D. Guo, Z. Li, H. Yan, C. Wang, J. Bi, C. Yu, X. Liu, et al., Photonic-aided w-band dual-vector rf signal generation and detection enabled by bandpass delta-sigma modulation and heterodyne detection. Opt. Lett. **48**(8), 2146–2149 (2023)
3. F. Tian, T. Wu, Y. Ji, C. Wang, Q. Zhang, R. Gao, Z. Li, Q. Tian, X. Xin, Design of probabilistic shaping 4d ultra high order modulation format with 8apsk pilot aided carrier phase recovery. J. Lightwave Technol. **40**(12), 3688–3697 (2022)
4. N. Cvijetic, D. Qian, J. Hu, 100 gb/s optical access based on optical orthogonal frequency-division multiplexing. IEEE Commun. Mag. **48**(7), 70–77 (2010)
5. D.O. Otuya, K. Kasai, T. Hirooka, M. Nakazawa, Single-channel 1.92 tbit/s, 64 qam coherent nyquist orthogonal tdm transmission with a spectral efficiency of 10.6 bit/s/hz. J. Lightwave Technol. **34**(2), 768–775 (2015)
6. G. Raybon, A. Adamiecki, P.J. Winzer, S. Randel, L. Salamanca, A. Konczykowska, F. Jorge, J.-Y. Dupuy, L.L. Buhl, S. Chandrashekhar, et al., High symbol rate coherent optical transmission systems: 80 and 107 gbaud. J. Lightwave Technol. **32**(4), 824–831 (2013)
7. X. Chen, G. Raybon, D. Che, J. Cho, K. Kim, Transmission of 200-gbaud pdm probabilistically shaped 64-qam signals modulated via a 100-ghz thin-film linbo3 i/q modulator, in *Optical Fiber Communication Conference* (Optica Publishing Group, 2021), pp. F3C–5
8. F. Li, J. Yu, Z. Cao, J. Zhang, M. Chen, X. Li, Experimental demonstration of four-channel wdm 560 gbit/s 128qam-dmt using im/dd for 2-km optical interconnect. J. Lightwave Technol. **35**(4), 941–948 (2017)
9. S.T. Le, H. Buelow, 64× 0.5 gbaud nonlinear frequency division multiplexed transmissions with high order modulation formats. J. Lightwave Technol. **35**(17), 3692–3698 (2017)
10. F. Klejs, E.P. Da Silva, M. Lillieholm, M.P. Yankov, T. Morioka, L.K. Oxenløwe, M. Galili, Spectrally efficient dp-1024qam 640 gb/s long haul transmission using a frequency comb, in *Optical Fiber Communication Conference* (Optica Publishing Group, 2020), pp. M4K–4
11. X. Liu, A. Chraplyvy, P. Winzer, R. Tkach, S. Chandrasekhar, Phase-conjugated twin waves for communication beyond the kerr nonlinearity limit. Nature Photon. **7**(7), 560–568 (2013)
12. D. Hillerkuss, R. Schmogrow, T. Schellinger, M. Jordan, M. Winter, G. Huber, T. Vallaitis, R. Bonk, P. Kleinow, F. Frey, et al., 26 tbit s- 1 line-rate super-channel transmission utilizing all-optical fast fourier transform processing. Nature Photon. **5**(6), 364–371 (2011)
13. S.T. Le, V. Aref, H. Buelow, Nonlinear signal multiplexing for communication beyond the kerr nonlinearity limit. Nature Photon. **11**(9), 570–576 (2017)
14. P. Guan, F. Da Ros, M. Lillieholm, N.-K. Kjøller, H. Hu, K.M. Røge, M. Galili, T. Morioka, L.K. Oxenløwe, Scalable wdm phase regeneration in a single phase-sensitive amplifier through optical time lenses. Nature Commun. **9**(1), 1049 (2018)
15. L.-W. Luo, N. Ophir, C.P. Chen, L.H. Gabrielli, C.B. Poitras, K. Bergmen, M. Lipson, Wdm-compatible mode-division multiplexing on a silicon chip. Nature Commun. **5**(1), 1–7 (2014)
16. D.H. Sim, H. Kim, Y.-C. Chung, Direct-detection receiver for polarization-division-multiplexed ook signals. IEEE Photon. Technol. Lett. **27**(21), 2238–2241 (2015)

17. X. Liu, F. Buchali, R.W. Tkach, Improving the nonlinear tolerance of polarization-division-multiplexed co-ofdm in long-haul fiber transmission. J. Lightwave Technol. **27**(16), 3632–3640 (2009)
18. C. Xie, Interchannel nonlinearities in coherent polarization-division-multiplexed quadrature-phase-shift-keying systems. IEEE Photon. Technol. Lett. **21**(5), 274–276 (2009)
19. K. Igarashi, D. Soma, Y. Wakayama, K. Takeshima, Y. Kawaguchi, N. Yoshikane, T. Tsuritani, I. Morita, M. Suzuki, Ultra-dense spatial-division-multiplexed optical fiber transmission over 6-mode 19-core fibers. Opt. Express **24**(10), 10213–10231 (2016)
20. G. Rademacher, R. Ryf, N.K. Fontaine, H. Chen, R.-J. Essiambre, B.J. Puttnam, R.S. Luís, Y. Awaji, N. Wada, S. Gross, et al., Long-haul transmission over few-mode fibers with space-division multiplexing. J. Lightwave Technol. **36**(6), 1382–1388 (2018)
21. J. Sakaguchi, Y. Awaji, N. Wada, A. Kanno, T. Kawanishi, T. Hayashi, T. Taru, T. Kobayashi, M. Watanabe, Space division multiplexed transmission of 109-tb/s data signals using homogeneous seven-core fiber. J. Lightwave Technol. **30**(4), 658–665 (2012)
22. M. Nakajima, K. Suzuki, K. Yamaguchi, H. Ono, T. Goh, M. Fukutoku, Y. Miyamoto, T. Hashimoto, Multilane photonic spectral processor integrated in a spatial and planar optical circuit for a space-division multiplexing network. J. Lightwave Technol. **36**(2), 309–317 (2017)
23. D.J. Richardson, J.M. Fini, L.E. Nelson, Space-division multiplexing in optical fibres. Nature Photon. **7**(5), 354–362 (2013)
24. A. Li, X. Chen, A. Al Amin, J. Ye, W. Shieh, Space-division multiplexed high-speed superchannel transmission over few-mode fiber. J. Lightwave Technol. **30**(24), 3953–3964 (2012)
25. J. Wang, J.-Y. Yang, I.M. Fazal, N. Ahmed, Y. Yan, H. Huang, Y. Ren, Y. Yue, S. Dolinar, M. Tur, et al., Terabit free-space data transmission employing orbital angular momentum multiplexing. Nature Photon. **6**(7), 488–496 (2012)
26. H. He, Y. Chen, Q. Xiao, L. Zhou, Z. Dong, Ldpc-coded probabilistic shaping pam4 based on many-to-one mapping in wdm-pon. J. Lightwave Technol. **38**(15), 3918–3925 (2020)
27. H. Sun, M. Torbatian, M. Karimi, R. Maher, S. Thomson, M. Tehrani, Y. Gao, A. Kumpera, G. Soliman, A. Kakkar, et al., 800g dsp asic design using probabilistic shaping and digital sub-carrier multiplexing. J. Lightwave Technol. **38**(17), 4744–4756 (2020)
28. Z. Cao, Y. Chen, X. Zhao, L. Zhou, Z. Dong, Probabilistic shaping 44qam based on many-to-one mapping in dmt-wdm-pon. IEEE Photon. Technol. Lett. **32**(11), 639–642 (2020)
29. M.P. Yankov, D. Zibar, K.J. Larsen, L.P. Christensen, S. Forchhammer, Constellation shaping for fiber-optic channels with qam and high spectral efficiency. IEEE Photon. Technol. Lett. **26**(23), 2407–2410 (2014)
30. M.P. Yankov, F. Da Ros, E.P. da Silva, S. Forchhammer, K.J. Larsen, L.K. Oxenløwe, M. Galili, D. Zibar, Constellation shaping for wdm systems using 256qam/1024qam with probabilistic optimization. J. Lightwave Technol. **34**(22), 5146–5156 (2016)
31. S. Iqbal, P.M. Kaminski, F. Klejs, M.P. Yankov, E.P. Da Silva, F. Da Ros, L.K. Oxenløwe, S. Forchhammer, Probabilistically shaped rate-adaptive polar-coded 256-qam wdm optical transmission system. J. Lightwave Technol. **38**(7), 1800–1808 (2019)
32. S. Karabetsos, E. Pikasis, T. Nikas, A. Nassiopoulos, D. Syvridis, Dft-spread dmt modulation for 1-gb/s transmission rate over 100 m of 1-mm si-pof. IEEE Photon. Technol. Lett. **24**(10), 836–838 (2012)
33. D. Zou, Y. Chen, F. Li, Z. Li, Y. Sun, L. Ding, J. Li, X. Yi, L. Li, Z. Li, Comparison of bit-loading dmt and pre-equalized dft-spread dmt for 2-km optical interconnect system. J. Lightwave Technol. **37**(10), 2194–2200 (2019)
34. X. Li, J.A. Ritcey, Bit-interleaved coded modulation with iterative decoding, in *1999 IEEE International Conference on Communications (Cat. No. 99CH36311)*, vol. 2. (IEEE, 1999), pp. 858–863
35. Y. Xu, G. Zhang, H. Ju, D. He, J. Sun, Y. Wu, W. Zhang, A low complexity decoding scheme for raptor-like ldpc codes. IEEE Transa. Broadcasting **65**(4), 770–776 (2019)
36. Y. Cai, M. Zhu, J. Zhang, M. Lei, B. Hua, Y. Zou, W. Luo, S. Xiang, L. Tian, J. Ding, et al., Real-time 100-gbe fiber-wireless seamless integration system using an electromagnetic dual-

polarized single-input single-output wireless link at the w band. Opt. Lett. **48**(4), 928–931 (2023)
37. W. Li, J. Yu, Y. Wang, F. Wang, B. Zhu, L. Zhao, W. Zhou, J. Yu, F. Zhao, Ofdm-ps-256qam signal delivery at 47.45 gb/s over 4.6-kilometers wireless distance at the w band. Opt. Lett. **47**(16), 4072–4075 (2022)
38. R. Deng, J. Yu, J. He, M. Chen, Y. Wei, L. Zhao, Q. Zhang, X. Xin, Twin-ssb-ofdm transmission over heterodyne w-band fiber-wireless system with real-time implementable blind carrier recovery. J. Lightwave Technol. **36**(23), 5562–5572 (2018)
39. J. Ding, L. Zhang, J. Liu, W. Li, Y. Wang, K. Wang, L. Zhao, W. Zhou, J. Zhang, M. Zhu, et al., Thz-over-fiber transmission with a net rate of 5.12 tbps in an 80 channel wdm system. Opt. Lett. **47**(12), 3103–3106 (2022)
40. J. Ding, M. Wang, W. Li, B. Zhu, F. Wang, C. Wang, Y. Wang, W. Zhou, J. Zhang, M. Zhu, et al., Wireless transmission of a 200-m ps-64qam thz-wave signal using a likelihood-based selection radius-directed equalizer. Opt. Lett. **47**(15), 3904–3907 (2022)
41. J. Wang, Z. Jia, L.A. Campos, C. Knittle, Real-time demonstration of 5-gsa/s delta-sigma digitization for ultra-wide-bandwidth lte and 5g signals in next generation fronthaul interface, in *2018 European Conference on Optical Communication (ECOC)* (IEEE, 2018), pp. 1–3
42. Y. Zhu, X. Fang, L. Yin, F. Zhang, W. Hu, Radio-over-fiber transmission supporting 65536-qam at 25ghz band with high-pass delta-sigma modulation and rf fading mitigation, in *Optical Fiber Communication Conference* (Optica Publishing Group, 2022), pp. Tu2G–2
43. D. Tan, L. Zhong, Y. Zou, J. Zhang, W. Lu, X. Dai, Q. Yang, S. Fu, M. Cheng, L. Deng, et al., A novel structure design method of delta-sigma modulator based on genetic algorithm for mobile fronthaul, in *2022 Optical Fiber Communications Conference and Exhibition (OFC)* (IEEE, 2022), pp. 1–3
44. F. Zhao, X. Yang, L. Zhao, Y. Wei, J. Yu, Demonstration of 4096qam thz mimo wireless delivery employing one-bit delta-sigma modulation. Opt. Lett. **47**(24), 6361–6364 (2022)
45. X. Li, J. Yu, 2× 2 multiple-input multiple-output optical–wireless integration system based on optical independent-sideband modulation enabled by an in-phase/quadrature modulator. Opt. Lett. **41**(13), 3138–3141 (2016)
46. X. Pan, X. Liu, H. Zhang, K. Wang, Y. Zhang, D. Ran, X. Wang, C. Wang, Independent dual single-sideband vector millimeter-wave signal generation by one single i/q modulator. Opt. Exp. **27**(14), 19906–19914 (2019)
47. X. Pan, B. Zhang, X. Liu, H. Zhang, K. Wang, Y. Zhang, D. Ran, X. Wang, C. Wang, Mitigation of crosstalk between rsb and lsb signals generated by one i/q modulator. Opt. Exp. **28**(2), 2238–2250 (2020)
48. L. Zhao, H. Guo, Y. Liu, J. Xiao, T. Wu, S. Song, L. Guo, Independent dual-single-sideband qpsk signal detection based on a single photodetector. Opt. Exp. **30**(13), 22946–22956 (2022)
49. J. Ming, J. Xiao, D. Wu, L. Wang, Y. Zhou, L. Zhao, Synthesis of a gs-16qam signal for a simplified optical isb system and direct detection with the k-means clustering algorithm. Opt. Exp. **30**(22), 39663–39678 (2022)
50. C. Yu, X. Sheng, R. Gao, F. Wang, Z. Li, H. Chang, L. Huang, Q. Zhang, X. Xin, X. Huang, et al., Experimental demonstration of twin-single-sideband signal detection system based on a single photodetector and without optical bandpass filter. Opt. Exp. **30**(21), 37341–37349 (2022)
51. Y. Koizumi, K. Toyoda, M. Yoshida, M. Nakazawa, 1024 qam (60 gbit/s) single-carrier coherent optical transmission over 150 km. Opt. Exp. **20**(11), 12508–12514 (2012)
52. S. Beppu, K. Kasai, M. Yoshida, M. Nakazawa, 2048 qam (66 gbit/s) single-carrier coherent optical transmission over 150 km with a potential se of 15.3 bit/s/hz. Opt. Exp. **23**(4), 4960–4969 (2015)
53. R. Maher, K. Croussore, M. Lauermann, R. Going, X. Xu, J. Rahn, Constellation shaped 66 gbd dp-1024qam transceiver with 400 km transmission over standard smf, in *2017 European Conference on Optical Communication (ECOC)* (IEEE, 2017), pp. 1–3

54. S. Okamoto, M. Terayama, M. Yoshida, K. Kasai, T. Hirooka, M. Nakazawa, Experimental and numerical comparison of probabilistically shaped 4096 qam and a uniformly shaped 1024 qam in all-raman amplified 160 km transmission. Opt. Exp. **26**(3), 3535–3543 (2018)
55. S.L. Olsson, J. Cho, S. Chandrasekhar, X. Chen, E.C. Burrows, P.J. Winzer, Record-high 17.3-bit/s/hz spectral efficiency transmission over 50 km using probabilistically shaped pdm 4096-qam, in *2018 Optical Fiber Communications Conference and Exposition (OFC)* (IEEE, 2018), pp. 1–3
56. S.L. Olsson, J. Cho, S. Chandrasekhar, X. Chen, P.J. Winzer, S. Makovejs, Probabilistically shaped pdm 4096-qam transmission over up to 200 km of fiber using standard intradyne detection. Opt. Exp. **26**(4), 4522–4530 (2018)
57. M. Terayama, S. Okamoto, K. Kasai, M. Yoshida, M. Nakazawa, 4096 qam (72 gbit/s) single-carrier coherent optical transmission with a potential se of 15.8 bit/s/hz in all-raman amplified 160 km fiber link, in *Optical Fiber Communication Conference* (Optica Publishing Group, 2018), pp. Th1F–2
58. Y. Wang, S. Okamoto, K. Kasai, M. Yoshida, M. Nakazawa, Single-channel 200 gbit/s, 10 gsymbol/s-1024 qam injection-locked coherent transmission over 160 km with a pilot-assisted adaptive equalizer. Opt. Exp. **26**(13), 17015–17024 (2018)
59. X. Chen, S. Chandrasekhar, J. Cho, P. Winzer, Transmission of 30-gbd polarization-multiplexed probabilistically shaped 4096-qam over 50.9-km ssmf. Opt. Exp. **27**(21), 29916–29923 (2019)
60. X. Chen, J. Cho, A. Adamiecki, P. Winzer, 16384-qam transmission at 10 gbd over 25-km ssmf using polarization-multiplexed probabilistic constellation shaping, in *45th European Conference on Optical Communication (ECOC 2019)* (IET, 2019), pp. 1–4
61. T.A. Eriksson, M. Sjödin, P. Johannisson, P.A. Andrekson, M. Karlsson, Comparison of 128-sp-qam and pm-16qam in long-haul wdm transmission. Opt. Exp. **21**(16), 19269–19279 (2013)
62. J.K. Fischer, C. Schmidt-Langhorst, S. Alreesh, R. Elschner, F. Frey, P.W. Berenguer, L. Molle, M. Nölle, C. Schubert, Generation, transmission, and detection of 4-d set-partitioning qam signals. J. Lightwave Technol. **33**(7), 1445–1451 (2014)
63. J. Lu, S. Fu, Z. Hu, L. Deng, M. Tang, D. Liu, C.C.-K. Chan, Carrier phase recovery for set-partitioning qam formats. J. Lightwave Technol. **36**(18), 4129–4137 (2018)
64. P.J. Winzer, High-spectral-efficiency optical modulation formats. J. Lightwave Technol. **30**(24), 3824–3835 (2012)
65. F. Steiner, F. Da Ros, M.P. Yankov, G. Böcherer, P. Schulte, S. Forchhammer, G. Kramer, Experimental verification of rate flexibility and probabilistic shaping by 4d signaling, in *Optical Fiber Communication Conference* (Optica Publishing Group, 2018), pp. M4E–3
66. Q. Guo, W.-R. Peng, Y. Cui, Y. Bai, Multi-dimensional probabilistic shaping for higher fibre nonlinearity tolerance, in *45th European Conference on Optical Communication (ECOC 2019)* (IET, 2019), pp. 1–3
67. J. Hu, Z. Sun, X. Xu, Y. Lu, Y. Qiao, Carrier phase recovery exploiting recursive blind phase search algorithm for probabilistic shaping 64qam systems, in *Asia Communications and Photonics Conference* (Optica Publishing Group, 2020), pp. M4A–318
68. I. Fatadin, D. Ives, S.J. Savory, Laser linewidth tolerance for 16-qam coherent optical systems using qpsk partitioning. IEEE Photon. Technol. Lett. **22**(9), 631–633 (2010)
69. Y. Gao, A.P.T. Lau, C. Lu, J. Wu, Y. Li, K. Xu, W. Li, J. Lin, Low-complexity two-stage carrier phase estimation for 16-qam systems using qpsk partitioning and maximum likelihood detection, in *Optical Fiber Communication Conference* (Optica Publishing Group, 2011), p. OMJ6
70. Q. Zhang, C. Shu, Viterbi and viterbi algorithm based phase recovery for probabilistically shaped signals. J. Lightwave Technol. **39**(5), 1364–1370 (2021)
71. J. Lu, S. Fu, M. Tang, M. Xiang, P. Shum, D. Liu, Low-complexity carrier phase estimation for m-ary qam based on blind phase search using simplified measurement, in *2016 15th International Conference on Optical Communications and Networks (ICOCN)* (IEEE, 2016), pp. 1–3

72. D.A. Mello, F.A. Barbosa, J.D. Reis, Interplay of probabilistic shaping and the blind phase search algorithm. J. Lightwave Technol. **36**(22), 5096–5105 (2018)
73. J. Cho, X. Chen, S. Chandrasekhar, P. Winzer, On line rates, information rates, and spectral efficiencies in probabilistically shaped qam systems. Opt. Exp. **26**(8), 9784–9791 (2018)
74. G. Böcherer, F. Steiner, P. Schulte, Bandwidth efficient and rate-matched low-density parity-check coded modulation. IEEE Trans. Commun. **63**(12), 4651–4665 (2015)
75. M. Morsy-Osman, Q. Zhuge, L.R. Chen, D.V. Plant, Feedforward carrier recovery via pilot-aided transmission for single-carrier systems with arbitrary m-qam constellations. Opt. Exp. **19**(24), 24331–24343 (2011)
76. F.R. Kschischang, S. Pasupathy, Optimal nonuniform signaling for gaussian channels. IEEE Trans. Inf. Theory **39**(3), 913–929 (1993)
77. G. Böcherer, Probabilistic signal shaping for bit-metric decoding, in *2014 IEEE International Symposium on Information Theory* (IEEE, 2014), pp. 431–435
78. H. Cui, X. Xu, S. Liu, Y. Lu, Y. Qiao, Qpsk pilot-aided carrier phase recovery algorithm for 800-gb/s/λ dp-256qam transmission, in *Asia Communications and Photonics Conference* (Optica Publishing Group, 2020), pp. M4A–217
79. A. Sheikh, A.G.I. Amat, G. Liva, Achievable information rates for coded modulation with hard decision decoding for coherent fiber-optic systems. J. Lightwave Technol. **35**(23), 5069–5078 (2017)

Chapter 4
Raman Amplifier

Stimulated Raman scattering (SRS) plays a vital role as a nonlinear process in optical communication systems. As the optical signal power increases, the SRS effect grows, causing severe inter-channel crosstalk, thereby limiting the maximum power of optical signal transmission. However, judicious utilization of SRS can transform an optical fiber into a broadband Raman amplifier, offering power amplification capabilities. This chapter deeply explores into a comprehensive exploration of SRS effects in optical fibers. Firstly, the fundamental principles of Raman scattering are analyzed, with particular emphasis on the threshold conditions for Raman amplification and the Raman gain spectrum [1]. Secondly, the optical sources used for pump light in Raman amplifiers are introduced [2–4]. Finally, recent research developments in novel Raman amplifiers are examined, and the applications of Raman amplifiers across a wide range of domains are systematically summarized [5–7].

4.1 Fundamental

When a beam of light enters a medium, a portion of the light scatters in different directions. The majority of this scattered light maintains the same frequency as the incident light, this phenomenon identified as Rayleigh scattering. However, about 1% of this scattered light has frequencies different from the incident light, and this type of scattered light is known as Raman scattering. From a quantum perspective, Rayleigh scattering involves molecules in their ground state absorbing a radiative photon to reach a virtual energy level, then transmitted back to the initial energy level and emitted a photon with the same frequency as the incident radiation. However, when a photon is absorbed by molecules to reach a virtual energy level, Raman scattering, then transmitted to a different energy level and emitted a photon with a frequency different from the incident radiation. Rayleigh scattering solely

X. Xin, *Advanced Optical Fiber Transmission Systems*, Wireless Networks,
https://doi.org/10.1007/978-3-032-09516-9_4

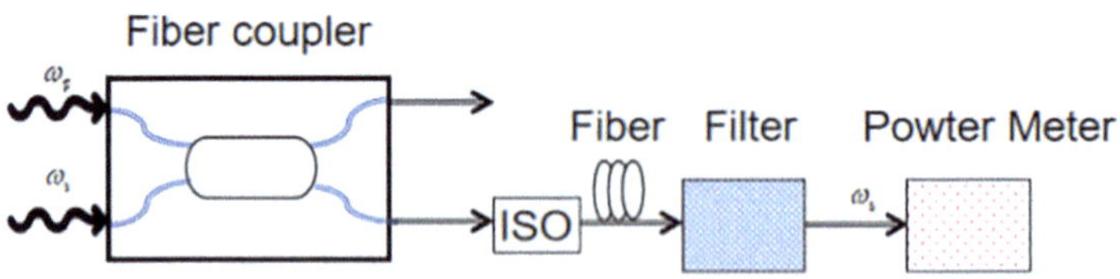

Fig. 4.1 Fiber Raman Amplifier system. The pump and signal light are coupled through a fiber coupler and introduced into the optical fiber to amplify the signal light power. Subsequently, a filter is used to eliminate non-signal light

alters the direction of motion without any exchange of energy, whereas Raman scattering involves both a change in motion direction and an exchange of energy.

Raman scattering primarily falls into two categories: spontaneous Raman scattering and stimulated Raman scattering. In 1928, Raman discovered that within any molecular medium (solid, liquid, transparent gas, etc.), spontaneous Raman scattering could transfer a portion of the incident light's energy from one optical field to another. The scattered photons' frequency differs from the incident photons', influenced by the vibrational modes within the medium. This phenomenon is known as Raman scattering [8]. However, in optical fiber transmission, the effect of spontaneous Raman scattering is exceedingly weak, so much so that it can be neglected. In 1962, researchers observed that under a strong pump light field, significant Raman scattering effects could occur, with most of the energy transferring to the scattered light, leading to its rapid growth within the medium (exponential). This process is termed stimulated Raman scattering [9].

Practically, observing SRS in optical fibers via continuous-wave (CW) pump beams proves challenging due to its relatively high threshold ($\sim$1 W). However, co-launching a Stokes beam at the appropriate frequency with the pump beam (Fig. 4.1) enables significant amplification even with CW pump powers as low as 100 mW. The near-isotropic nature of SRS permits counter-propagating pump and signal beams. In fact, the backward-pumping configuration is often preferred in practical implementations,owing to its superior performance in Raman amplifiers. While fiber-based Raman amplifiers garnered substantial interest in the 1980s, their maturation for telecommunications awaited the development of suitable pump lasers in the late 1990s [10–12].

This section primarily introduces the fundamental principles of Raman and discusses its gain spectrum and threshold.

4.1.1 Raman Scattering

Raman scattering refers to the phenomenon where incident light interacts with molecules within a medium, causing a frequency shift in the scattered light. Macroscopically, Raman scattering involves transferring energy from a portion of high-frequency photons to lower-frequency photons, facilitating the conversion of power between different-frequency photons. From a quantum mechanics perspective, Raman scattering is like a tow-photon process. A molecule scatters an incident photon of high frequency and energy into another photon of lower frequency. Simultaneously, the molecule undergoes a transition from one vibrational state to another. The resulting lower-frequency photon is referred to as the Stokes wave, where the incident light is considered the pump light wave. Operating on the principle of nonlinear optical effects, fiber Raman amplifiers exploit stimulated Raman scattering generated during the propagation of intense pump light through optical fibers. Consequently, vibrations in the silicon molecules constituting the optical fiber interact with the pump light, generating scattered light at wavelengths longer than that of the pump light ω_R (ω_{P-QR} Stokes light). The scattered light overlaps with the signal light W_S to be amplified, thereby amplifying the weak signal light and resulting in Raman gain. Assuming the energy of the incident light is h_{v_p} , in the Stokes scattering process, a molecule initially in state 'a' absorbs the energy of an incident photon, transitioning to a higher virtual energy level 'c'. However, due to the molecule's instability at this virtual level, it radiates a photon approximately picoseconds later, known as the shifted photon. Consequently, the original molecule shifts to an intermediate energy level 'b'. On this intermediate level, after a relatively longer time, molecular vibration relaxation occurs within the medium, or anti-Stokes scattering takes place. The process of anti-Stokes scattering is depicted in figure 'b'. While in the intermediate level 'b', molecules get excited by an incident photon, transitioning them to a higher virtual energy level 'd'. Similarly, as the molecule is unstable at this virtual energy level, after about picoseconds, it scatters a photon, concurrently transitioning the molecule back to its initial state. h_{v_s} represents the energy of Stokes photons, h_{v_ω} signifies the energy of anti-Stokes photons, and h_{v_m} represents the energy difference between energy levels, specifically between levels 'd' and 'c'. According to the law of energy conservation, the frequencies of the scattered light are symmetrical around the frequency of the incident light. As show in Fig. 4.2, there have tow Stokes lines, when the frequency of a scattered photon is lower than that of the incident photon, Stokes lines appear in the left region of the Raman spectrum. Conversely, if the emitted photon's frequency exceeds the incident photon's frequency, anti-Stokes lines are observed in the right region [13] (Fig. 4.3).

4.1.2 Raman Gain Spectrum

The Raman gain coefficient g_R is a critical parameter characterizing a Raman amplifier. g_R is a function of Ω, $\Omega=\omega_p - \omega_s$ where represents the Raman shift or Stokes shift, with ω_p being the pump light frequency and ω_s being the Stokes light frequency, g_R delineates how the pump light power is transferred to the Stokes light power under the SRS effect, illustrating the trend in the growth of Stokes light power. For simplicity, the anti-Stokes wave is neglected, and the analysis is focused

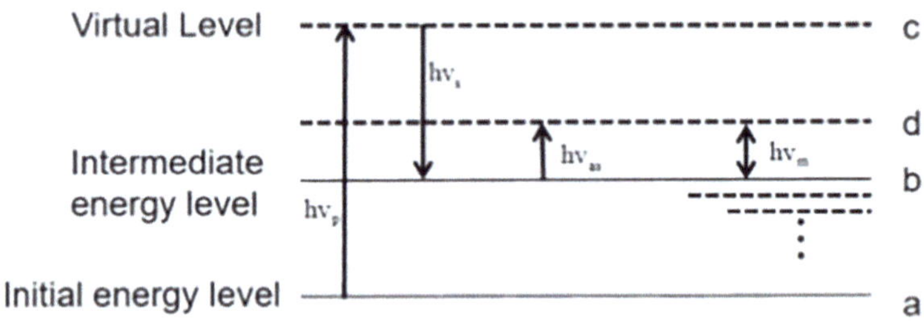

Fig. 4.2 Stimulated Raman scattering

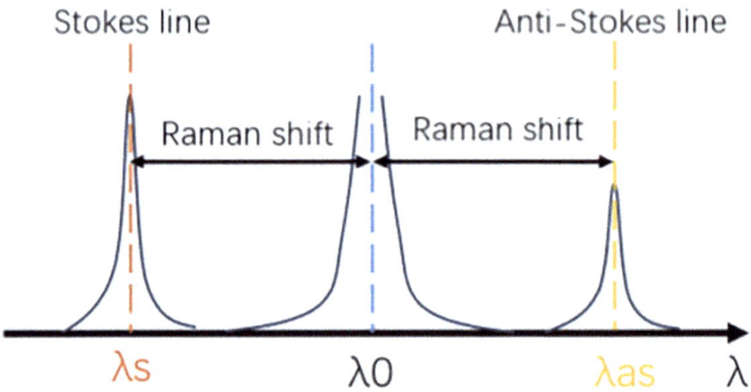

Fig. 4.3 Stokes lines and anti-Stokes lines

on fundamental Raman gain under continuous wave (CW) or quasi-CW conditions [14]. The initial growth of a weak optical signal is governed by:

$$\frac{dI_s}{dz} = \gamma_R(\omega_p, \omega_s)I_pI_s \tag{4.1}$$

$$\mathrm{I_p} = \frac{\mathrm{P_{pump}}}{\mathrm{A_{eff}}} \tag{4.2}$$

where γ_R (Ω) is related to (Ω), I_P and I_S are the optical intensities associated with the pump and signal fields having frequency ωp and ω_s , respectively. Pump is the optical power of pump light, A_{eff} is the effective core area of the fiber, which can be defined as 1.2.3.

$$A_{\text{eff}} = \frac{\left(\int\int_{-\infty}^{\infty} I_P\,(x, y, z)\,dxdy\right)\left(\int\int_{-\infty}^{\infty} I_s\,(x, y, z)\,dxdy\right)}{\int\int_{-\infty}^{\infty} I_P\,(x, y, z)\,I_s\,(x, y, z)\,dxdy} \tag{4.3}$$

We assume that the mode field distributions of pump light and Stokes light are approximately the same, and further use Gaussian function to simulate the mode field distribution function, the above expression will be greatly simplified, and we obtain the approximate relationship of effective mode field diameter: $A_{\text{eff}} \approx \pi\omega^2$, ω is the field-mode radius, and for any fiber, A_{eff} is a known parameter with values ranging from 10 to 100 μm2, depending on the fiber design [15].

The formula (4.1) indicates that the growth in signal light intensity is related to the Raman gain coefficient g_R and the pump light intensity I_P . g_R is influenced by the fiber core material and structure; different materials possess distinct molecular vibration modes and optical characteristics, directly impacting the efficiency of Raman scattering. In smaller fiber cores, the light field's intensity is higher, leading to a stronger Raman scattering efficiency. Impurities, defects, and dopants in the optical fiber also affect g_R . Some dopants, such as erbium, can enhance the Raman amplification coefficient. It also depends on the polarization state between the signal and pump light (orthogonal polarization or co-polarization), which will be elaborated on in Sect. 4.4. High pump light power is a necessary condition for significant SRS. Generally, as the pump light intensity increases, the SRS effect becomes more pronounced. Section 4.3 will study the threshold conditions for Raman amplification.

The Raman gain spectrum in silica-based fiber has long been calculated by researchers. The following Fig. 4.4 shows the normalized Raman gain spectrum of standard silica fiber. Under pump light at 1500 nm wavelength, its gain bandwidth can reach more than 30 THz (close to 40 THz), in other words, it can provide a gain bandwidth of nearly 100 nm. As can be seen from Fig. 4.1, due to the amorphous nature of silica glass, the molecular vibration frequency diffuses into overlapping energy bands and generates a continuous spectrum, and there are multiple intermediate energy levels in molecular vibration. Electrons in the virtual energy level can jump back to different energy levels and emit photons of different

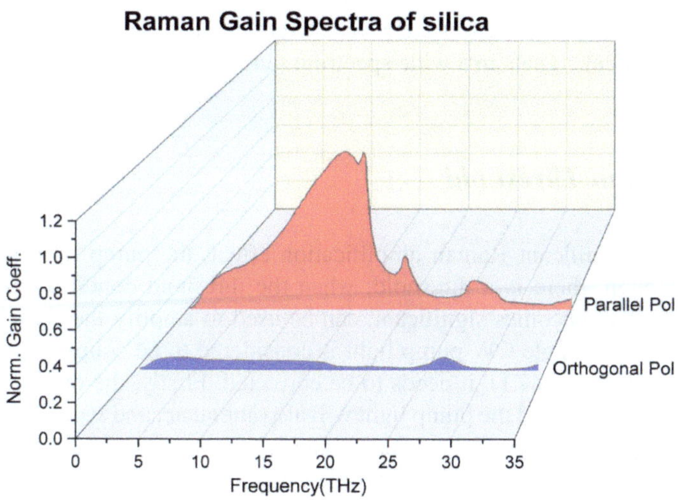

Fig. 4.4 Normalized Raman gain coefficients in a typical silica fiber

wavelengths, and the scattered light (Stokes wave) has a large frequency range, so stimulated Raman amplification has a wide gain spectrum. The amplification of signal light is primarily determined by the pump-signal frequency difference. Maximum gain occurs at a 13-THz frequency difference. Below this value, the gain coefficient scales roughly proportionally with the frequency difference, whereas above 13 THz, it becomes a pproximately inversely proportional. The resulting flat gain spectrum peaks near 13 THz, exhibiting a full width at half maximum (FWHM) of nearly 7 THz. Obviously, the normalized gain coefficient is also related to the polarization direction, the gain coefficient of orthogonal polarization is very small, and the gain coefficient of co-polarization is very large, so it is necessary to pay attention to the influence of polarization direction on the gain when pumping.

When the difference between the pump optical frequency ω_p and the signal optical frequency ω_S Is within the Stokes shift range, the signal optical I_S is stimulated by the pump optical I_p with larger optical power, and the pump optical energy is transformed into the signal optical energy, and the signal optical power is amplified. Because the Raman gain spectrum has gains over the entire bandwidth range, signal light at all frequencies is amplified as long as the pumping power exceeds the threshold of stimulated Raman scattering. The only difference is that the gain of the signal light with different frequencies is different, especially when the frequency difference is near 13.2 THz, the signal light will reach the maximum gain state first.

If the pump frequency is suitable and the Raman shift is within the gain spectrum, the signal optical power can be amplified effectively. That is to say, according to the signal light frequency, choose the appropriate pump light source, the O-band, C-band and S-band signal can be amplified, not limited by the band. This characteristic of fiber Raman amplifier is very different from erbium-doped fiber amplifier, the signal light does not need to be in a specific band to be amplified, and it is widely used in long-distance wavelength division multiplexing systems for amplifying optical signals in a wide spectrum range.

4.1.3 Raman Threshold

To achieve a significant Raman amplification effect, the pump light power must be large enough, there is a threshold, when the threshold condition is exceeded, the Raman effect becomes significant, can be used to amplify the signal light. For simplicity, only a single CW pump light is considered to be coupled into the fiber. Before using formula (4.1), it needs to be corrected. Firstly, the optical signal will attenuate in the fiber, and the pump light will also attenuate, and secondly, the optical power of the pump light will also change due to the nonlinear interaction of energy exchange between the pump light and Stokes wave. Hence, the formula (4.1) can be rewritten as:

$$\begin{aligned} \frac{\mathrm{d}I_s}{\mathrm{d}z} &= g_{\mathrm{R}} I_p I_s - \alpha_s I_s \quad \text{(a)} \\ \frac{\mathrm{d}I_p}{\mathrm{d}z} &= -\frac{\omega_p}{\omega_s} g_{\mathrm{R}} I_p I_s - \alpha_p I_p \quad \text{(b)} \end{aligned} \tag{4.4}$$

where, $\aleph_S$ and $\aleph_p$ represent the fiber loss at Stokes and pump frequencies, respectively. The first term on the right of formula (4.4) represents the light gain generated by Raman amplification, and the second term represents the dispersion of signal light in the fiber. The first term on the right of (b) represents the optical attenuation caused by the conversion of the pump optical power to the signal optical power, and the second term represents the dispersion of the pump light in the fiber [16].

According to the SRS process, the number of photons in the pump light and Stokes beam remains constant, and the following expression exists:

$$\frac{\mathrm{d}}{\mathrm{d}z}\left(\frac{I_s}{\omega_s} + \frac{I_p}{\omega_p}\right) = 0 \tag{4.5}$$

Because dispersion is the main cause of pump light loss, when estimating the Raman threshold, we ignore the first term to the right of equation (b) and bring it into equation (a) to get:

$$\begin{aligned} &\mathrm{d}I_s/\mathrm{d}c = g_{\mathrm{R}} I_0 \exp(-\alpha_p z) I_s - \alpha_s I_s \quad \text{(a)} \\ &I_s(L) = I_s(0) \exp(g_{\mathrm{R}} I_0 L_{\mathrm{eff}} - \alpha_s L) \quad \text{(b)} \end{aligned} \tag{4.6}$$

where I_0 is the incident pump intensity at $z = 0$, $L_{\mathrm{eff}} = [1 - \exp(-\alpha_p L)]/\alpha_p$ is the effective fiber length. $I_s(L)$ is integrated in the entire gain spectrum bandwidth, and with reasonable approximate calculation we can be obtained:

$$\begin{aligned} &P_s(L) = P_{s0}^{\mathrm{eff}} \exp[g_{\mathrm{R}}(\Omega_{\mathrm{R}}) I_0 L_{\mathrm{eff}} - \alpha_s L] \\ &P_{s0}^{\mathrm{eff}} = \hbar\omega_s \left(\frac{2\pi}{I_0 L_{\mathrm{eff}}}\right)^{1/2} \left|\frac{\partial^2 g_{\mathrm{R}}}{\partial\omega^2}\right|_{\omega=\omega_s}^{-1/2} \end{aligned} \tag{4.7}$$

According to the definition of pump threshold power, the input pump power is:

$$P_{s0}^{\mathrm{eff}} \exp(g_{\mathrm{R}} P_0 L_{\mathrm{eff}}/A_{\mathrm{eff}}) = P_0 \tag{4.8}$$

The formula gives the lowest pump optical power for Raman amplification. When the pump light with a wavelength of 1443 nm enters a standard fiber, different pump light powers will produce different gains. When the pump optical power is 200 mW, the maximum gain is 7.78 dB; When the pump optical power is 100 mW,

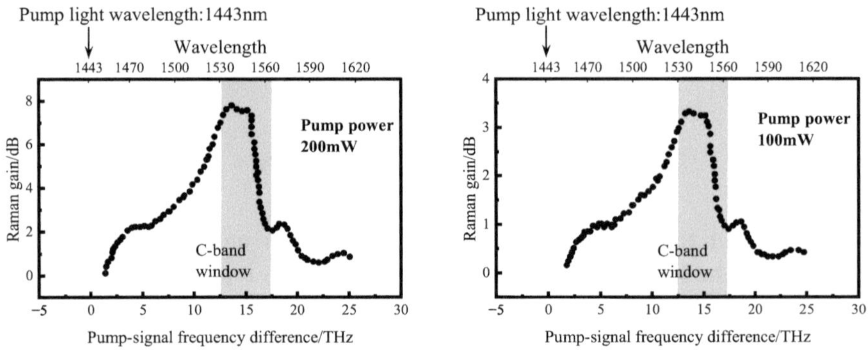

Fig. 4.5 Effect of pump power on Raman gain spectrum

the maximum gain is 3.6 dB, and the gain bandwidth near the peak gain is 7–8 THz [17] (Fig. 4.5).

Four-wave mixing will also affect the SRS effect in the medium. When analyzing the SRS principle, Raman scattering will be converted into photons of lower frequencies and phonons related to the molecular vibration mode. However, under certain energy and momentum, phonons and pump photons may combine to produce photons of higher frequencies, that is, anti-Stokes waves, whose frequency is $\omega_{\omega} = \omega_p + \omega$. Coexisting with Stokes light with frequency $\omega = \omega_p - \omega$. Because $2\omega_p = \omega_x + \omega_s$, a process in which two pump photons are simultaneously annihilate to generate Stokes and anti-Stokes photons can occur under the condition of total momentum conservation. In other words, as long as the phase matching condition is met, that is, $\Delta k = 2k(\omega_p) - k(\omega_\infty) - k(\omega_s) = 0$, four-wave mixing will be generated. To obtain a more accurate Raman threshold, consideration should be given to the propagation of anti-Stokes waves and the coupling effects caused by four-wave mixing with Stokes waves [1].

4.1.4 *Polarization Effects*

The polarization state of the pump and signal light also affects Raman amplification. As seen in Fig. 4.4, if the signal light aligns its polarization direction with the pump light (co-polarization), significant Raman gain is achieved. Conversely, when their polarization directions are orthogonal (cross-polarization), Raman gain is nearly zero. Throughout the analysis, we assumed co-polarization between the pump and signal lights. However, during signal transmission, the signal undergoes birefringence, causing random changes in the light field's polarization state. This phenomenon induces polarization mode dispersion, consequently impacting Raman amplification efficiency. Consider the transmission coupling equation under the birefringent effect:

$$\begin{cases} \frac{dP_s}{dz} = -\alpha_p P_s + \frac{1+\frac{\hat{p}\cdot\hat{s}}{2}}{g_R} P_s P_p \\ \pm\frac{dP_p}{dz} = -\alpha_p P_p + \frac{w_p}{w_s} \cdot \frac{1+\hat{p}\cdot\hat{s}}{2} g_R P_s P_p \end{cases} \tag{4.9}$$

where $\hat{s}(z)$ and $\hat{p}\,(z)$ represent the polarization states of the signal and pump lights, respectively, at position Z. The term $\frac{1+\hat{p}\cdot\hat{s}}{2} \doteq \rho(z)$ denotes the polarization overlap factor (POF). Due to the stochastic nature of birefringence effects, the polarization state also undergoes random variations. Therefore, the following study investigates the statistical characteristics of gain concerning the polarization state under forward pumping. The gain can be rewritten as follow:

$$G_{db} = \int_0^l (10 \log e) g_R \cdot \rho(z) P(0) e^{-\alpha_p z} dz \tag{4.10}$$

The polarization state of the signal at position Z can be represented as:

$$\begin{aligned} d\hat{s}(z) &= R_s(0, z)\hat{s}(0) \\ d\hat{p}(z) &= R_p(0, z)\hat{p}(0) \end{aligned} \tag{4.11}$$

where $R_s(0, z)$ and $R_p(0, z)$ represents the rotation matrix for the birefringent optical fiber. After a series of mathematical operations, we obtain the relationship between G_{dB} and $\hat{s}(0)/\hat{p}(0)$:

$$\begin{aligned} G_{dB}^2 &= K^2 \cdot L_{eff}^2 \\ &+ 2 \cdot K^2 \cdot L_{eff} \cdot \int_0^L \exp[-\alpha_p z - N_0 \frac{\Delta w^2}{3} z] \cdot (\hat{s}(0)\hat{p}(0)) dz \\ &+ K^2 \cdot (\int_0^L \exp[-\alpha_p z - N_0 \frac{\Delta w^2}{3} z] dz)^2 \cdot (\hat{s}(0)\hat{p}(0))^2 \end{aligned} \tag{4.12}$$

If in the case of orthogonal polarization, where $\hat{s}(0) * \hat{p}(0)$ equals 0, the gain is very low. However, in the case of co-polarization, where $\hat{s}(0) * \hat{p}(0)$ equals 1, the gain is maximal. FRA polarization-dependent gain is closely related to the polarization states of the signal and pump light. Employing a polarization multiplexer to depolarize the outgoing pump light is the most direct and effective method to eliminate polarization-dependent gain in FRA. Currently, most commercially available Raman amplifiers utilizing multi-wavelength pumps adopt this approach.

4.2 Pump

According to the relationship between the direction of pump light transmission and the direction of optical signal transmission, Raman amplifiers are categorized into forward Raman amplifiers, backward Raman amplifiers and bidirectionally pumped Raman amplifiers. Forward Raman amplifiers are used at the transmitter side and backward Raman amplifiers are used at the receiver side (Fig. 4.6).

Forward Raman amplifiers amplify bands that cannot be amplified by erbium-doped fiber amplifiers, improving the gain flatness of the signal. Backward Raman amplifier prevents pump noise from cross-talking into the signal, resulting in a lower noise amplifier.

Distributed and lumped configurations represent the two main categories of Raman amplifiers. It is also possible to combine these approaches into a hybrid amplifier. In this hybrid arrangement, an erbium-doped fiber amplifier (EDFA) can replace the discrete fiber Raman amplifier (FRA).

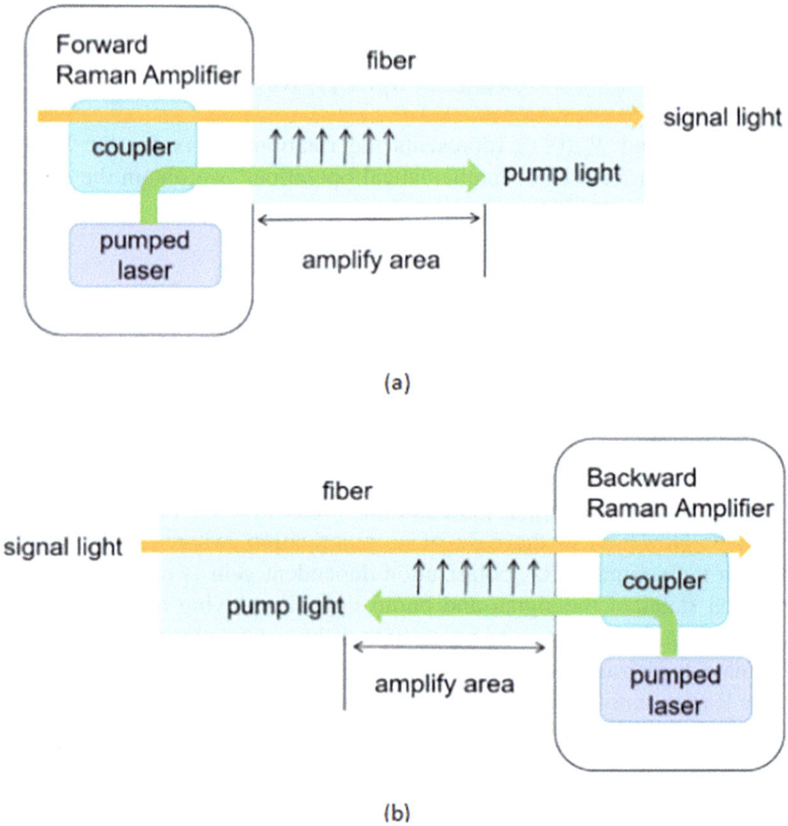

Fig. 4.6 (**a**) Forward Raman amplifier (**b**) backward Raman amplifier

The application of back-pumped Raman fiber amplifier (RFA) in dense wavelength division multiplexing (DWDM) has been verified by many experiments. It is generally believed that the optical signal-to-noise ratio (OSNR) of the backward pumped RFA is better than that of the forward pumped RFA due to the noise effect of the pump source. It is this view that makes the research and development of forward pump RFA seriously lag behind, and there are few reports on the detailed research results of forward pump RFA. When pump source noise is neglected, the signal-to-noise ratio (SNR) achieved with forward pumping typically surpasses that of backward pumping. This superiority arises because, in the forward pumping configuration, signal amplification occurs prior to fiber attenuation. Consequently, the subsequently generated amplified spontaneous emission (ASE) noise experiences suppression. Therefore, if the pump source can work under low noise, and then the pump source module is reasonably designed, the performance of the forward pump RFA is also superior, and even better than the backward pump RFA in some aspects. The low noise pump source is now readily available because of the rapid development of high-power semiconductor lasers. Therefore, it can be concluded that forward RFA will have a good application prospect in DWDM transmission system [2].

Forward pumped and backward pumped Raman fiber amplifiers (RFAs) are themselves limited by Rayleigh scattering and two-way Rayleigh scattering, and the gain provided in long-distance transmission systems is difficult to compensate for line losses, so it is usually different from other amplifiers. The application of optical amplifiers—including erbium-doped fiber amplifiers (EDFA), semiconductor optical amplifiers (SOA), and hybrid types—achieves two key objectives: compensation of transmission path attenuation and subsequent improvement in overall system performance. Nevertheless, the integration inherent in hybrid amplifiers impedes the full exploitation of Raman fiber amplifier (RFA) benefits. Crucially, the low noise figure and broad bandwidth achievable with RFA technology—distinct advantages unavailable from erbium-doped fiber amplifiers (EDFA) or semiconductor optical amplifiers (SOA)—are diminished in such combined systems.

Bidirectionally pumping is a better and more promising application of Raman fiber amplifiers. It can fully compensate the line loss, improve the system performance, but also can really realize broadband amplification, so that the advantages of Raman fiber amplifier can be fully played [3].

Figure 4.7 illustrates the standard configuration for bidirectional pumping. This architecture comprises several core elements: high-power laser diodes serving as pump sources, a pump combiner, optical isolators for both pump and signal paths, a pump-signal combiner, and wavelength multiplexer/demultiplexer components. The most important component is the high-power laser diode with appropriate wavelength. Only recently, its technique begins to mature, which greatly propels the development of RFA. Ch1 to Ch40 are typical DWDM signals. OR1 to OR40 are signal receivers. The function of each component can be understood by corresponding names. Redundant introduction is omitted. Figure 4.7 can be transformed to forward pumping RFA by removing backward pumps and to backward pumping RFA by removing forward pumps [18].

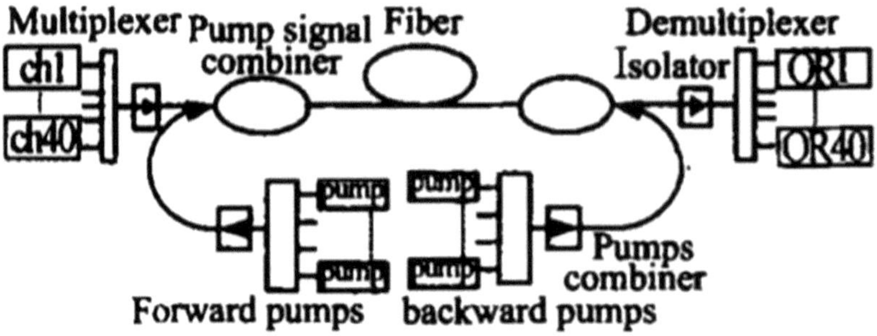

Fig. 4.7 Layout of bidirectional pumping

Table 4.1 Pumps initial power and propagating direction corresponding to four cases

		Case			
Wavelength/nm	Power/mW	a	b	c	d
1421	400	B	F	F	B
1436	140	B	F	F	B
1451	110	B	F	B	F
1466	160	B	F	B	F

While Raman fiber amplifiers (RFA) support multiple operational configurations, their performance characteristics exhibit significant variations. These distinctions primarily manifest in the evolution of signal/pump power distribution along transmission fibers, optical signal-to-noise ratio (OSNR), and gain profiles. Consequently, practical optical transmission systems demand tailored performance specifications based on application requirements [18].

Simulations are performed according to four cases, which are shown in Table 4.1. In these simulations, the average power of each channel in 40-channel DWDM signals is 0dBm. The C-band spectrum spanning 192.10–196.00 THz conforms to ITU-T channel specifications with 100 GHz spacing. For case (a), a 90-km standard single-mode fiber facilitates signal transmission achieving net-zero loss through appropriate gain compensation. In Table 4.1, B denotes that the pump propagates backward and F denotes forward. The powers of four pump sources are kept constant for four cases, but their propagating direction is different in case a, b, c and d. Flat gain can be achieved in case a.

Figure 4.8 presents the measured signal output power distribution. Backward pumping implementation enables gain variation control within 1.2 dB, confirming superior amplitude stability across channels.

Figure 4.9 shows the OSNR without considering pumps noise. The OSNR with forward pumping is the best, the worst with backward pumping, and medium with bidirectional pumping. The reason is mainly because of the ASE at signals output end. The ASE of backward pumping is the strongest because of the highest level of pumps sources. The case is contrary to forward pumping. This is consistent with the result published. In bidirectional pumping configuration, two curves show different trend for case c and d. The reason is similar to the above two cases. The OSNR

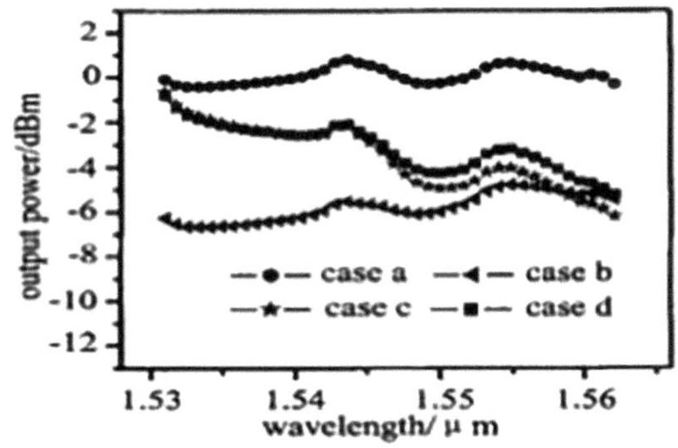

Fig. 4.8 Output signals power

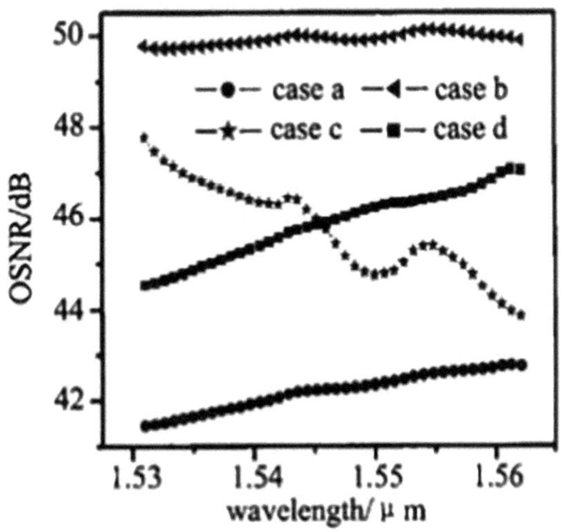

Fig. 4.9 OSNR of 40 channels

is mainly determined by the corresponding backward propagating pumps. Pump sources operating at shorter wavelengths preferentially amplify signals in corresponding spectral regions. This wavelength-dependent amplification characteristic similarly applies to longer-wavelength pump-signal interactions. Consequently, the above trend is reasonable corresponding to case c and d. Although the OSNR is bad in the backward pumping configuration as compared with the other two cases, but the smallest OSNR of each channel is still higher than 41 dB. The figure is higher than 20–22 dB defined by ITU-T.

The results show that the powers of output signals with backward pumping are larger comparatively in long haul transmission. Under conditions excluding pump ripple, forward pumping yields optimal optical signal-to-noise ratio (OSNR). All pumping configurations maintain OSNR above 41 dB, establishing a fundamental performance baseline. Implementation of backward pumping achieves gain flatness enabling uncompensated 90-km transmission for 40-channel DWDM systems (10 Gbit/s per channel). This configuration supports spectral coverage from 192.05 THz (channel 1) to 195.95 THz (channel 40) across the C-band, with gain variations constrained below 1.2 dB. The conclusion of this paper can be referred in the designing of RFA.

4.2.1 Distributed Fiber Raman Amplifiers

Raman amplification technology primarily serves three functional domains: distributed amplifiers, discrete amplifiers, and fiber lasers. Distributed configurations constitute the predominant implementation in optical transmission systems, utilizing the transmission fiber itself as the gain medium. Conversely, the latter two application categories typically employ specialty fibers engineered as Raman gain media to maximize optical conversion efficiency [19].

Raman-gain-based distributed feedback random fiber lasers exhibit an ultracompact configuration, comprising solely a pump source and passive fiber functioning as the gain medium. Such lasers fundamentally operate through wavelength conversion processes, where pump radiation is shifted to spectral regions dictated by the pump wavelength and Stokes shift characteristics of the Raman medium. The requisite random distributed feedback originates from Rayleigh scattering (RS) induced by refractive index inhomogeneities within the fiber core. Cumulative enhancement of the weak RS effect is achievable through strategic fiber length extension. Concurrently, stimulated Raman scattering (SRS) amplification enables generation of stabilized random lasing outputs with distributed feedback characteristics [20].

Distributed fiber Raman amplifiers (FRA) co-propagate pump radiation at frequency ω_p and optical signals within a common transmission fiber. Raman amplification occurs when the frequency separation $\Omega_s = \omega_p - \omega_\nu$ aligns with the Raman-gain spectrum bandwidth. This amplifier architecture maintains single-mode propagation for both wavelengths to optimize modal field overlap. A typical distributed FRA module integrates 14xx nm pumping sources without discrete gain media, comprising laser diodes, pump combiners, isolators, monitoring taps, and wavelength-division multiplexing (WDM) couplers for signal-pump integration. Polarization scramblers are essential when employing polarized pumps to mitigate polarization-dependent gain variations [21].

The reason why Raman amplifier has such great application potential is that it has advantages that EDFA does not have in the WDM transmission system, EDFA is usually a discrete amplifier. Distributed EDFA has also been reported, with the reported longest single transmission distance of only 65 km, and distributed doped fiber is relatively expensive, large spontaneous radiation noise, and other limitations, distributed EDFA ultimately did not get commercial application.

Functioning as a discrete amplification unit, the erbium-doped fiber amplifier (EDFA) delivers optical gain at specific discrete locations within the transmission infrastructure. Conversely, distributed amplification schemes generate continuous gain profiles along the entire propagation path, thereby sustaining optimal signal power levels through the fiber link. This distributed amplification approach consequently enhances overall system performance, particularly regarding noise

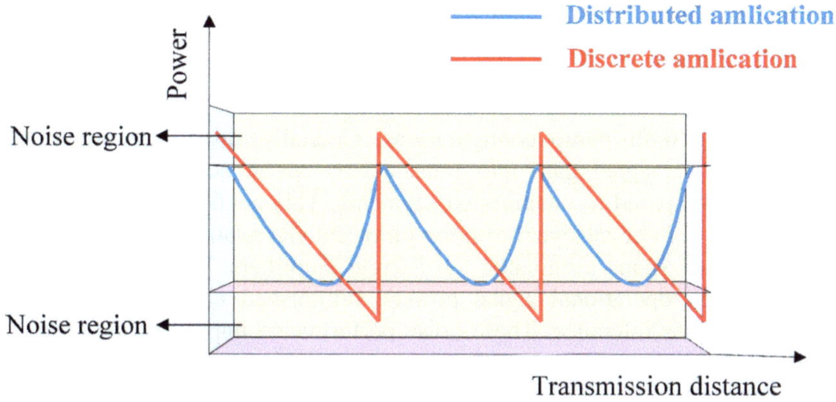

Fig. 4.10 The gain of distributed and discrete amplifiers varies along distance

suppression efficacy. Critical to system design is optimization of the optical power evolution profile, wherein signal power must remain constrained between dual operational boundaries: an upper threshold dictated by nonlinear effects and a lower limit defined by detection sensitivity. In the representative power level diagram 1-2, highlighted regions above the nominal trajectory signify power excursions exceeding maximum permissible values, while sub-nominal segments indicate noncompliant attenuation. When the signal light power exceeds the upper limit horizontal line, then the nonlinear effect of the light will have an obvious effect, when below the lower limit, because the signal light is too weak, it will produce a large noise in the next amplifier. In discrete amplification systems, it is difficult to balance power levels, so that the distribution of signal light is between the upper and lower limits, while distributed amplification is easy to do as shown in Fig. 4.10 [4].

As previously established, distributed fiber Raman amplifiers (FRA) support bidirectional pumping configurations. Under forward-pumped operation, amplification occurs proximal to the fiber input interface where signal intensity remains comparatively elevated. This configuration functionally emulates a booster amplifier stage. Conversely, backward pumping amplifies optical signals near the receiver terminal prior to photodetection—a regime where signal attenuation may render power levels commensurate with noise-floor radiation. By enhancing photodetector sensitivity and reducing minimum detectable power thresholds, this architecture effectively implements a preamplification stage for detection systems [22].

Raman amplification exhibits near-instantaneous dynamics, necessitating minimal intensity noise in forward-propagating pumps to prevent noise transfer to signal channels. Such noise coupling would elevate bit-error-rate (BER) in transmission systems. Backward pumping configurations are therefore predominantly employed, leveraging counter-propagating signal-pump geometry to average out pump-intensity fluctuations and suppress undesired crosstalk. Standard implementations integrate high-power counter-propagating Raman pumps with discrete amplification modules, notably erbium-doped fiber amplifiers (EDFAs) [23].

Distributed fiber Raman amplifiers (FRAs) demonstrate three principal advantages: structural simplicity through direct in-line signal amplification within transmission fibers; spectral flexibility enabled by Raman gain spectra shifting with pump wavelength rather than dopant emission profiles; and broad gain bandwidth accommodating multi-pump configurations. Crucially, distributed amplification uniquely elevates signal gain into transmission spans, substantially mitigating signal decay compared to unamplified segments. This configuration consequently maintains superior optical signal-to-noise ratio (OSNR) relative to systems employing passive fiber spans terminated by discrete amplifiers. The resultant benefits include reduced operational signal powers, diminished nonlinear impairments, and enhanced loss tolerance. These noise performance improvements offer three implementation pathways: extended repeater spacing, increased total system reach, or augmented transmission capacity [24].

Distributed FRAs transform transmission fibers into amplification media, conferring significant system benefits while introducing dual-role optimization challenges. Simultaneous optimization for signal propagation and amplification performance complicates fiber design. Comprehensive FRA evaluation requires multidimensional parameter analysis: accumulated linear noise, group-velocity-dispersion-induced pulse distortion, Rayleigh backscattering effects, nonlinear pump-signal interactions, and inter-channel crosstalk. These noise mechanisms attain critical importance given the extended gain distribution across kilometer-scale spans coupled with the femtosecond-order response time inherent to Raman processes.

4.2.2 *Lumped/Discrete Fiber Raman Amplifier*

Discrete fiber Raman amplifiers (FRA) employ specialty optical fibers exhibiting elevated Raman gain coefficients—typically highly germanium-doped variants—with lengths generally confined to several kilometers. This configuration necessitates substantial pump power requirements. Capable of achieving gain values exceeding 40 dB, discrete FRAs perform lumped signal amplification functionally analogous to erbium-doped fiber amplifiers (EDFA). Consequently, they primarily extend amplification coverage to spectral bands inaccessible to EDFA technology.

Lumped fiber Raman amplifiers (FRA) employ localized gain media positioned before or after transmission segments to provide full or partial attenuation compensation. Distinct from distributed configurations, these discrete units confine total pump power exclusively within the amplification module (Fig. 4.11). Specifically for counter-propagating pump schemes, optical isolators circumscribe pump radiation within the amplifier assembly. This design ensures zero residual pump radiation propagates into the transmission fiber.

Discrete fiber Raman amplifiers (FRA) employ significantly shorter gain fibers (tens versus hundreds of kilometers) compared to distributed configurations. Given the length-dependent accumulation of Raman gain, distributed architectures demonstrate superior amplification efficacy. Consequently, discrete FRAs primarily

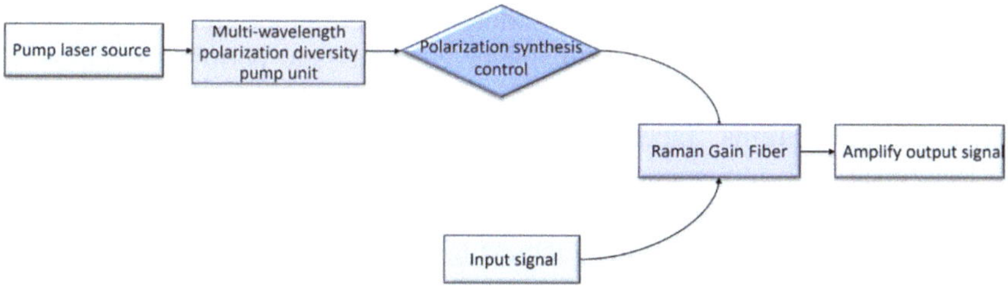

Fig. 4.11 Discrete Fiber Raman amplifier (FRA) scheme which pump power is limited to lumped elements

enhance optical network capacity by accessing spectral regions beyond conventional erbium-doped fiber amplifier (EDFA) coverage. The underlying mechanism involves signal amplification at wavelengths determined by the pump-Stokes frequency shift, enabling arbitrary gain spectral positioning through pump wavelength selection. Furthermore, multi-wavelength pump schemes facilitate customized gain profile engineering: wavelength-multiplexed pump sources enable ultrawide-bandwidth flat-gain spectra [25].

Achieving spectral gain flatness represents a critical objective in wavelength-division multiplexing (WDM) systems, ensuring uniform channel amplification. Consequently, broadband Raman amplifier optimization prioritizes gain profile equalization across operational bands. This is accomplished through simultaneous deployment of spectrally offset pump sources, whose partially overlapping Raman gain profiles generate a composite response. Precise control of individual pump wavelengths and intensities enables ultrawide-bandwidth gain uniformity. The resultant amplification bandwidth, dictated by the Stokes frequency shift magnitude, necessitates large shifts for extensive spectral coverage [15]. In silica-based fiber operating at 1550 nm, multi-pump configurations achieve approximately 100 nm amplification bandwidth—sufficient for near-complete coverage of WDM signal bands and particularly advantageous for high-density channel systems.

Broadband pumping methodologies enable demonstration of amplifiers exceeding 100 nm gain bandwidths. Standard single-mode fiber (SMF) achieves 2 dB average net gain, whereas dispersion-shifted and reverse dispersion fibers attain 6.5 dB. Gain bandwidth expansion employs a 12-channel wavelength-division multiplexed laser diode array (1405–1510 nm spectral range), integrated via planar lightwave circuit technology with an Il-MZI combiner. Nevertheless, such broadband fiber Raman amplifiers exhibit intensified noise characteristics due to pump-pump interactions—particularly demanding elevated power at shorter-wavelength pumps that amplify longer-wavelength counterparts.

Dispersion-compensating Raman amplifiers perform dual-function integration: chromatic dispersion correction and discrete amplification within a unified component. Conventional dispersion-compensating fiber (DCF) modules, operating at kilometer-scale lengths due to modal dispersion constraints, exhibit significant intrinsic insertion loss. Hansen et al. established that integrating Raman gain into DCF enables loss-compensated modules without compromising system functionality. The reduced mode-field diameter characteristic permits operation with semiconductor-laser-compatible pump powers. Resultant net-positive gain expands system design margins, facilitating integration of optical add-drop multiplexers. Reference [17] implemented a polarization-multiplexed four-channel 14xx nm pump array, demonstrating broadband amplification in DCF with ±0.5 dB gain variation across 50 nm bandwidth—achieving complete loss compensation [12].

Single-Stage Discrete Raman Amplifier (SS-DRA)

Discrete Raman amplifiers (DRAs) are distinguished from distributed configurations by confining the gain medium to discrete amplification nodes—analogous to erbium-doped fiber amplifiers (EDFAs)—rather than utilizing the transmission span itself. Consequently, DRAs forgo the significant noise figure enhancement (typically several decibels) inherent to distributed implementations. However, they leverage specialty nonlinear fibers with elevated Raman gain coefficients to minimize pump power requirements, or incorporate multi-segment fiber architectures to augment gain while mitigating detrimental nonlinear phenomena. The inherent quantum efficiency limitations necessitate kilometer-scale gain fiber lengths in DRAs, rendering nonlinear signal distortions substantially more pronounced than in EDFAs. The latter employ sub-10-meter gain media wherein nonlinear effects remain negligible [24].

Dual-Stage Discrete Raman Amplifier (DS-DRA)

A dual-stage discrete Raman amplifier (DS-DRA) configuration is schematically presented, featuring duplicated core components from single-stage designs within an identical secondary module. For simulation parameters herein, the second-stage gain medium employs either 7.5 km of inverse dispersion fiber (IDF) or 10 km of standard single-mode fiber (SMF). Backward pumping of this stage occurs through a wavelength-division multiplexing (WDM) coupler at the Fiber-2 output terminal. An inter-stage isolator provides critical module decoupling, concurrently mitigating optical signal-to-noise ratio (OSNR) degradation induced by Rayleigh backscattering. Although simulated second-stage pump wavelengths replicate primary-stage values for simplicity, empirical evidence indicates that spectral interleaving of pump sources—for instance, alternating wavelengths between Fiber-1 and Fiber-2—enhances overall gain spectral uniformity.

4.3 Raman Fiber Amplifier Improvement

According to the Raman amplification principle and gain spectrum curve, the Raman fiber amplifier (RFA) suffers from uneven gain issues. To achieve a more even gain, larger gain bandwidth, and greater gain efficiency, researchers have explored various methods involving light sources, pumping schemes, filtering, and more. This section will delve into theoretical research on RFA, methods for spectral flattening, and conclude by discussing relevant applications of RFA.

4.3.1 The Research in Different Conditions

The gain spectrum of Raman amplification demonstrates different Raman gain coefficients for various frequencies. In Wavelength Division Multiplexing (WDM) systems that transmit optical signals at different wavelengths, Raman Fiber Amplifiers (RFAs) are often used to boost signal power. It is crucial to ensure uniform amplification across all wavelengths to maintain signal quality and system performance. This approach helps in reducing signal distortion and enhances system stability.

Raman Fiber Amplifiers are one of the key technologies used to extend the reach and capacity of Dense Wavelength Division Multiplexing (DWDM) optical fiber communication systems. Ensuring flat gain across the bandwidth range is a critical focus area in its research. The flatness of gain primarily depends on factors such as pump spacing, pump power, pump wavelength, interactions between the pump and signal, as well as fiber losses. Figure 4.12 depicts the influence of different pump configurations and signal powers on the Raman gain curve. Studying different pump arrangements along with signal power provides valuable insights for optimizing the flatness profile of the Raman amplifier's gain [25].

In [5], we investigated the characteristics of single-sideband (SSB) optical signals generated through the four-wave mixing (FWM) effect in dispersion-shifted fiber (DSF). The study revealed several key factors affecting optimization, including signal and pump power levels, fiber length, relative positions of the pump and signal, dispersion slope, and the zero-dispersion wavelength of the fiber. Figure 4.13 shows the experimental setup: the generated SSB signals, after amplification by an Erbium-Doped Fiber Amplifier (EDFA), are combined with the pump light via an optical coupler. The signals are then transmitted through DSF, and the optical power is measured using a power meter.

Our study focused on investigating the impact of fiber length, pump power, and wavelength on the Q-value and FWM efficiency. We determined the optimal length range for highly nonlinear DFS (Fig. 4.14), the optimal power range for the pump light (Fig. 4.15), and the influence of pump wavelength. Similarly, we examined the impact of differences between the pump wavelength and the zero-dispersion wavelength of the fiber on FWM efficiency. FWM efficiency is higher

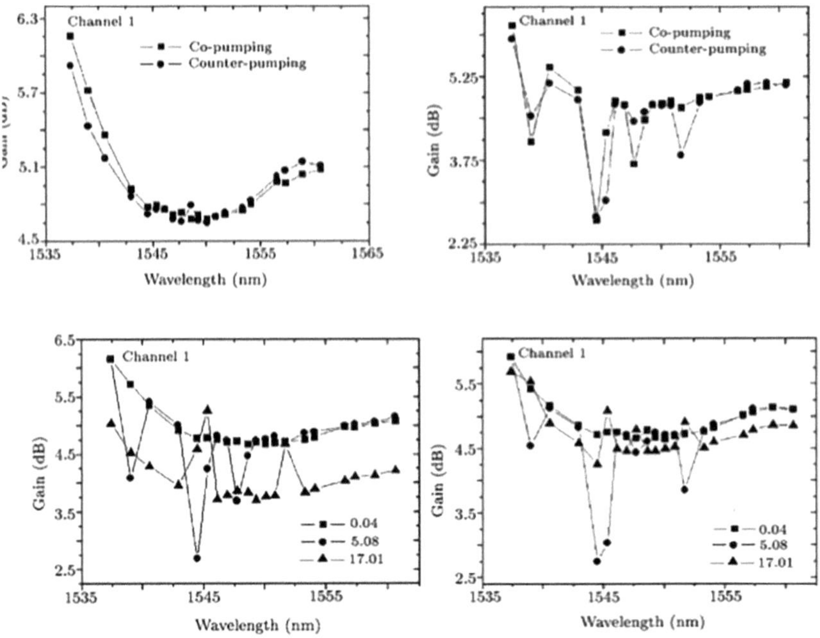

Fig. 4.12 The influence of pump direction and signal power on Raman gain. The above two images illustrate the impact of different signal powers on gain, where the signal launch power is 0.04 dBm (left) and 5.08 dBm (right), respectively. The following two images show the effects of co-pumping (left) and counter-pumping (right) on gain

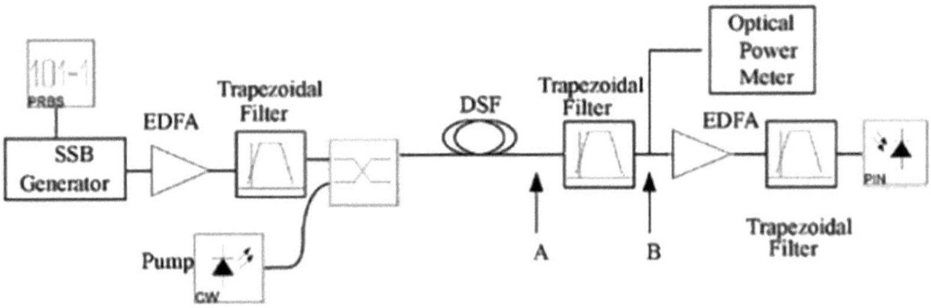

Fig. 4.13 Experimental setup for 10 Gb/s SSB optical signal wavelength conversion using FWM in DSF

for positive deviation than for negative deviation of pump (Fig. 4.16). This work providing guidelines to the utilization of these very promising technologies in the implementation of the future high debit all-optical networks [5, 26].

The channel power and pump configuration significantly influence the gain distribution of distributed Raman fiber amplifiers (DRFAs), as demonstrated in [6]. Experimental results confirm that both factors play a crucial role in determining the

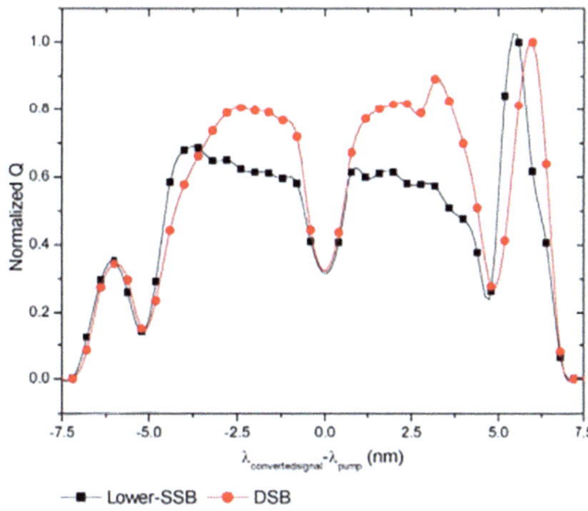

Fig. 4.14 The influence of the discrepancy between signal and pump wavelengths on the Q-factor

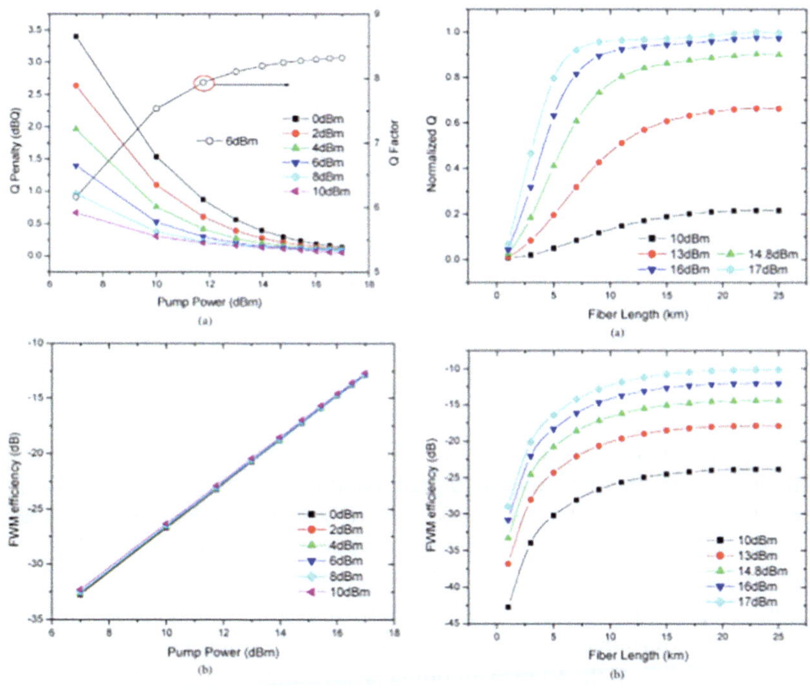

Fig. 4.15 The experimental results of pump power and fiber length on Q factor and FWM efficiency

amplifier's gain. Therefore, optimizing the gain flatness of DRFAs requires careful consideration of both channel power and the pumping scheme.

In our experiments, we validated the impact of signal power and pump configuration on the gain performance of the Raman amplifier during the transmission of 20

Fig. 4.16 The experimental results of pump wavelength on FWM efficiency

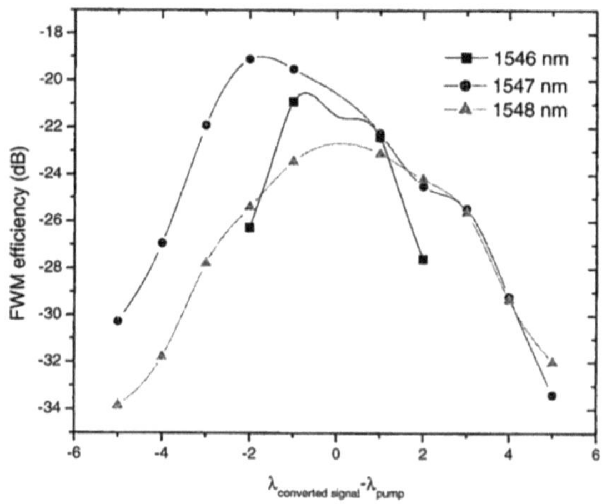

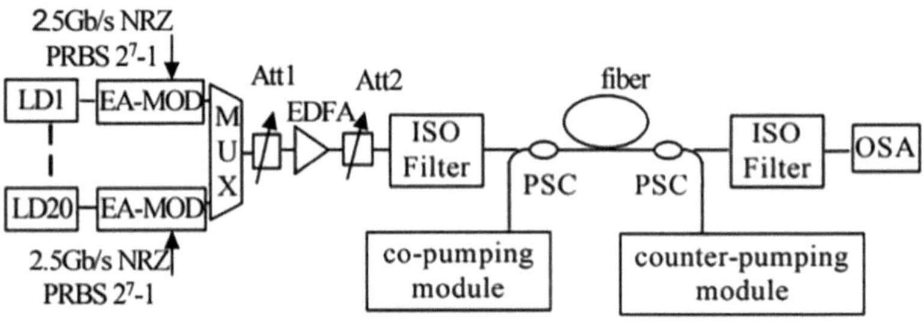

Fig. 4.17 The bidirecotionally pumped Raman fiber amplifier experimental system with function of measuring the gain saturation and the pump power conversion efficiency

independent 2.5 Gbps non-return-to-zero (NRZ) signal channels over a 75-kilometer standard single-mode fiber (SMF) link. The experimental outcomes indicate that optimizing the pump power and signal parameters enhances gain uniformity. Figure 4.17 depicts the system setup where 20 independent NRZ signals are transmitted via a multiplexer into the fiber. An Isolator (ISO) is employed to separate the pump light. The testing system utilizes both co-pumping and counter-pumping bidirectional pump methods, with the pump wavelength being fixed while the pump power remains adjustable. Figure 4.18 demonstrates the impact of different pumping methods on the gain of the Raman amplifier. The results show that the optical signal-to-noise ratio (OSNR) achieved with bidirectional pumping lies between those of co-pumping and counter-pumping, and can be further improved by appropriately increasing the co-propagating gain. Gain saturation is influenced by the ratio of co-gain to counter-gain: as this ratio increases, the saturation power decreases and the gain saturation becomes more pronounced. These characteristics of bidirectional pumping offer greater flexibility in the design of Raman Fiber Amplifiers (RFAs). If OSNR is a primary concern in the transmission system, the co-gain should be

wavelength (nm)	power (mw)							
	Case 1		Case 2		case 3		co-pumping only	counter-pumping only
	co	counter	co	counter	Co	counter		
1426	110	70	110	90	141	139	139	141
1426	110	70	110	90	141	139	139	141
1456	90	50	90	80	145	130	130	145
1456	90	50	90	80	145	130	130	145

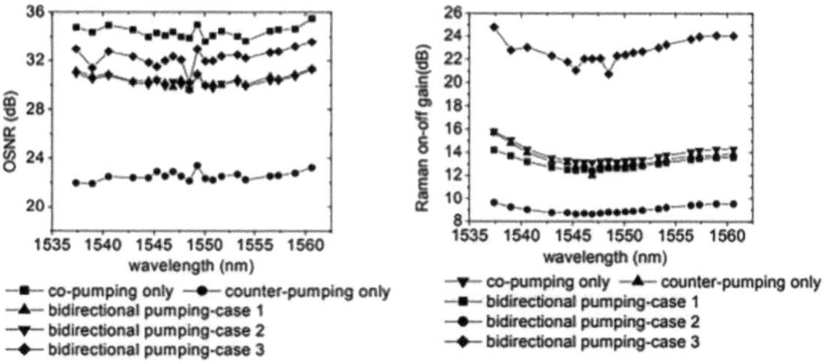

Fig. 4.18 The experiment condition and result. The upper table clarify the different pump conditions (pump power and configuration) and the lower figure display the result with the conditions in the table

moderately increased. Conversely, if gain saturation or dynamic performance is prioritized, the co-gain should be reduced accordingly [6, 7].

4.3.2 Raman Gain Spectrum Flatten

Assuming the fixed utilization of a certain wavelength for the pump light, the gain obtained at different wavelengths varies according to the gain spectrum. To avoid this scenario, a simple approach involves employing multiple pump lights with different wavelengths and controlling their wavelengths and powers to achieve bandwidth gain flatness. For example, by employing a broadband Amplified Spontaneous Emission (ASE) light source (Fig. 4.19), the powers of three different pump wavelengths can be individually controlled using a pump controller. An optical isolator (ISO) is used to prevent residual pump light from reflecting back into the ASE source, thereby protecting it from potential damage. This setup allows for uniform power amplification across different wavelengths, achieving gain flatness of <1 dB in the C-band amplification [27].

Another common method for achieving gain flatness involves the utilization of specialized optical fibers, such as highly nonlinear fibers (HNLFs) and dispersion-shifted fibers (DFS). These fibers possess high Raman gain and significant dispersion, facilitating control over Raman amplification. Moreover, HNLFs generate

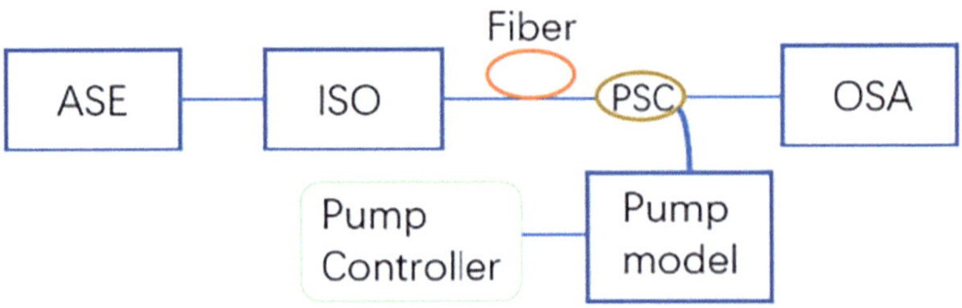

Fig. 4.19 The schematic diagram illustrating the principle of ASE (Amplified Spontaneous Emission) controlled gain spectrum

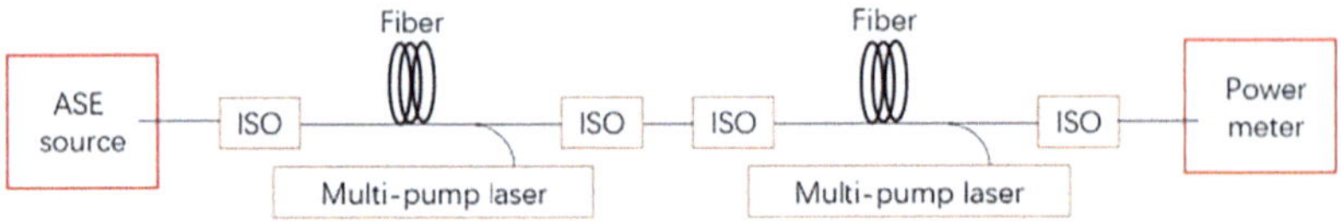

Fig. 4.20 Experimental setup of the broadband dual-stage RFA

substantial Raman gain over shorter distances, thereby reducing the required fiber length and significantly minimizing the impact of Rayleigh scattering on the signal. Consequently, using HNLF is a prevalent technique for improving gain flatness.

For instance, employing ASE as the light source, adopting a second-order pumping technique, and utilizing HNLFs, a system can be designed with four wavelengths of pump light. By implementing spectral gain control in each amplifier stage, the system can achieve both high overall gain and a flat noise coefficient. This setup allows for flat gain within the wavelength range of 1529–1609 nm, with gain fluctuations of <1 dB. The schematic diagram of the system is depicted in Fig. 4.20. The signal light undergoes optical power amplification through two Raman fiber amplification systems. Employing backward pumping for pump light, and integrating isolated pump light coupling with the signal light, these improvements enable the broadband lumped fiber Raman amplifier to simultaneously achieve high gain and low noise performance while maintaining a relatively compact design [28].

We also propose a method and device for controlling the gain spectrum of a fiber Raman amplifier. This method uses extended Kalman filtering to estimate the current state of the gain spectrum and calculates the offset between this estimate and the ideal gain profile. Using this offset and an initial correlation matrix, it determines the necessary adjustments to pump wavelength and pump power, which are then translated into driving voltage changes for the flexible grid optical network. This approach enables dynamic gain spectrum adjustment and significantly improves gain flatness. By applying perturbation theory, it models the relationship between pump power adjustments and gain spectrum variations under multiple pump wavelength combinations, constructing the correlation matrix to facilitate efficient and rapid feedback control.

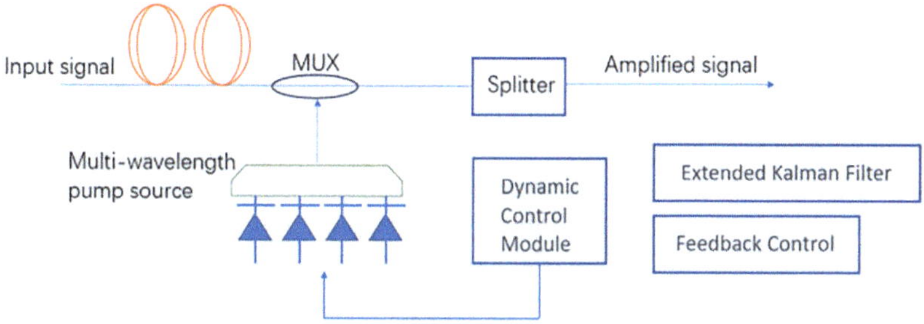

Fig. 4.21 Principle diagram of dynamic control for Raman Fiber Amplifier gain

Fig. 4.22 Results of gain spectrum before and after optimization

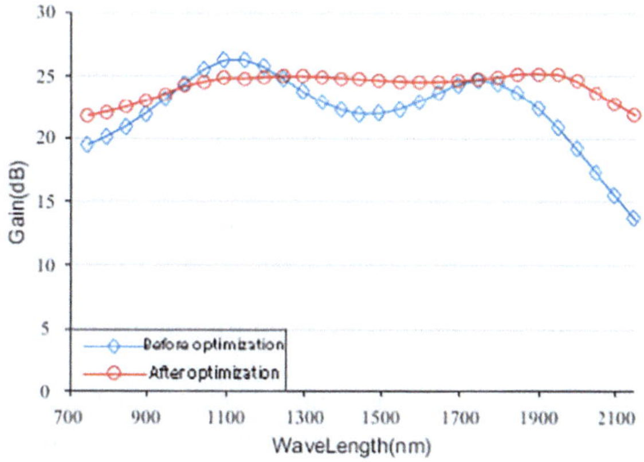

As shown in Fig. 4.21, the process begins by obtaining the initial correlation matrix relating pump power adjustments to changes in the gain spectrum of the fiber Raman amplifier under various pump wavelength combinations, based on perturbation theory. Next, measurements of multiple state variables of the amplifier are collected. Using these measurements from the previous state, extended Kalman filtering is applied to estimate the optimal current state variables. Then, based on the current estimated pump wavelength and pump power, the correlation matrix of the gain spectrum for the current state is computed. The offset from the ideal gain spectrum is calculated, and using this offset along with the correlation matrix and estimated pump parameters, the necessary pump wavelength and pump power offsets are determined. Finally, these offsets are used to calculate the adjustment needed in the driving voltage of the flexible grid optical network to correct the pump wavelength and power deviations [29] (Fig. 4.22).

4.3.3 Neural Network Model

Researchers have made improvements to Raman amplifiers in order to achieve higher transmission rates, lower power losses, and wider gain bandwidths. Various techniques such as multi-wavelength pump Raman fiber amplifiers, forward-pumped Raman fiber amplifiers, and hybrid amplifiers combining erbium-doped fiber amplifiers have been investigated and applied to create high-gain, wide-bandwidth, and high signal-to-noise ratio amplification systems. In recent years, with the advancement of computational capabilities and the development of neural networks, optimization and compensation methods for Raman fiber amplifiers based on neural network algorithms have gradually emerged and demonstrated significant improvement [30–33].

Raman fiber amplifiers are complex optical systems influenced by multiple parameters. Neural networks effectively model and predict the behavior of these nonlinear systems, including nonlinear effects, pump power, and wavelength. By training and optimizing with extensive input data, neural networks identify intricate relationships between various parameters. Using neural networks for training and optimization, they identify the optimal amplifier configurations to achieve desired performance requirements. Neural networks can be employed for real-time control of gain and gain flatness in Raman fiber amplifiers. They can automatically adjust the power and wavelength distribution of pump light based on input signal characteristics and real-time system conditions to attain the desired gain and flatness. Based on real-time monitoring data, neural networks adapt the amplifier's parameters to accommodate dynamic changes in the system. For instance, when environmental conditions or input signal characteristics change, neural networks dynamically adjust the amplifier's operation to optimize its performance to the maximum extent possible [34].

The primary optimization direction for neural network application in Raman fiber amplifiers involves integrating various neural network models, continuously learning the Raman characteristics in the channel, and providing feedback to optimize the wavelength and power of the pump lasers. Figure 4.23 represents a dynamically controlled Raman Fiber Amplifier (RFA) based on neural network algorithms. A signal processing unit dynamically controls multiple pump source wavelengths and output powers. The output detector receives the amplified optical signal, and a neural network signal processing unit optimizes it to achieve flexible, grid-based, dynamic gain control [35].

4.4 Summary

Raman amplification, vital in optical fiber systems, relies on stimulated scattering to amplify signal light. Pump light interacts with molecular vibrations, governed by factors like fiber material and pump intensity, to determine amplification efficiency.

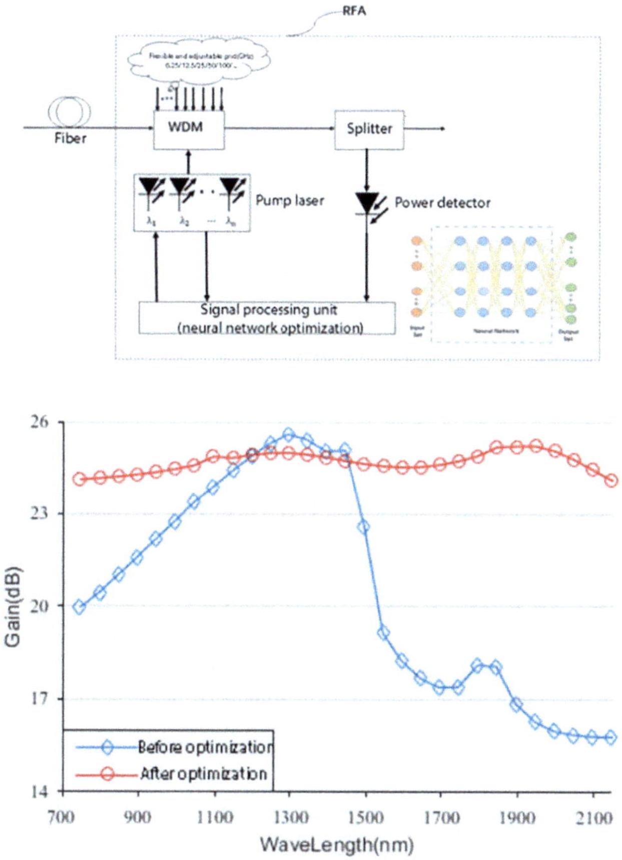

Fig. 4.23 A dynamically controlled optical fiber Raman amplifier integrating neural network algorithms

Polarization alignment enhances Raman gain, crucial for efficient amplification, while depolarization techniques ensure polarization-independent gain. Forward and backward pumping configurations offer advantages such as improved signal-to-noise ratio and gain flatness. Bidirectional pumping combines these benefits, compensating for line losses and enhancing system performance. Distributed Raman amplifiers offer simplicity and flexibility, while discrete Raman amplifiers provide localized gain compensation. Researchers explore methods like optimizing pump configurations, spectral gain control, and dynamic adjustment for Raman Fiber Amplifiers (RFAs), with neural network models emerging for real-time optimization.

References

1. X. Xin, C. Yu, J. Ren, A. Li, Q. Wu, Z. Wang, X. Yu, X. Zheng, R. Chen, G. Du, et al., Research of the performance of forward-pumped raman fiber amplifier. Study Opt. Commun. **121**, 64–67 (2004)
2. X. Xin, C. Yu, P.S. Andre, A.L.J. Teixeira, Q. Zhang, A. Li, P.M. Monteiro, Detailed research of the performance of bidirectionally pumped raman fiber amplifier. *Holography, Diffractive Optics, and Applications II*, vol. 5636 (SPIE, Bellingham, 2005), pp. 744–748
3. X. Xiangjun, Y. Chongxiu, Z. Ru, R. Jianhua, W. Qiang, X. Yu, W. Chuanlin, Z. Wenxiao, L. Anjian, R. Shuangchen, et al., The influence of configuration of raman fiber amplifier on performance. Acta Photonica Sinica **32**(2), 140–143 (2003)
4. X. Xin, C. Yu, J. Ren, A. Li, Q. Wu, C. Wang, W. Zheng, S. Ruan, G. Du, X. Yang, et al., Influence of pumping configuration in hyrid raman fiber amplifier on its performance. Guangxue Xuebao/Acta Optica Sinica **23**(11), 1311–1314 (2003)
5. X. Xin, P. André, A. Teixeira, A. Ferreira, T. Silveira, P. Monteiro, F. da Rocha, J. Pinto, Detailed numerical analysis of a four-wave mixing in dispersion-shifted fiber based all-optical wavelength converter of 10 gb/s single sideband optical signal. Opt. Fiber Technol. **12**(3), 288–295 (2006)
6. X. Xiang-Jun, Y. Chong-Xiu, Z. Ru, R. Jian-Hua, L. An-Jian, W. Qiang, X. Yu, W. Chuan-Lin, Z. Wen-Xiao, R. Shuang-Chen, et al., Influence of channel power and pumping configuration on the gain profile of distributed raman fibre amplifier. Chin. Phys. Lett. **20**(2), 237 (2003)
7. X. Xiangjun, Y. Chongxiu, R. Jianhua, L. Anjian, W. Qiang, X. Yu, W. Chuanlin, Z. Wenxiao, R. Shuangchen, D. Geguo, et al., Experimental investigation of the performance of co-and counter-pumped raman fiber amplifiers. Microwave Opt. Technol. Lett. **37**(3), 190–194 (2003)
8. C.V. Raman, A new radiation. Indian J. Phys. **2**, 387–398 (1928)
9. E. Woodbury, W. Nag, First demonstration of stimulated raman scattering. Proc. IRE **50**(6), 2347 (1962)
10. D. Gapontsev, S. Chernikov, J. Taylor, Fibre raman amplifiers for broadband operation at 1.3 μm. Opt. Commun. **166**(1–6), 85–88 (1999)
11. B. Zhu, L. Leng, L. Nelson, Y. Qian, L. Cowsar, S. Stulz, C. Doerr, L. Stulz, S. Chandrasekhar, S. Radic, et al., 3.08 tbit/s (77× 42.7 gbit/s) wdm transmission over 1200 km fibre with 100 km repeater spacing using dual c-and l-band hybrid raman/erbium-doped inline amplifiers. Electron. Lett. **37**(13), 1 (2001)
12. D. Grosz, A. Agarwal, S. Banerjee, D. Maywar, A. Kung, All-raman ultralong-haul single-wideband dwdm transmission systems with oadm capability. J. Lightwave Technol. **22**(2), 423–432 (2004)
13. G.P. Agrawal, Nonlinear fiber optics, in *Nonlinear Science at the Dawn of the 21st Century* (Springer, Berlin, 2000), pp. 195–211
14. Y.-R. Shen, Principles of nonlinear optics (1984)
15. G.P. Agrawal, *Applications of Nonlinear Fiber Optics*, vol. 10 (Academic Press, Cambridge, 2008)
16. R.G. Smith, Optical power handling capacity of low loss optical fibers as determined by stimulated raman and brillouin scattering. Appl. Opt. **11**(11), 2489–2494 (1972)
17. G. Keiser, *Optical Fiber Communications*, vol. 2 (McGraw-Hill New York, 2000)
18. X. Zhang, L. Shao, Z. Yu, D. Cheng, R. Jin, S. He, Automatic real-time control for gain-flattened fiber raman amplifiers. Opt. Commun. **239**(1–3), 79–84 (2004)
19. Y. Cai, J. Wu, Z. Deng, J. Liu, Y. Chen, Short-cavity random distributed feedback fiber laser with ultra-low threshold. Appl. Phys. Exp. **13**(1), 012008 (2019)
20. L. Sirleto, M.A. Ferrara, Fiber amplifiers and fiber lasers based on stimulated raman scattering: a review. Micromachines **11**(3), 247 (2020)
21. Y. Aoki, Properties of fiber raman amplifiers and their applicability to digital optical communication systems. J. Lightwave Technol. **6**(7), 1225–1239 (1988)

22. A.H. Gnauck, G. Charlet, P. Tran, P.J. Winzer, C.R. Doerr, J. Centanni, E. Burrows, T. Kawanishi, T. Sakamoto, K. Higuma, 25.6-tb/s wdm transmission of polarization-multiplexed rz-dqpsk signals. J. Lightwave Technol. **26**(1), 79–84 (2008)
23. S. Kobtsev, A. Pustovskikh, Improvement of raman amplifier gain flatness by broadband pumping sources. Laser Phys. **14**(12), 1488–1492 (2004)
24. Y. Emori, Y. Akasaka, S. Namiki, Broadband lossless dcf using raman amplification pumped by multichannel wdm laser diodes. Electron. Lett. **34**(22), 2145–2146 (1998)
25. C. Headley, G.P. Agrawal, *Raman Amplification in Fiber Optical Communication Systems* (Elsevier, Amsterdam, 2005)
26. X. Xiangjun, P. André, A. Teixeira, A. Ferreira, T. Silveira, P. Monteiro, F. Da Rocha, J. Pinto, The influence of pump wavelength on the performance of wavelength conversion of 10-gb/s single-side-band optical signal based on four-wave mixing in dispersion-shifted fiber. Microwave Opt. Technol. Lett. **46**(5), 493–495 (2005)
27. X. Xin, C. Yu, J. Ren, Q. Wu, Y. Xin, C. Wang, W. Zheng, A. Li, S. Ruan, G. Du, et al., The influence of signals power and pumping configuration on the gain of raman amplifier. Study Opt. Commun. **04**, 61–63 (2003)
28. S. Liang, S. Jain, L. Xu, K.R. Bottrill, N. Taengnoi, M. Guasoni, P. Zhang, M. Xiao, Q. Kang, Y. Jung, et al., High gain, low noise, spectral-gain-controlled, broadband lumped fiber raman amplifier. J. Lightwave Technol. **39**(5), 1458–1463 (2021)
29. J. Chen, X. Liu, C. Lu, Y. Wang, Z. Li, Design of multistage gain-flattened fiber raman amplifiers. J. Lightwave Technol. **24**(2), 935 (2006)
30. D. Zibar, A.M. Rosa Brusin, U.C. de Moura, F. Da Ros, V. Curri, A. Carena, Inverse system design using machine learning: the raman amplifier case. J. Lightwave Technol. **38**(4), 736–753 (2019)
31. U.C. de Moura, F.D. Ros, A.M. R. Brusin, A. Carena, D. Zibar, Experimental characterization of raman amplifier optimization through inverse system design. J. Lightwave Technol. **39**(4), 1162–1170 (2021)
32. M. Soltani, F. Da Ros, A. Carena, D. Zibar, Spectral and spatial power evolution design with machine learning-enabled raman amplification. J. Lightwave Technol. **40**(12), 3546–3556 (2022)
33. U.C. de Moura, D. Zibar, A.M.R. Brusin, A. Carena, F. Da Ros, Fiber-agnostic machine learning-based raman amplifier models. J. Lightwave Technol. **41**(1), 83–95 (2022)
34. Y. Chen, J. Du, Y. Huang, K. Xu, Z. He, Intelligent gain flattening of fmf raman amplification by machine learning based inverse design, in *Optical Fiber Communication Conference* (Optica Publishing Group, 2020), pp. T4B–1
35. M. Tang, Y. Gong, P. Shum, Dynamic properties of double-pass discrete raman amplifier with fbg-based all-optical gain clamping techniques. IEEE Photon. Technol. Lett. **16**(3), 768–770 (2004)

Chapter 5
Advanced Modulation Format Recognition Technology

Future optical network is gradually evolving from the traditional fixed system to a flexible architecture, while meeting the requirements of large capacity, high speed and dynamic characteristics. The next generation optical network can adaptively change the modulation format and transmission rate according to the dynamic link conditions and user business requirements [1]. Intelligent identification and monitoring of network parameters is one of the core technologies of optical communication systems. Modulation Format Identification (MFI) automatically identifies modulation formats the modulation formats based on the characteristics of the received signal to provide information for subsequent digital signal processing algorithms. Optical Performance Monitoring (OPM) is used to monitor the physical state of the optical network and the quality of the transmitted signal [2, 3]. In this chapter, a novel blind MFI method based on principal component analysis (PCA) and singular value decomposition (SVD) is proposed [4]. Then, the features are extracted based on signal constellation diagrams and the support vector machine (SVM) is used to complete the modulation format recognition [5]. Finally, we propose a joint modulation format identification and OSNR monitoring scheme based on long short-term memory neural networks (LSTM) by extracting features from raw IQ data samples [6]. Transfer learning is also introduced to optimize the joint monitoring scheme, so that it can be adaptively adjusted for different channel conditions [7].

5.1 Principal Component Analysis and Singular Value Decomposition Method

Emerging applications such as cloud computing and streaming media bring significant challenges to optical networks in terms of bandwidth and flexibility [8]. To address these challenges, optical networks are evolving from traditional fixed-

X. Xin, *Advanced Optical Fiber Transmission Systems*, Wireless Networks,
https://doi.org/10.1007/978-3-032-09516-9_5

wavelength grid architectures to flexible adaptive architectures [9]. In elastic optical networks (EONs), a variety of modulation formats exist to match different data rates and bandwidth requirements. Therefore, modulation format recognition algorithms need to be deployed in digital receivers to achieve correct reception and demodulation.

Digital signal processing (DSP) flow of optical fiber communication digital receivers includes multiple steps to compensate for different signal impairments. Some of these steps, such as carrier frequency offset estimation and phase recovery, require a priori information on the modulation format to work properly or achieve excellent results. Therefore, it is necessary to determine the modulation format of signals before performing these steps to compensate for the signal.

According to technical characteristics, these methods can be classified into the following categories. Based on aided data, such as training sequence and radio frequency (RF), various MFI methods have recently been proposed [10–12]. An RF-pilot aided MFI method is proposed in [10] to enable a hitless flexible transceiver. Additionally aided by RF data, MFI based on frequency offset loading are proposed [11, 12]. These methods often achieve excellent modulation format independence identify performance. However, they require additional cost to aid identification, which results in a reduction in the effective rate or bandwidth utilization. To avoid the loss of efficiency, a large number of MFI methods are based on statistical features such as signal cumulants, amplitude histograms or peak-to-average power ratio (PAPR) [13–19]. These methods classify different modulation formats by setting the threshold of the statistical variable. For example, the fourth order cumulants of received signals are calculated in the literature [13] for the distinction of on-off keying (OOK), binary phase shift keying (BPSK), quadrature phase-shift keying (QPSK) and 16 Quadrature amplitude modulation (QAM). A scheme based on the intensity profile is proposed in [14] to identify QPSK and 8/16/32/64 QAM by amplitude features. For the identify of QPSK and 16/32/64 QAM, a method based on the information entropy is proposed in [15], which is calculated from the amplitude distribution. By using some transformations, more hidden features can be obtained. These transformations could be fourth-power [16], nonlinear power transformation [17] or others [18, 19].

Applying the Stokes transformation, a variety of MFI methods in Stokes space have been proposed [20–24]. By mapping the received signals into Stokes space and extracting the density distributions, a low complexity MFI method is proposed in [20]. When different energy level features of different modulation formats are considered, a density peak-based MFI method is proposed in [21]. To reduce clusters in Stokes space, the method in [22] uses square operation before mapping and PCA for improved performance with less complexity. For feature extraction, [23] proposes two schemes based on SVD and radon transformation in Stokes space analysis. For the clustering algorithm in Stokes space, an improved particle swarm optimization clustering algorithm combined with a two-dimension Stokes plane is proposed in [24]. Recently, many MFI methods based on machine learning have been proposed [25–37]. For example, an MFI method based on deep neural networks (DNNs) combined with amplitude histograms is proposed in [25]. The identification of four modulation formats is realized in [26], by combining the density distributions in Stokes axes. A lightweight CNNs-based MFI scheme in Stokes plane is proposed in [27] for the identification of QPSK, 8PSK and 16/32/64 QAM. Another method in [28] is based on asynchronous delay-tap plot, achieving the recognition of 16/32/64 QAM by using image processing. More MFI method-based CNNs are proposed in [29, 30]. For the application of the multi-task neural networks in MFI, various of MFI methods are proposed in [31–34]. These MFI methods can achieve fast and high-accuracy identification after good training.

In this section, a blind modulation format identification based on PCA and SVD is proposed for the identification of seven modulation formats of BPSK, QPSK, 8PSK and 8/16/32/64 QAM. In our methods, the square operation of received signals and PCA is used for the removal of phase noise, so that both amplitude and phase information can be available for the identification of seven modulation formats. After PCA, the density distribution matrix of the received constellation is calculated, and then SVD is applied to the matrix to achieve noise removal and trend smoothing. Finally, a support vector machine is used for identification according to the smoothed matrix. Due to the proposed correcting method based on PCA, the phase information is available to improve the accuracy of identification, and to extend our method to the scenario even with multiple PSK modulation formats, which will be unavailable in the MFI methods based on amplitude information. In order to avoid over-fitting in the machine learning-based MFI method, a distribution matrix instead of the image is utilized for identification.

5.1.1 Algorithm Principle

In this section, the proposed MFI method based on PCA and SVD is described, and the methodology is explained by mathematical equations and figure.

Principle of MFI Method Based on PCA and SVD

Figure 5.1a describes the receiver DSP flow used in an optical digital coherent receiver containing the proposed MFI method. All the algorithms (except MFI) in flow could be divided into a modulation format independent DSP and modulation format dependent DSP. First, the modulation format independent DSPs are used to compensate the impairment of signals. Then, based on the compensated signals, MFI is executed to get the modulation format. Lastly, the modulation format dependent DSPs are configured according to the MFI results.

During the stage of modulation format independent DSPs, the received signals are normalized to the standard power at first. Then, QI compensation is used to mitigate the mismatch within the I-branch and Q-branch signals caused by receiver. After that the chromatic dispersion (CD) compensation and the nonlinear (NL) compensation are performed to compensate for the channel impairment, and the

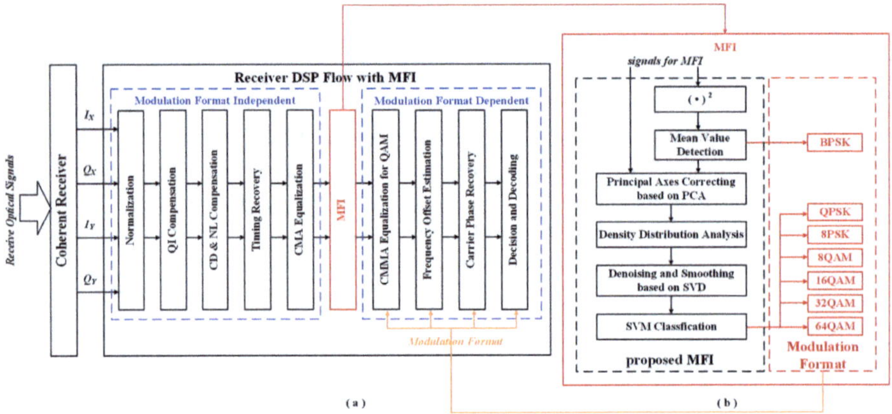

Fig. 5.1 The DSP flow in optical digital coherent receive; (**a**) Receiver DSP flow with MFI; (**b**) The proposed MFI method based on PCA and SVD

timing recovery is used to reduce timing error. Before MFI, the constant modulus algorithm (CMA) equalization is taken to compensate for residual impairment. After MFI, the modulation formats undergo cascaded multi-modulus algorithm (CMMA), frequency offset estimation, carrier phase recovery, decision and decoding, respectively.

Figure 5.1b shows the schematic of the proposed MFI method based on PCA and SVD, which consists of five crucial steps: square operation and average power detection; principal axes correction based on PCA; density distribution analysis (DDA); denoising and smoothing based on SVD; SVM classification.

Square Operation and Average Value Detection

The square operation is executed for the signals after CMA firstly because the symmetry of constellation will degenerate after this operation. For example, signals with a constellation rotational symmetry with a degree of 4 will only have a degree of 2 after the square operation.

To illustrate the changes after square operation, consider signals with the following form:

$$E_i = A_i \cdot e^{j(\theta_i + \phi)} \tag{5.1}$$

where E_i represents the i-th complex signal in the symbol sequence with the modulated amplitude of A_i and the modulated phase of θ_i, and ϕ donates the phase rotation caused by laser linewidth or initial phase difference between the laser sources at the transmitter and receiver.

Consider the rotational symmetry with the degree of n, which means:

$$\begin{aligned} E_i' &= E_i \cdot e^{j\frac{\pi}{n}} \\ &= A_i \cdot e^{j(\theta_i+\phi)} \cdot e^{j\frac{\pi}{n}} \\ &= A_i \cdot e^{j(\theta_i+\frac{\pi}{n}+\phi)} \\ &= A_k \cdot e^{j(\theta_k+\phi)} \\ &= E_k \end{aligned} \tag{5.2}$$

where E_k is another complex signal with the same modulation format as E_i.

After the square operation, the complex signal in Eq. (5.1) can be represented by:

$$\begin{aligned} E_i^2 &= A_i^2 \cdot e^{j(2\theta_i+2\phi)} \\ &= A_k^2 \cdot e^{j(2(\theta_k-\frac{\pi}{n})+2\phi)} \\ &= A_k^2 \cdot e^{j(2\theta_k-\frac{\pi}{n/2})+2\phi)} \end{aligned} \tag{5.3}$$

This means that only the rotational symmetry with the degree of $n/2$ is satisfied after the square operation. It should be noted that although the condition $\theta_k = \theta_i + \frac{\pi}{n}$ is assumed in Eq. (5.2), the same result can be deduced under the condition of $\theta_k = (\theta_i + \frac{\pi}{n}) \quad Mod2\pi$, where Mod means the modulo operation.

As shown in Fig. 5.2, the changes between constellations before and after the square operation of these seven modulation formats (BPSK, QPSK, 8PSK and 8/16/32/64 QAM). Using this property, it is easier to find the direction with the greatest variance, which will be useful in next step of principal axes correction.

Another property that should be noted is the average power after transformation. When considering the modulation format of BPSK in Eq. (5.2), the average power of 1 is obtained according to $E_i^2 = A_i^2 \cdot e^{j2\Phi}$, since $A_i = A_k = 1$ and $\theta_i, \theta_k \epsilon \{0, \pi\}$.

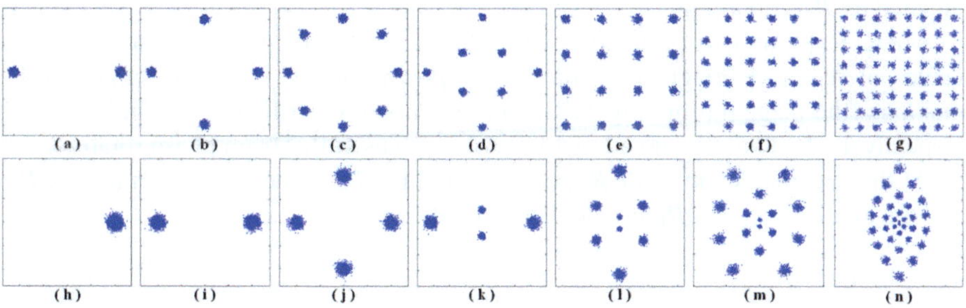

Fig. 5.2 Constellation diagrams before and after the square operation of signals in modulation formats of BPSK, QPSK, 8PSK, 8QAM, 16QAM, 32QAM and 64QAM, respectively; **(a)**–**(g)** Constellation diagrams before the square operation of signals; **(h)**–**(n)** Constellation diagrams after the square operation of signals

The average power of BPSK signals after the square operation is approximately 1, while others with an average power is approximately 0 as shown in Fig. 5.2h. Thus, the BPSK signals can be identified by detecting the average power after the square operation.

Principal Axes Correcting Based on PCA

Due to the non-uniformity of laser sources at the transmitting and receiving ends, the receiving constellation experiences an overall phase rotation. Previous MFI methods based on constellation diagrams required training with different rotation angles, significantly increasing implementation complexity. This paper proposes principal axes correction based on PCA to unify constellations with varying phase rotation angles. Thus, the phase rotation invariant properties to the constellation diagram are obtained without additional training on the phase rotation. In contrast to the application of PCA in data dimension reduction, the usage of PCA is to find the direction with the greatest variance. Furthermore, the phase rotation is removed by correcting this direction. For convenience, the coordinate axis composed of the direction with the largest variance and one of its orthogonal vectors is called the principal axis.

As shown in Fig. 5.2b–g, the distributions of the signals before the square operation are uniform, without any direction showing prominent variance. However, the distributions after the square operation exist in the desired directions, which is extracted and corrected through the process as follows.

For the received signals of length N after the square operation, a N×2 matrix

$$\mathbf{R}=\begin{bmatrix} r_{11} & r_{12} \\ r_{21} & r_{22} \\ \vdots & \vdots \\ r_{N1} & r_{N2} \end{bmatrix}=\begin{bmatrix} \mathrm{Re}\{E_1^2\} & \mathrm{Im}\{E_1^2\} \\ \mathrm{Re}\{E_2^2\} & \mathrm{Im}\{E_2^2\} \\ \vdots & \vdots \\ \mathrm{Re}\{E_N^2\} & \mathrm{Im}\{E_N^2\} \end{bmatrix} \tag{5.4}$$

donated as $\boldsymbol{R}$ is obtained:

where Re{·} and Im{·} denote the real and imaginary parts, respectively. The values of real and imaginary parts are the projections of complex signals to the real axis (the magenta line Fig. 5.3a) and imaginary axis (the red line in Fig. 5.3a), respectively. It should be noted that here our signals suffer from phase rotation.

At first the mean values of $\boldsymbol{R}$ are subtracted to obtain the zero-mean matrix $\boldsymbol{R_0}$, and the element $r_{0,ij}$ in $\boldsymbol{R_0}$ is calculated as follows:

$$r_{0,ij}=r_{ij}-\frac{1}{N}\sum_{k=1}^{N}r_{kj} \tag{5.5}$$

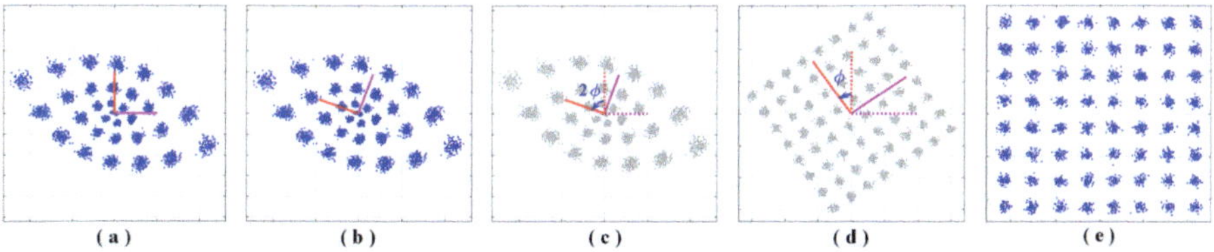

Fig. 5.3 The process diagram of principal axes corresponding based on PCA (64QAM for example). (**a**) The projections of signals on the real and imaginary axes after the square operation; (**b**) The principal axes formed by eigenvectors of covariance matrix; (**c**) The phase rotation of principal axes with respect to the original axes after the square operation; (**d**) The phase rotation of principal axes with respect to the original axes before the square operation; (**e**) Signals after principal axes correcting based on PCA

Next, the covariance matrix for the real and imaginary parts are calculated:

$$\mathbf{C} = \mathbf{R_0 R_0^{\mathit{T}}} \tag{5.6}$$

The largest eigenvector of the covariance matrix points in the direction of the greatest variance and the second largest eigenvector is always orthogonal to the largest one. The principal axes as shown in Fig. 5.3b can be attained by eigenvalue decomposition or singular value decomposition of the matrix.

The eigenvalue decomposition of covariance matrix can be expressed as:

$$\mathbf{C} = \mathbf{U \Lambda U^{\mathit{T}}} \tag{5.7}$$

where $\mathrm{U} = \left[\mathrm{u_1}\ \mathrm{u_2}\right]$ denotes the eigenvector matrix and $\Lambda = diag(\lambda_1, \lambda_2)$ is the diagonal matrix of eigenvalues. The eigenvectors $\mathrm{u_1}$ and $\mathrm{u_2}$ form the desired principal axes, which is shown in Fig. 5.3b. The eigenvector with larger eigenvalue is the direction with the greatest variance (the red line in Fig.5.3b).

In Fig. 5.2i–n, for original QPSK, 8PSK and 8QAM, the constellations after the square operation have the greatest variance in the real axis, and in the imaginary axis for 16/32/64 QAM. Since the modulation formats of received signals are unknown during the MFI process, the direction with the greatest variance is corrected to the imaginary axis consistently. The direction with greatest variance is given by the following equation:

$$\mathbf{u} = \mathbf{u}_{\mathrm{m}} = \begin{bmatrix} u_x \\ u_y \end{bmatrix}, m = \underset{i \in \{1,2\}}{\arg\max} \lambda_i \tag{5.8}$$

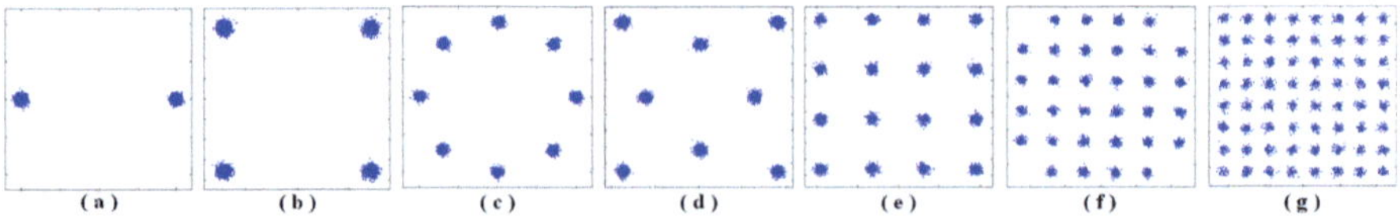

Fig. 5.4 Signals after the principal axes correcting based on PCA. (**a**)–(**g**) BPSK, QPSK, 8PSK and 8/16/32/64 QAM, respectively

The angle 2ϕ of phase rotation after the square operation shown in Fig. 5.3c is calculated as follows:

$$2\phi = \begin{cases} \arccos(u_y) & \text{, if } u_x \geq 0 \\ 2\pi - \arccos(u_y) & \text{, if } u_x < 0 \end{cases} \tag{5.9}$$

Note that this angle is between the direction with the greatest variance and the imaginary axis, so it is u_y instead of u_x in the arc-cosine function.

Equation (5.3) shows that the phase rotation after the square operation is twice as much as before. Therefore, the angle of phase rotation before the square operation can be attained by halving the 2ϕ.

Finally, the received data is projected onto the principal axes in Fig. 5.3d to obtain the data without phase rotation as shown in Fig. 5.3e. The above method can also be applied to the BPSK signals without square operation, thus, restoring its phase.

After the principal axes correcting based on PCA, the signals without phase rotation is obtained, which is shown in Fig. 5.4. Furthermore, according to the final MFI results, the constellations of QPSK and 8QAM can be rotated by 45°, so as to realize the rough phase recovery of all modulated signals and reduce the complexity of the subsequent carrier phase recovery algorithm.

Density Distribution Analysis

Figure 5.4 shows that the density distribution analysis is employed to obtain the distribution of constellations.

Considering the signals are normalized to the average energy of 1, the maximum level of the in-phase component and quadrate component is approximately 1. Thus, a range for analysis is set, and any signal outside this range is treated as noise.

This range is then divided into several intervals, both in real and imaginary axes, and these intervals are used to divide a constellation into several blocks. The number of signals in each block is then calculated.

Finally, the density is calculated according to the number of signals in each block and then a matrix of density distribution is obtained.

Denoising and Smoothing Based on SVD

Due to the presence of noise in the received signal, the constellation diagram exhibits numerous noise spots. In order to minimize the impact of noise on identification accuracy, denoising is performed using a density matrix based on SVD, and the low-value zeroing method is employed to emphasize the primary features.

The principle of it is to make use of the characteristic in which the large singular value of the density matrix contains the main information, while small singular values contain more noise. By SVD of the matrix, the part with large eigenvalues is extracted, and the denoised matrix is obtained.

For the density distribution matrix **M**, SVD is applied:

$$\mathbf{M} = \mathbf{USV}^T \tag{5.10}$$

where **U** and **V** are the left and right singular matrices, respectively. **S** is the singular value matrix of the following form:

$$\mathbf{S} = diag(\sigma_1, \sigma_2, \ldots) \tag{5.11}$$

where σ_i is the singular value. Selecting the first n larger singular values, the density distribution matrix is reconstructed as:

$$\mathbf{M}' = \mathbf{U}_{l\times n}\mathbf{S}_{m\times n}\mathbf{V}_{l\times n}^T = \begin{bmatrix} m_{11} & m_{12} & \cdots & m_{1l} \\ m_{21} & m_{22} & \cdots & m_{2l} \\ \vdots & \vdots & \ddots & \vdots \\ m_{l1} & m_{l2} & \cdots & m_{ll} \end{bmatrix} \tag{5.12}$$

where $\mathrm{U}_{(l\times n)}$ and $\mathrm{V}_{(l\times n)}$ are the first n columns of U and V with l rows, respectively. $\mathrm{S}_{(n\times n)}$ is the submatrix consisting of the first n rows and the first n columns of S.

After SVD, the density for blocks with low density is set to 0:

Here, m_{thr} is the threshold of low density.

$$m_{jk} = \begin{cases} m_{jk} & \text{, if } m_{jk} \geq m_{thr} \\ 0 & \text{, if } m_{jk} < m_{thr} \end{cases} \tag{5.13}$$

SVM Classification

BPSK has already been identified in the previous procedure, so only the remaining modulation formats of QPSK, 8PSK, and 8/16/32/64 QAM need to be classified in the SVM classifier. In order to apply the SVM classifier, the density distribution matrix is reconstructed into a vector as the feature vector.

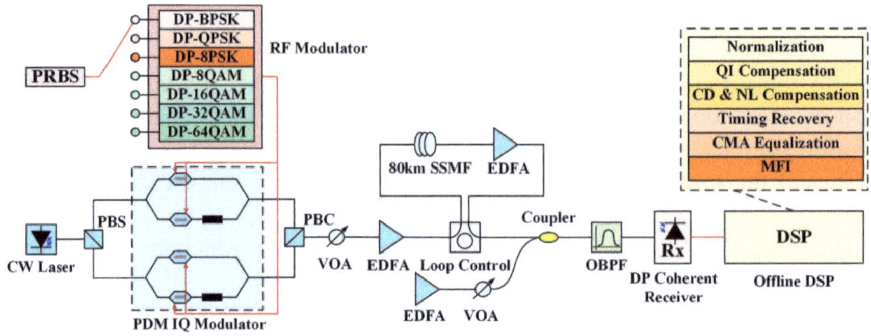

Fig. 5.5 Simulation setup of the proposed blind MFI method

5.1.2 Simulation Setup and Results

In the simulation system, the 28-GBaud signals of dual polarization (DP) BPSK, QPSK, 8PSK and 8/16/32/64 QAM is transmitted through the fiber channel. The MFI accuracy is obtained after coherent reception and modulation format independent DSP processing.

Simulation Setup

As shown in Fig. 5.5, we have built a simulation system to verify the blind MFI method. First, pseudo-random bit sequences (PRBS) are generated and used to generate different modulation signals in the RF modulator at a symbol rate of 28-GBaud. In the polarization division multiplexing (PDM) IQ modulator, four channel RF signals are modulated to two optical carriers of different polarization states, respectively. The optical carriers are generated by a continuous-wave (CW) laser with a central wavelength of 1550 nm and linewidth of 10 kHz and are split into different polarization states by a polarization beam splitter (PBS). Then, the modulated optical signals in two polarization states are combined into one by using of polarization beam combiner (PBC). A combination of a tunable optical attenuator (VOA) and an erbium-doped fiber amplifier (EDFA) is used for the power control, and the optical power is set to 0 dBm before entering the transmission link. The transmission link consists of multiple loops, and each of them contain an 80 km standard single mode fiber (SSMF) and an EDFA whose gain is 16 dB to compensate for fiber attenuation. The attenuation, chromatic dispersion, differential group delay and nonlinear refractive index coefficient of SSMF are set as 0.2 dB/km, 16.75 ps/(nm·km), 0.2 ps/km and 26×10–21 m2/W, respectively. The adjustment of different OSNR is achieved by coupling the amplified spontaneous emission (ASE) noise of different powers into the fiber at the receiving end using couplers. For this purpose, an EDFA is employed as an ASE noise source and a VOA is used to adjust the noise power. After the filtering out of out-of-band noise by an optical bandpass filter (OBPF) with a central wavelength of 1550 nm and bandwidth of 0.4 nm, the received signals are obtained in the DP coherent receiver. Finally, processed by the offline DSP in the figure, the modulation format information is obtained. It should be noted that the noise figure (NF) of the EDFA employed in the simulation system are all set as 4 dB.

Simulation Results

As shown in Fig. 5.5, OSNR varies from 5 to 35 dB with a step value of 1dB. In each OSNR, 1000 data sets are generated for each modulation format, and each of them contains 4096 samples of received signals. Then, the proposed MFI method is applied to the data sets to obtain the SVM classification feature sets. From each OSNR, 70% of the 1000 feature sets for each modulation format are mixed together as training features and the remaining 30% as testing features. Since features under different OSNR are mixed together for training and testing, the proposed method is blind and do not require prior information of OSNR.

Figure 5.6 shows the average power of signals of each modulation format after square operation. The lines of different modulation formats in Fig. 5.6 represent the mean value of the average power of 1000 data sets under each OSNR and the regions in light-color represent the range of the average power of all data sets. BPSK is obviously different from other modulation formats in the average power after square operation, so the identification of BPSK can be realized only by setting a threshold value. In our work, the threshold shown by the dotted purple line is set to 0.0554.

First of all, the signal power is squared and the mean value is detected, and then the remaining signal is calibrated based on PCA. Take the 16QAM signals for example, the constellation before and after correcting are shown, respectively, in Fig. 5.7a, b. The density distribution matrix is then obtained by the processing of density distribution analysis. For visualization and comparison, the distribution of

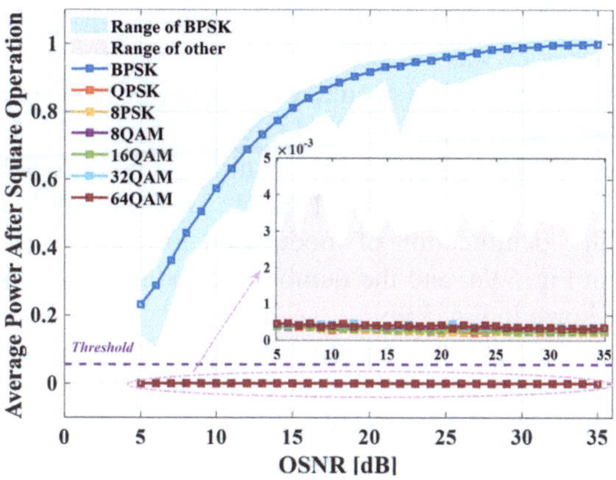

Fig. 5.6 The average power of signals after the square operation

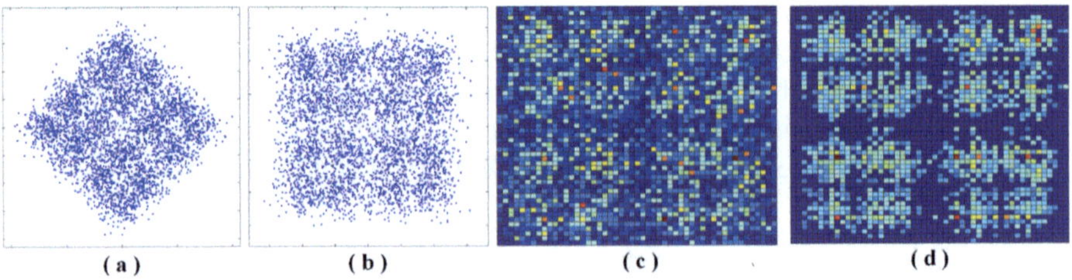

Fig. 5.7 The process and effect of the proposed MFI method after the square operation, take the 16QAM signals for example. (**a**) Constellation before principal axes correcting; (**b**) Constellation after principal axes correcting; (**c**) Density matrix obtained by DDA; (**d**) Density distribution matrix after denoising and smoothing based on SVD

Fig. 5.8 The identification accuracy of the proposed MFI

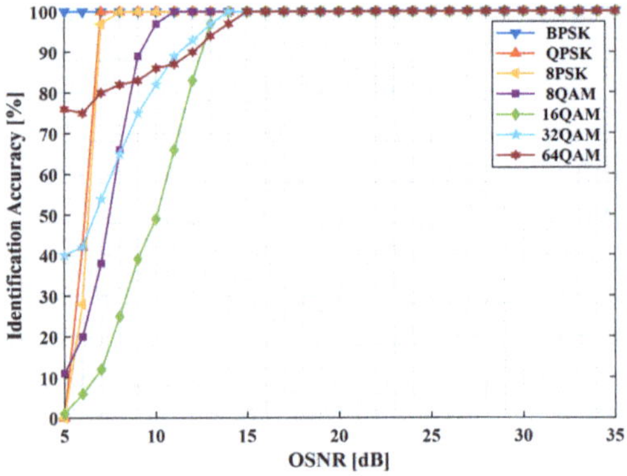

the matrix is shown in Fig. 5.7c. It can be seen that there are many noise points in the original matrix, which is not conducive to identification. In order to improve the identification accuracy, denoising and smoothing based on SVD is applied to the matrix. After the matrix is reconstructed with the first five largest eigenvalues, the distribution, as shown in Fig. 5.7d, is obtained.

The density distribution matrix is converted into a vector input to the SVM classifier, and features were extracted for training or testing. The recognition accuracy was obtained, as shown in Fig. 5.8. The minimum OSNR values to achieve an identification accuracy of 100% for these seven modulation formats of BPSK, QPSK, 8PSK, 8/16/32/64 QAM are 5, 7, 8, 11, 14, 14, and 15 dB, respectively.

We have compared the minimum OSNR required for 100% recognition accuracy with other methods in [18, 21, 22, 24, 26, 30]. The minimum OSNR required for the identification of modulation formats under different MFI methods is shown in Fig. 5.9a, and the number of modulation formats identified by these methods is shown in Fig. 5.9b.

The minimum required OSNR for the proposed MFI method to achieve 100% accuracy are 5, 7, 8, and 15 dB for BPSK, QPSK, 8PSK and 64QAM. In other methods, the best achievable performance is 8, 7.5, 13, and 21 dB, respectively. Therefore, we achieved gains of OSNR of approximately 3, 0.5, 5, and 6 dB,

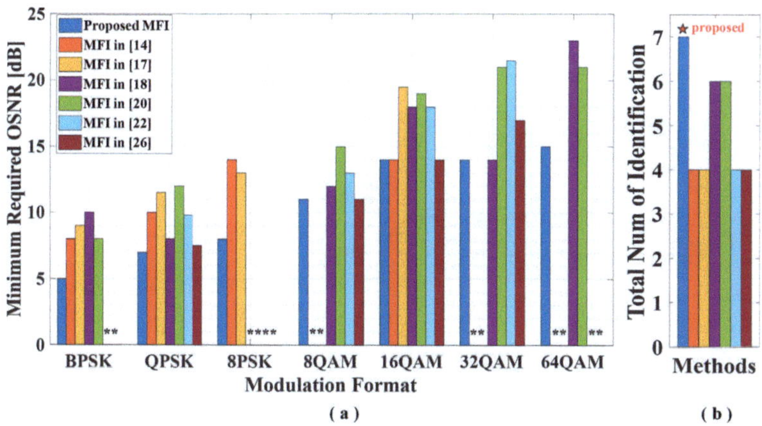

Fig. 5.9 Performance comparison between the proposed method and the methods in the other literature, where * indicates that there is no identification result at the corresponding position. (**a**) The minimum required OSNR for 100% accurate identification; (**b**) Total number of identified modulation formats

respectively, in the identification of these modulation formats. For the remaining modulation formats of 8QAM, 16QAM and 32QAM, the minimum required OSNR for the proposed MFI are 11, 14, and 14 dB, respectively, which is comparable to the best performance among other schemes. In terms of the number of identified modulation formats, the identification of most modulation formats is realized by the proposed method, with the total number of seven, while in other methods, they are four, four, six, six, four and four, respectively.

The more modulation formats an MFI method can identify, the more difficult it is to improve the accuracy because the increase in modulation formats means that more distinguishing features are required. However, as shown in Fig. 5.9, the proposed method requires lower OSNR for the identification of most modulation formats, even though it needs to complete the identification of more modulation formats, which demonstrates the superior performance of the proposed method.

The recognition performance under different sample sizes is alse tested. The average power after the square operation of different modulation formats with sample numbers of 2048, 1024 and 512 are shown in Fig. 5.10a–c, respectively. As can be seen from these Fig. 5.10a–c, there are still significant differences between BPSK and other modulation formats after the square operation, which means that BPSK can still be easily separated from other modulation formats. This can be reflected in the identification accuracy in Fig. 5.10d–f: the identification accuracy of BPSK does not deteriorate with the decrease of the number of samples. Except for BPSK, the recognition performance of the other modulation formats decreases with the decrease in the number of samples.

In addition, the minimum OSNR values required by the proposed method and the other two methods in literature [18, 22] to achieve 100% recognition accuracy under different sample numbers are shown in Fig. 5.11. The horizontal axis is divided into

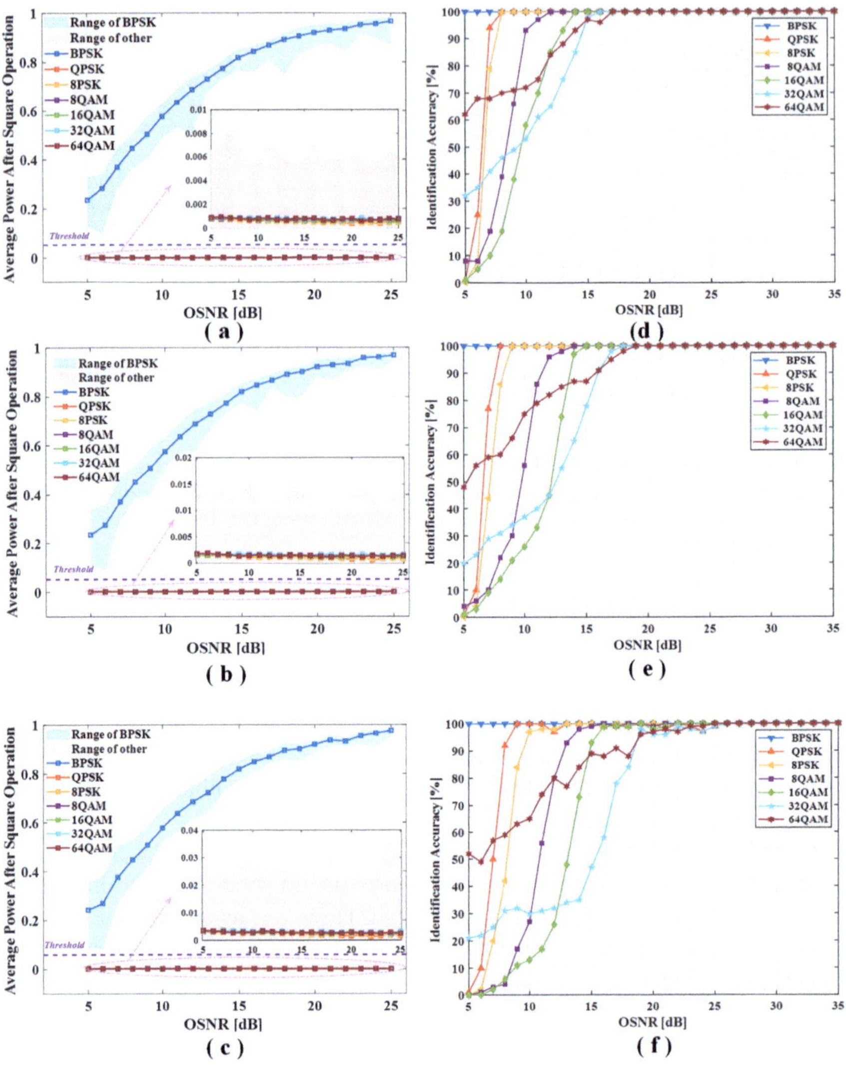

Fig. 5.10 The identification performance under a different number of samples. (**a**)–(**c**) The average power of signals after square operation with the number of samples of 2048, 1024 and 512, respectively. (**d**)–(**f**) The identification accuracy under 2048, 1024 and 512 samples, respectively

seven equal parts according to the modulation formats, and under each modulation format the bars are grouped according to the MFI method. When the number of samples is 4096, the proposed method has obvious advantages in the identification of BPSK, QPSK, 8PSK and 16QAM, and achieved OSNR gains of at least 3, 1, 10, 1, and 2 dB, respectively. Under the condition of 2048 samples, the best identification performances of BPSK, 8PSK,16QAM and 64QAM are achieved

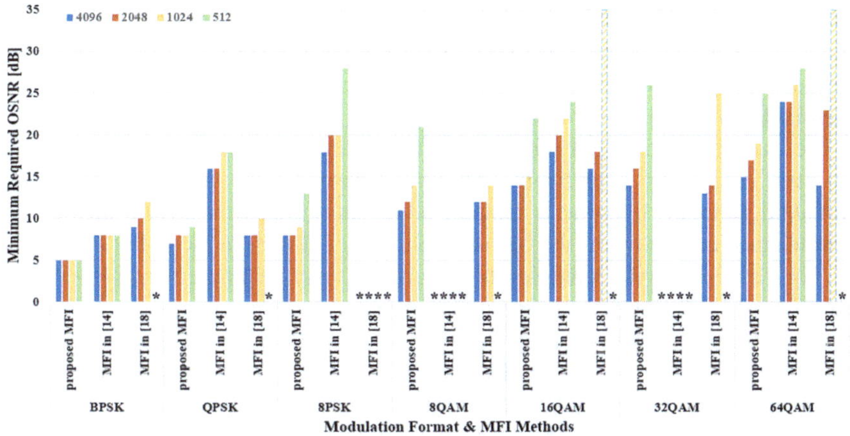

Fig. 5.11 The comparison of the minimum required OSNR to achieve 100% accuracy for different MFI methods, under a different number of samples

by the proposed method, with the advantages of 3, 12, 4, and 6 dB. For BPSK and 8QAM, the same OSNR requirements for the best performances in the other methods are required for the proposed method. As for the sample number of 1024, the best performances are achieved by the proposed method for all modulation formats, and the OSNR requirements are reduced by at least 3, 2, 11, 0, 7, 7, and 7 dB. It should be noted that the bars with slashes in Fig. 5.11 indicates that 100% identification accuracy cannot be achieved until the corresponding OSNR (35 dB). When the number of samples is 512, only the comparison between the proposed method and the method in [18] is carried out. The results show that advantages in all modulation format identifications exist in the proposed scheme. For BPSK, QPSK, 8PSK, 16QAM and 64QAM, the advantages are 3, 9, 15, 2, and 3 dB. Simulation results show that our method works well even when the number of samples is reduced, which is conducive to the implementation of low complexity.

5.2 Signal Constellation Diagrams and Support Vector Machine Method

As services like 5G, cloud computing, high-definition video, and cloud conferencing continue to advance, there is a growing need for more bandwidth and spectrum usage. This is leading to an exponential increase in channel capacity in optical communication systems [35–39]. The future of fiber optic communication systems is expected to be flexible and varied, capable of accommodating a wide range of signals and utilizing different modulation formats to meet the diverse needs of users [40–44]. Optical signals are susceptible to various transmission impairments that change dynamically in time, which places new demands on optical receivers. At the digital receiver, demodulate the transmitted signal; it is necessary to know the type of modulation format. Therefore, the correct identification of modulation formats is essential for high-quality communication [45–47].

The existing modulation format recognition algorithms can be roughly divided into two categories. One is the likelihood-based (LB) method based on the likelihood function hypothesis testing, which minimizes the probability of false identification and provides the optimal solution of Bayesian ideal, but this method has high computational complexity and is not easy to implement [48–50]. The feature-based (FB) recognition method must extract the important features from the received signal and then identify the signal modulation format. Although FB is a suboptimal approach, it is generally simpler and easier to implement, and can also provide near-optimal recognition performance if properly designed [27, 51–53].

For example, an MFI scheme based on lightweight convolutional neural networks (CNN) in the 2D-Stokes planes is proposed in [27]. In [51], an MFI scheme that depends on using CNNs on constellation diagrams is proposed. A blind optical MFI scheme based on the Hough transform of the constellation diagram is proposed in [52], which is used to display phase and magnitude information. In [53], the authors proposed a modulation classification scheme based on signal constellation diagram and deep learning to identify different modulated signals. Most of the above MFI techniques can identify modulation formats such as QPSK, 16QAM, and 64QAM.

In this section, an MFI scheme based on signal constellation diagrams and support vector machine (SVM) is proposed. Firstly, the signal constellation diagrams are divided by the fractal dimension of the weighted linear least squares (WLS-FD) algorithm, and each region's fractal dimension (FD) is calculated, regarded as one of the image features. Then, the feature values of the image in different directions are extracted by the gray-level cooccurrence matrix (GLCM), and their mean and variance are calculated, which is regarded as another feature. Finally, the two features are input into the modulation format classifier constructed by the SVM to achieve MFI in coherent optical communication systems.

5.2.1 Theory and Principle

Figure 5.12 shows the architecture of the proposed MFI scheme. Firstly, the signal constellation diagrams are divided by the WLS-FD algorithm, and the FD in each region is calculated, which is regarded as one of the image features. Then, the feature values of the image in different directions are extracted by the GLCM, and their mean and variance are calculated, which is regarded as another feature. Finally, the two features are input into the modulation format classifier constructed by the SVM. Meanwhile, to facilitate the learning of feature data by the SVM learning algorithm, it is necessary to change the feature data into one-dimensional data. The training feature data and test feature data are obtained by randomly extracting the feature set

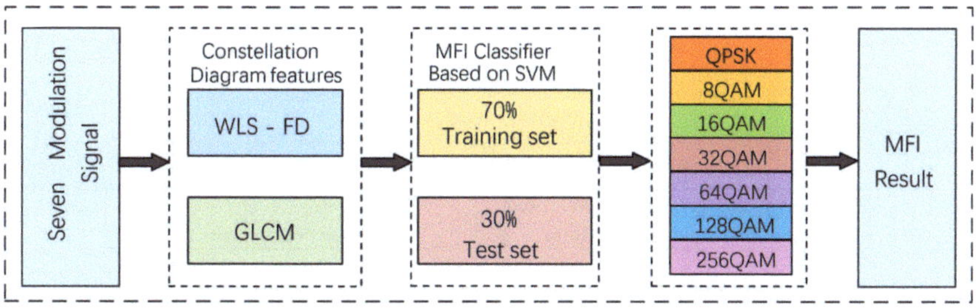

Fig. 5.12 The architecture of the MFI scheme

several times according to the ratio of 7:3, and the number of samples is 6370 and 2730, respectively.

Algorithm for Extracting Signal Constellation Diagram Features

FD can be used to describe the features of image surface [54, 55]. A larger fractal size values represent more complex image surfaces and usually correspond to roughness texture. In contrast, images of smooth surfaces typically have relatively small fractal dimensions. Therefore, image FD is chosen as one of the features to measure image quality.

The box-counting (BC) method proposed by Sarkar and Chaudhari is used to calculate the fractal dimension of the images [56]. Although the BC method can estimate the BC of the fractal image, there are certain defects, and it can not reflect the inhomogeneity of this graph, so the differential box-counting (DBC) method is used. Most researchers employed linear least squares (LLS) regression to calculate the FD of the images; however, it is susceptible to outliers. Furthermore, it provides equal weight to all points while drawing a line, which is impracticable; therefore, in this paper, the FD of the images is obtained using WLS regression, and each fitting point is given a distinct weight using the trapezoidal membership function (TMF). As a mature statistical image analysis method [57], GLCM has excellent flexibility and robustness, and the technique is simple, which is extensively utilized in the field of statistical analysis. The co-occurrence matrix is acquired by calculating the grayscale image and its different features in different directions to obtain the feature values. Finally, the mean and variance of these feature values are calculated as the image features.

Figure 5.13 shows the algorithm flow for extracting the features of signal constellation. When the fractal dimension of constellation map is calculated as the input image, because the gray value of the vector point concentration area is usually higher than that of other blank areas, the calculated fractal dimension is bound to be very different. The constellation map is divided into 16*16 regions, and subsequent dimensional judgments and fractal box dimensions are calculated separately for

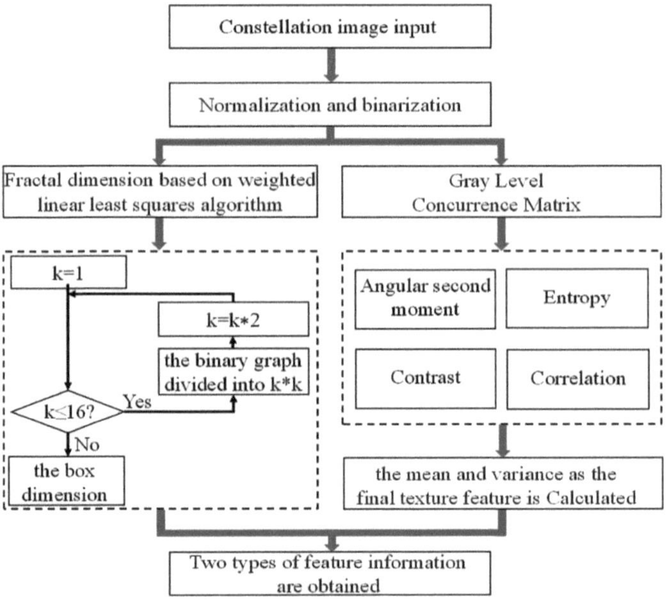

Fig. 5.13 Algorithm flow for extracting signal constellation diagram features

each region. WLS-FD sorts the fractal box dimensions calculated from each small region according to the location of the region as one of the image features.

Calculating the constellation diagrams is similar to constructing a three-dimensional surface on which pixels are located along (x, y), and their grayscale values represent the third coordinate z. The fractal image of pixel size $M * M$ is divided into small pieces of size $s * s$, where $\frac{M}{2} \geq s > 1$ and s is an integer. The scale of the grid is r, which is calculated by $\frac{s}{M}$. Each small block has a column of $s{*}s{*}h$ boxes. Where G is the total amount of gray levels in an 8-bit grayscale image, which is 256, and the height h of the box can be formulated as [52] Eq. (5.14):

$$\mathrm{h} = \frac{s}{\mathrm{M}}\mathrm{G} \tag{5.14}$$

The image's highest and lowest gray levels at position (i, j) fall in the uth and vth boxes, respectively, and the total amount of boxes that are size $s * s \times h$ comes to n_r, which is obtained by Eq. (5.15):

$$\mathrm{n_r(i, j) = u - v + 1} \tag{5.15}$$

The value of all grids is N_r, which can be obtained by Eq. (5.16):

$$\mathrm{N_r(i, j) = u - v + 1} \tag{5.16}$$

The weight for each grid size is assigned by WLS-FD using the rules of the TMF. Each point (x_d, y_d) on the fitting line corresponds to a different grid size s, which is assigned the weight $W(s)$. Where $x_d = \log\frac{1}{r}$, $y_d = \log N_r$, the weight of the point (x_d, y_d) is written as $W(s)$, which can be used with w_s instead of each other. The weight for each data point is assigned by TMF using Eq. (5.17):

$$\mathrm{W(s)} = \begin{cases} 0, & s \le a_t \\ \frac{s-a_t}{b_t-a_t}, & a_t < s < b_t \\ 1, & b_t \le s \le c_t \\ \frac{d_t-s}{d_t-c_t}, & c_t < s < d_t \\ 0, & s \ge d_t. \end{cases} \tag{5.17}$$

TMF divides all data points into five parts with intervals of $(-\infty, a_t]$, (a_t, b_t), $[b_t, c_t]$, (c_t, d_t), and $[d_t, +\infty)$, respectively. The points in the first and fifth parts have a weight of 0, while the points in the third part have a weight of 1. The points in the second and fourth parts have weights in the range (0, 1). In this paper, the values of a_t, b_t, c_t, and d_t are estimated by Eq. (5.18):

$$\begin{cases} \mathrm{a_t = s_{min} - 1} \\ \mathrm{b_t = |\sqrt[3]{M}|} \\ \mathrm{c_t = max}\left\{\mathrm{s}\left[\left(\frac{\mathrm{M}}{\mathrm{s}}\right) + 1 \le \left(\frac{\mathrm{M}}{\mathrm{s-1}}\right)\right]\right\} \\ \mathrm{d_t = s_{max} + 1} \end{cases} \tag{5.18}$$

Since the size of all the images we use is 512 × 512 pixels, we can obtain $a_t = 1$, $b_t = 8$, $c_t = 27$, and $d_t = 257$ by Eq. (5.18). After the weight of the points (x_d, y_d) are obtained. Fitting these points to the regression line $y = a_x + b$ via WLS. The FD of the image is calculated by Eq. (5.19):

$$\mathrm{D_A} = \frac{(\sum_r \mathrm{w_s x_d y_d})(\sum_r \mathrm{w_s}) - (\sum_r \mathrm{w_s y_d})(\sum_r \mathrm{w_s x_d})}{(\sum_r \mathrm{w_s x_d^2})(\sum_r \mathrm{w_s}) - (\sum_r \mathrm{w_s x_d})^2} \tag{5.19}$$

The GLCM is an $n * n$ square matrix, where n is the number of gray-level categories in the image. The (i, j) elements in the matrix represent the number of pixel pairs. By calculating the GLCM in four directions (0°, 45°, 90° and 135°), GLCM$[p(i, j)]$is normalized by Eq. (5.20):

$$\frac{\mathrm{p(i,j)}}{\varepsilon} \Rightarrow \hat{\mathrm{p}}(\mathrm{i,j}) \tag{5.20}$$

where ϵ is the constant of normalization.

Angular second moment m_1, entropy m_2, contrast m_3, and correlation m_4 are used to perform image block retrieval, which can be calculated by Li et al. [57]

Eq. (5.21):

$$\begin{cases} \mathrm{m_1} = \sum_\mathrm{i} \sum_\mathrm{j} \mathrm{P(i,j)}^2 \\ \mathrm{m_2} \ \& = -\sum_\mathrm{i} \ \sum_\mathrm{j} \ \mathrm{P(i,j)log\,P(i,j)} \\ \mathrm{m_3} \ \& = \sum_\mathrm{i} \ \sum_\mathrm{j} \ \mathrm{(i-j)^2P(i,j)} \\ \mathrm{m_4} = \frac{\left[\sum_\mathrm{i} \sum_\mathrm{j} \ ((\mathrm{i*j})*\mathrm{P(i,j)})-\mu_\mathrm{x}\mu_\mathrm{y}\right]}{\delta_\mathrm{x}\delta_\mathrm{y}} \end{cases} \tag{5.21}$$

where: $\mu_x = \sum_i \sum_j i * P(i,j)$, $\mu_y = \sum_i \sum_j j * P(i,j)$, $\delta_x = \sum_i \sum_j (i-\mu_x)^2 * P(i,j)$, $\delta_y = \sum_i \sum_j \left(j-\mu_y\right)^2 * P(i,j)$; The final texture features in this paper are calculated using the mean and variance of m_1, m_2, m_3, and m_4.

MFI Classifier Based on SVM

SVM is used to construct the MFI classifier, which is used to process the extracted features of the signal constellation diagram [29]. The robustness of the classification algorithm is improved by constructing a linear segmentation hyperplane with maximum boundary to classify all data correctly.

A training dataset $D = \{(x_1, y_1), (x_2, y_2), \cdots, (x_M, y_M)\}$ containing M samples is given. The samples' feature vectors and labels are x_i and y_i, respectively, where the feature vector satisfies $x_i = (x_i^{(1)}, x_i^{(2)}, \cdots, x_i^{(m)})$ with m features and the labels satisfy $y_i \in \{-1, +1\}$. To train the maximum margin model with parameters $w \in R^D$ and $b \in R$, it solves the following optimization problem:

$$\begin{aligned} &\min_{\mathrm{w,b}} \frac{1}{2}\|\mathrm{w}\|^2 \\ &\mathrm{s.t.y_i(w^T x_i + b)} \geq 1, i = 1, \ldots, \mathrm{N} \end{aligned} \tag{5.22}$$

According to the feature vector x_i and Lagrange multiplier pairs λ_i,λ_i^*, the classification decision function is calculated by Eq. (5.23):

$$f(x) = \sum_{m=1}^{M} (\lambda_i - \lambda_i^*)\langle x_i, x\rangle + b \tag{5.23}$$

The classifier's performance is measured by comparing the predicted results with y_i, and the classification accuracy is expressed by Eq. (5.24):

$$P = \frac{1}{M} \sum_{m=1}^{M} \mathrm{II}(f(x_i) - y_i) \tag{5.24}$$

where II denotes the indicative function, which is 1 when the condition inside the brackets holds and 0 otherwise. According to Eq. (5.24), the accuracy of the classifier is obtained by adding the number of predicted samples with the same labels as the original ones and dividing them by the total number of samples M.

5.2.2 Simulation System Setup

To verify the feasibility and superiority of the scheme, we compare it with the MFI scheme based on higher-order statistical (HOS) features, GLCM features, and FD features, respectively. Further, we built a 30 GBaud coherent optical communication system with fiber lengths of 80 and 120 km, where the optical signal-to-noise ratio (OSNR) ranges from 0 to 30 dB. The proposed MFI scheme identifies seven modulation formats: QPSK, 8QAM, 16QAM, 32QAM, 64QAM, 128QAM, and 256QAM.

The setup of the simulation system is described in detail, as shown in Fig. 5.14. First, the pseudorandom bit sequence (PRBS) is generated and used to generate different modulated signals in the arbitrary waveform generator (AWG), where the digital-to-analog converter (DAC) module turns the digital signal into an analog signal. In the Mach–Zehnder modulator (MZM), four signals are modulated to two optical carriers with different polarization states. The optical carriers are generated by a continuous-wave (CW) laser with a center frequency of 193.1 THz and divided into different polarization states by a polarization beam splitter (PBS). Then, the two polarization states of the modulated optical signal are combined into one using a polarization beam combiner (PBC). The transmission links are 80 km and 120 km single-mode fiber (SMF) with an attenuation coefficient of 0.2 dB/km and a dispersion coefficient is 16.75 ps/nm/km. The amplification gains of the erbium-doped fiber amplifier (EDFA) are 16 and 24 dB, respectively. Then, the optical signals are transmitted through SMF and amplified by EDFA, which can completely compensate for the fiber attenuation loss in the process of optical signal

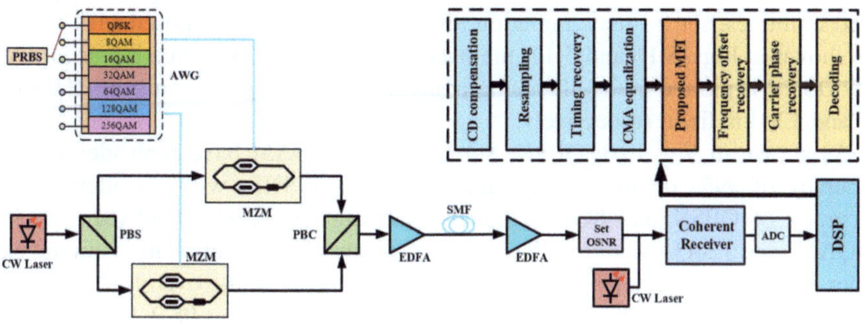

Fig. 5.14 Simulation system setup

Table 5.1 Coherent optical fiber communication system parameters

Parameter type	Value
Central wavelength	193.1 THZ
Signal rate	30 Gbaud
Fiber input power	10 dBm
Optical amplifier	EDFA
EDFA gain	16, 24 dB
EDFA noise	4 dB
Fiber	SMF
Fiber attenuation coefficient	0.2 dB/km
Fiber nonlinear coefficient	$1.31(\text{W} \cdot \text{km})^{-1}$
Fiber dispersion coefficient	$16.75\text{ps}/(\text{nm} \cdot \text{km})$
Fiber distance	80, 120 km
OSNR range	0~30 dB

transmission. Two optical signals, I and Q, are obtained at the receiver, and the photodiode is used to modulate the optical signals into analog signals. Next, after the coherent receiver, the analog signal is converted into a digital signal by the analog-to-digital converter (ADC) module. Further, dispersion compensation (DC), resampling, timing recovery, and constant mode algorithm (CMA) equalization are performed sequentially at the DSP. Finally, the required constellation diagrams are generated. The parameters of the coherent optical communication system are shown in Table 5.1.

5.2.3 Results and Discussion

The simulation system generated 9100 constellation diagrams in PNG format, which have a pixel size of 512 × 512. The sample constellation diagram obtained after 80 km SMF transmission is shown in Fig. 5.15.

The results show that compared with the other schemes, our proposed scheme has a better identification accuracy at low OSNR. Seven modulation signal samples are used in the test set for the simulation test, according to the proposed MFI scheme. As shown in Figs. 5.16 and 5.17, the overall and average identification accuracy at low OSNR is plotted in the 3D histograms for the seven signals transmitted through 80 and 120 km fiber, respectively.

As shown in Fig. 5.16, QPSK, 8QAM, and 256QAM have high identification accuracy of 99% in the overall case and at low OSNR, with the identification accuracy of QPSK and 8QAM signals reaching 100%. The 16QAM, 32QAM, 64QAM, and 128QAM signals have the possibility of misidentification among each other. However, in the MFI scheme proposed in this paper, the overall identification accuracy of 16QAM, 32QAM, 64QAM, and 128QAM reaches 98.2, 98.5, 95.1, and 98.2%, respectively, and the identification accuracy at low OSNR reaches 95.5, 96, 87.3, and 95.3%, respectively.

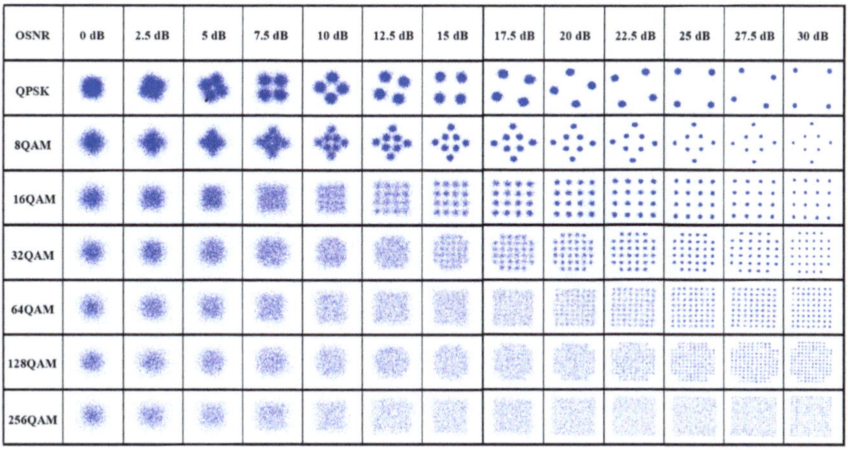

Fig. 5.15 The sample constellation diagram obtained after 80 km SMF transmission

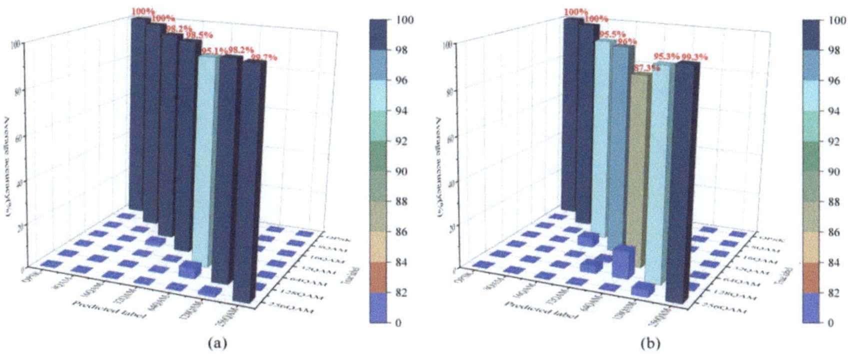

Fig. 5.16 The average identification accuracy after 80 km fiber transmission (**a**) OSNR range is 0~30 dB (**b**) OSNR range is 0~10 dB

In Fig. 5.17, QPSK, 8QAM, and 32QAM have high identification accuracy of 98% in the overall case and at low OSNR, with the identification accuracy of QPSK and 8QAM signals reaching 100%. The 16QAM, 64QAM, 128QAM, and 256QAM signals have the possibility of misidentification among each other. However, in the MFI scheme proposed in this paper, the overall identification accuracy of 16QAM, 64QAM, 128QAM, and 256QAM reaches 97.4, 95.9, 97.4, and 98.2%, respectively, and the identification accuracy at low OSNR reaches 93.3, 89.3, 93.3, and 95.3%, respectively.

In Fig. 5.18, the identification accuracy of the seven types of signals gradually increases as the OSNR increases until they all reach 100%. Among them, the identification accuracy of QPSK and 8QAM is always 100%. In Fig. 5.18a, the OSNR values of 16QAM, 32QAM, 64QAM, 128QAM, and 256QAM when reaching 100% identification accuracy are around 5, 2.5, 7.5, 10, and 2.5 dB, respectively.

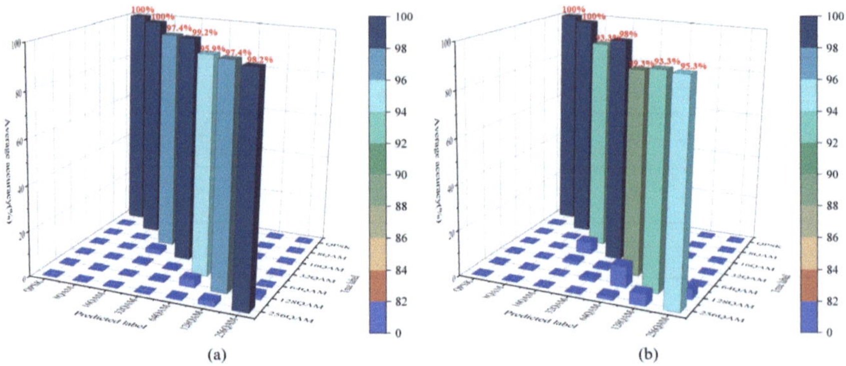

Fig. 5.17 The average identification accuracy after 120 km fiber transmission (**a**) OSNR range is 0~30 dB (**b**) OSNR range is 0~10 dB

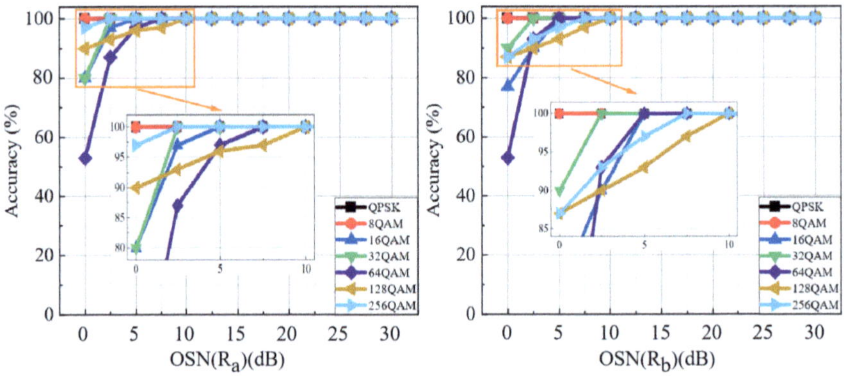

Fig. 5.18 Comparison of seven types of signal identification accuracy. The fiber distance is set as (**a**) 80 km, (**b**) 120 km

In Fig. 5.18b, the OSNR values of 16QAM, 32QAM, 64QAM, 128QAM, and 256QAM at 100% identification accuracy are around 5, 2.5, 5, 10, and 7.5 dB, respectively. The results show that the proposed MFI scheme can significantly improve the modulation format recognition accuracy and thus improve the performance of the classifier.

In addition, as shown in Fig. 5.19, different MFI methods are compared at fiber lengths of (a) 80 km and (b) 120 km. The results show that the proposed MFI scheme quickly achieves 100% overall identification accuracy and remains stable when OSNR $\geq$ 10 dB. High overall identification accuracy is also achieved at low OSNR. Because of the influence of high-order QAM signals in seven kinds of signals, such as 64, 128, and 256QAM, the overall identification accuracy of GLCM features and FD features is not good. Although the overall identification accuracy of MFI based on HOS features is high, it fails to achieve 100% identification accuracy within the set OSNR. The results show that compared with the other three schemes,

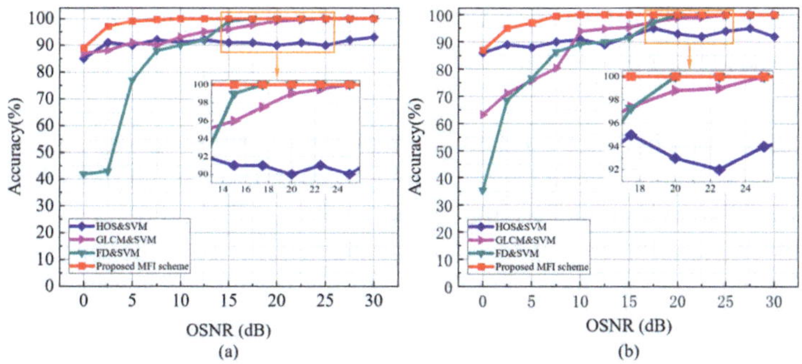

Fig. 5.19 Comparison of the overall identification accuracy of the seven signals with different MFI schemes. The fiber distance is set as (**a**) 80 km, (**b**) 120 km

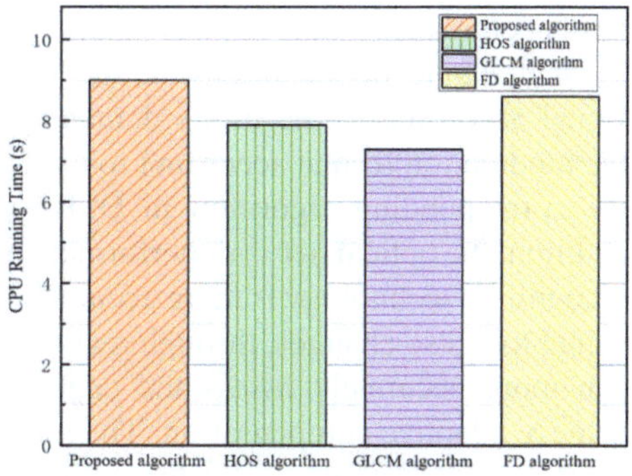

Fig. 5.20 Comparison of CPU running time of different algorithms for extracting constellation diagram features

our proposed scheme has a better identification accuracy at low OSNR. In addition, the identification accuracy of this scheme can reach 100% when the OSNR $\geq$ 10 dB.

As shown in Fig. 5.20, in order to understand the computational complexity of the proposed MFI algorithm, we calculated the CPU run time required to extract the features of a single constellation map and compared it with the other three algorithms. CPU running time tests were conducted on an Intel Personal Computer with Processor Core i5-10210U CPU at 1.60 GHz, 16 GB Random Access Memory, and Windows 10 Home Edition operating system. It can be seen that the proposed MFI scheme achieves a significant improvement in identification accuracy and stability at the cost of minor computational complexity and is the result of balancing accuracy and stability optimally.

5.3 Joint Modulation Format Identification and OSNR Monitoring Based on LSTM

With the development of various new services, optical networks show the characteristics of flexible bandwidth and variable rate. In order to achieve hitless receiver, it is necessary to extract and identify the signal features. MFI techniques automatically identify the modulation format based on the characteristics of the received signal to provide information for subsequent DSP algorithms, and OSNR reflects the degree of signal interference by noise during transmission and is directly related to the BER. Therefore, modulation format identification and OSNR estimation play an important role in monitoring optical transmission quality and developing a more intelligent optical communication system [58]. In order to reduce monitoring cost, how to achieve high accuracy joint monitoring of real-time network performance has become a current research hotspot.

In dynamic optical network, the transmitter dynamically adjusts the type of signal according to the user's demand, and the receiver automatically recognizes the modulation format according to the characteristics of the signal, and re-matches the corresponding algorithm in DSP to achieve the best demodulation [59]. In addition, the optical network performance monitoring assisted by modulation format recognition technology is more accurate, because the transmission path changes, the signal loss also changes, through real-time optical network performance monitoring can more accurately assess the degree of optical signal deterioration, so as to achieve adaptive compensation. The modulation format information provides more flexibility and autonomy for the optical network, and the performance parameters such as OSNR play an important role in the configuration and optimization of the optical network, monitoring the operation of the system and the transmission status of the signal, realizing real-time monitoring of the network, and ensuring the normal and reliable operation of the communication system.

Traditional optical network performance monitoring methods such as modulation format identification and OSNR monitoring based on data processing are costly and have poor real-time performance. With the continuous development of machine learning, neural network-based joint optical performance monitoring techniques have been improved. For example, a two-layer artificial neural network (ANN) is used for joint optical performance monitoring, where ANN1 is used for modulation format recognition and ANN2 selects different models to monitor OSNR according to the results of ANN1[2]. Four deep neural networks (DNNS) are used to extract the amplitude histograms (AHs) features equalized by constant module algorithm (CMA) in a hierarchical manner for OSNR monitoring and modulation format identification [60]. For signal amplitude histograms, features can also be extracted using multi-task artificial neural networks [61] or transfer learning simplified multi-task deep neural networks [62]. Convolutional neural network (CNN) extracts features of eye diagram [51], constellation diagram [63], and asynchronous delayed-tap sampling phase diagram [64] for joint modulation format identification and OSNR monitoring from the image perspective. The methods mentioned above require DSP compensation or image pre-processing of the signal, which increases the extra workload.

An optical link modulation format method combining Stokes spatial analysis and higher-order statistics is proposed to identify OOK, M-PSK, M-PAM and M-QAM signals, which is insensitive to residual carrier LO frequency and phase offset [65]. A method for identifying BPSK, QPSK, 8PSK and 16QAM based on nonlinear power transformation and peak detection is proposed, and the influence of key parameters such as FFT length and laser phase noise tolerance is discussed [17]. The blind modulation format recognition scheme based on the singular value decomposition of constellation maps and Radon transform achieves 100% recognition accuracy by utilizing constellation maps with different modulation formats [18]. An automatic modulation format recognition algorithm based on histogram features and convolutional neural networks is proposed to evaluate the classification performance considering the presence of channel interference, dispersion, phase noise and first polarization mode dispersion (PMD) under different optical signal-to-noise ratios [66].

In this section, we propose a joint modulation format identification and OSNR monitoring scheme based on long short-term memory neural network (LSTM) for optical communication systems, which directly generate combined data sets of four time-varying electrical signals types under different OSNR, modulation format recognition and OSNR monitoring are transformed into classification problems and regression problems in the field of machine learning, respectively, using the raw IQ data and LSTM to complete the construction and training of network models. The network has a shared model layer whose weights are trained jointly by the loss function of each task to minimize, and an output layer with separate weights for different tasks.

5.3.1 Algorithm Principle

Working Principle of Joint Monitoring Scheme

As shown in Fig. 5.21, the IQ data collected by coherent receiver from different modulation format signals have different amplitude and phase characteristics. The proposed scheme first trains the LSTM network using IQ raw data samples by learning mapping models from input data and corresponding labels to extract internal features. The loss function is calculated by comparing the predicted results with the true labels, and the optimal LSTM network model is finally obtained by constantly adjusting the weight parameters of the network using the optimization algorithm. And then the modulation format identification and OSNR monitoring for high-order format signals in the testing process can be completed in a very short time, which can meet the real-time requirements in practical applications.

Deep learning can obtain the internal features of input datasets and corresponding output results by classification or regression. For joint modulation format identi-

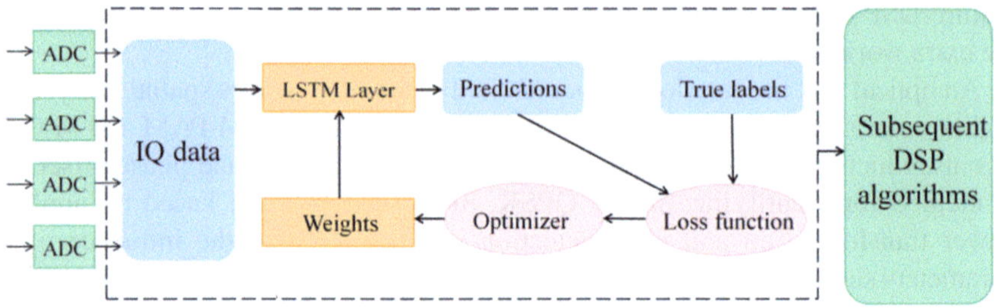

Fig. 5.21 Working principle of proposed scheme based on LSTM

fication and OSNR monitoring, it is necessary to generate combined datasets of multiple signal types in different OSNRs. The signals collected under different modulation formats have different amplitude and phase characteristics, and different OSNR values have different interference degrees of amplitude and phase. The corresponding modulation format and OSNR value of the signal are used as labels at the same time to realize the joint monitoring of optical performance. The commonly used feature extraction methods include data statistics (peak value, mean value, variance, number of bins) and images (eye diagram, constellation diagram, amplitude histogram), but these methods require additional processing and computation. In this paper, the proposed scheme directly uses the original signals XI, XQ, YI, and YQ of the coherent receiver output to construct the datasets for training the neural network model.

Principle of LSTM

As shown in Fig. 5.22, LSTM is a special type of recurrent neural network (RNN) with an internal structure composed of three gates: the forget gate, the input gate, and the output gate. These gates can effectively address the issues of exploding and vanishing gradients in the training of time-series data. In a forget gate in the LSTM cell, information from a current input and a previous hidden state is passed through a sigmoid function. The sigmoid function outputs a value in a range between 0 and 1. Numbers close to 0 mean to forget, whereas numbers close to 1 mean to keep the information.

An input gate updates the information in the cell. It passes the current input and the previous hidden state through a sigmoid and a tanh function and multiplies their outputs. The output of the sigmoid function close to 1 means the values are important and the values close to 0 are not. The tanh function helps with regulating the network and outputs values between -1 and 1.

The calculation of the cell state is done by pointwise multiplication of the previous cell state and the output of the forget gate. This is followed by pointwise addition with an output of the input cell so that the cell state can be updated with new values relevant to the neural network.

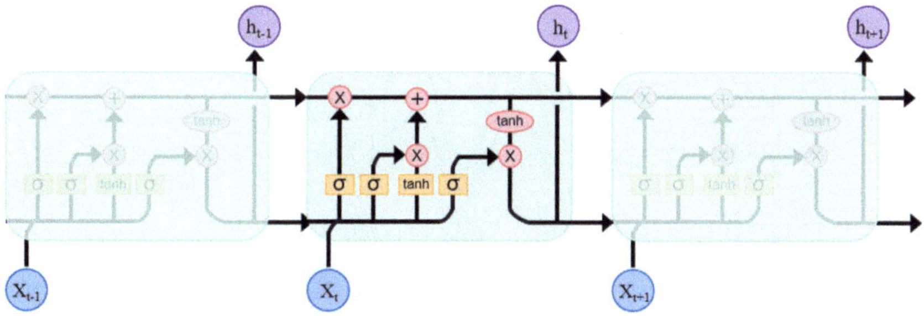

Fig. 5.22 LSTM unit structure

At last, the output gates decide about the new hidden state. It holds the information on previous inputs and is used for predictions. The values are given by passing the previous hidden state and current input into a sigmoid function and are multiplied with the new cell state passed through a tanh function. Both the new cell state and new hidden state are passed into a further cell.

$$Sigmoid(x) = \frac{1}{1 + e^{-x}} \tag{5.25}$$

$$tanh(x) = \frac{e^x - e^{-x}}{e^x + e^{-x}} \tag{5.26}$$

where x is the input value, the sigmoid function outputs a value between 0 and 1, indicating the size of information to be forgotten, and the tanh function outputs a value between -1 and 1 to adjust the network weight.

The calculation process of LSTM is as follows:

$$f_t = \sigma_g(W_f[h_{t-1}, x_t] + b_f) \tag{5.27}$$

$$i_t = \sigma_g(W_i[h_{t-1}, x_t] + b_i) \tag{5.28}$$

$$\tilde{C}_t = tanh(W_C[h_{t-1}, x_t] + b_C) \tag{5.29}$$

$$C_t = f_t * C_{t-1} + i_t * \tilde{C}_t \tag{5.30}$$

$$o_t = \sigma_g(W_o[h_{t-1}, x_t] + b_o) \tag{5.31}$$

$$h_t = o_t * tanh(C_t) \tag{5.32}$$

$$y_t = \sigma_y(W_{hy}h_t + b_y) \tag{5.33}$$

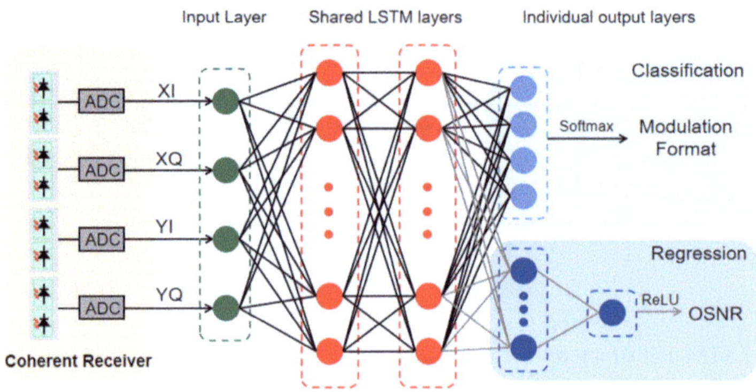

Fig. 5.23 Scheme of joint MFI and OSNR monitoring

where x is the input matrix, f, i, o, and C denote the forget gate, input gate, output gate, and cell state, respectively, t denotes the corresponding time, W is the weight matrix, and b is the bias vector. LSTM completes the cyclic calculation from time 1 to time t according to the above formula and finally obtains the output matrix $\mathrm{Y_t} = [\mathrm{y_1}, \mathrm{y_2}, \ldots\ldots, \mathrm{y_t}]$.

Neural Network Model

Compared to the output of a single parameter, the joint output of multiple parameters has emerged as a novel trend in the field of machine learning. Multi-task learning, which involves accomplishing multiple tasks by sharing the parameters of the neural network model, effectively enhances data utilization, reduces overfitting through weight sharing, and facilitates rapid learning. Figure 5.23 illustrates the need to construct a combined dataset that encompasses multiple signal types at different OSNRs, followed by the construction and training of a network model based on LSTM architecture. The monitoring of MFI and OSNR is transformed into classification and regression problems, respectively. The innovation of Multitask learning (MTL) in comparison to conventional network structures lies in its shared model layers, wherein the weight is jointly trained by the loss function of each task to minimize error. Furthermore, distinct tasks possess their own output layer with separate weights and biases to produce prediction results.

In the MFI section, the output neurons are determined by the number of modulation formats, and these neurons are transformed into a distribution range of [0,1] using the Softmax activation function. The probabilities of all modulation formats sum up to 1, and the modulation format with the highest probability signifies the corresponding signal. The loss function employed is cross entropy, while accuracy serves as the measurement index.

$$Softmax(y_i) = \frac{exp(y_i)}{\sum_{i=1}^{n} exp(y_i)} \tag{5.34}$$

where y_i denotes the value of the ithelement in the output array of the neural network multi-classification problem.

In the OSNR monitoring section, the ReLU activation function is employed to output the predicted value, while the MSE function serves as the loss function. The MAE function quantifies the discrepancy between the predicted and actual values.

$$MAE(X, h) = \frac{1}{m} \sum_{i=1}^{m} |h(x_i) - z_i| \tag{5.35}$$

where X denotes the dataset input to the neural network, hrefers to the set of prediction labels, $h(x_i)$ represents the predicted label corresponding to the ith sample in the dataset, and z_i denotes the true label.

5.3.2 Simulation Setup and Results

Simulation Setup

We carried out a 32 GBaud polarization multiplexing coherent optical simulation system, transmitting QPSK, 16QAM, 32QAM and 64QAM respectively, setting the OSNR range to 5–30 dB, and adjusting it in 1 dB steps. As shown in Fig. 5.24, the symbols generated by PRBS are mapped into signals with different modulation formats. After transmission through 1000 km optical fiber, ASE noise is added to adjust the OSNR value, and coherent receivers are used for analog-to-digital conversion. Finally, signal samples are collected and input to the LSTM network for joint modulation format identification and OSNR monitoring. The

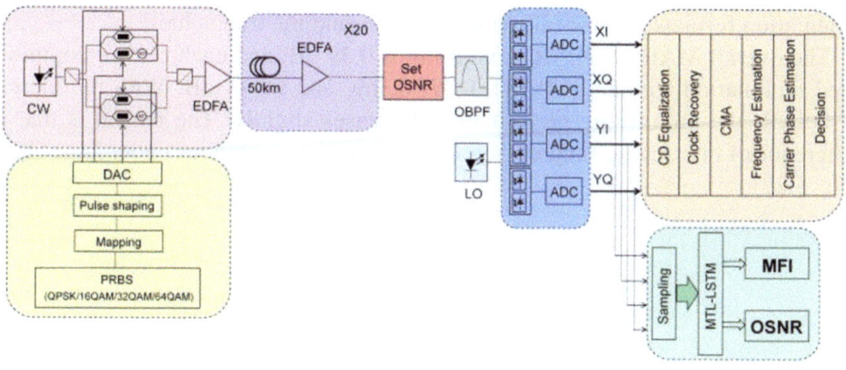

Fig. 5.24 Simulation setup for joint MFI and OSNR monitoring

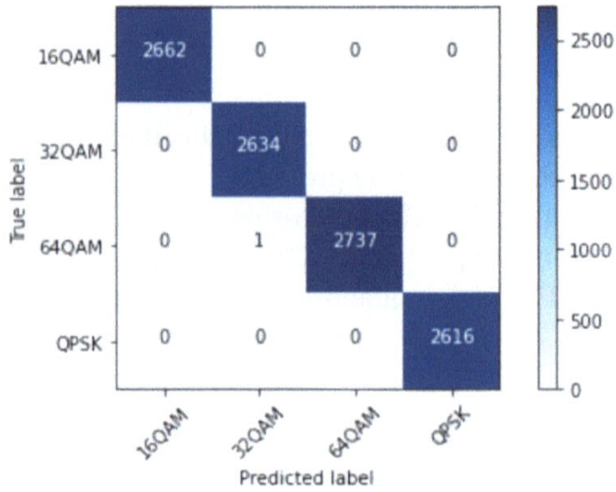

Fig. 5.25 Modulation scheme Identification confusion matrix

results of joint monitoring are used to provide necessary information for subsequent DSP processing to CD equalization, clock recovery, CMA equalization, frequency estimation, carrier phase estimation and symbol decision.

The collected signal data are divided into a group of 512 lengths, and a 512×4 matrix dataset was constructed, with the corresponding modulation format and OSNR value as labels. The LSTM network model in Fig. 5.22 is used for training. For the two output layers, cross entropy and MSE are used as loss functions, Adam optimization algorithm is used to minimize loss functions, and dropout and early-stopping are used to prevent overfitting. The initial learning rate is set to 0.01, batch-size is set to 64, and the data set is about 210,000 pieces in total. About 10,000 pieces are randomly selected as the test set, and the rest are used as the training set.

Results and Discussions

As shown in Figs. 5.25 and 5.26, for QPSK/16QAM/32QAM/64QAM, the MFI accuracy of the proposed scheme is close to 100% within the whole range of 5–30 dB, and only one sample is misjudged. The predicted OSNR values of the four modulation formats are basically distributed around the true values.

The overall MAE of OSNR monitoring is 0.39 dB, and each of four commonly used modulation formats is 0.384, 0.47, 0.369, and 0.344 dB. With the increase of OSNR, the monitoring performance decreases slightly. The reason is that the difference of interference degree of noise to signal is not obvious under large OSNR (Fig. 5.27).

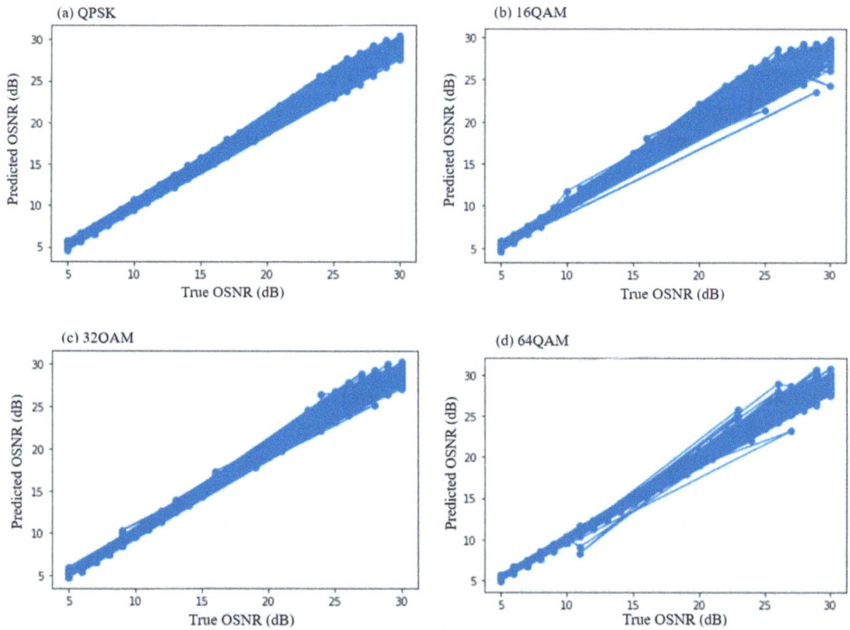

Fig. 5.26 Predicted OSNR values vs. true OSNR values

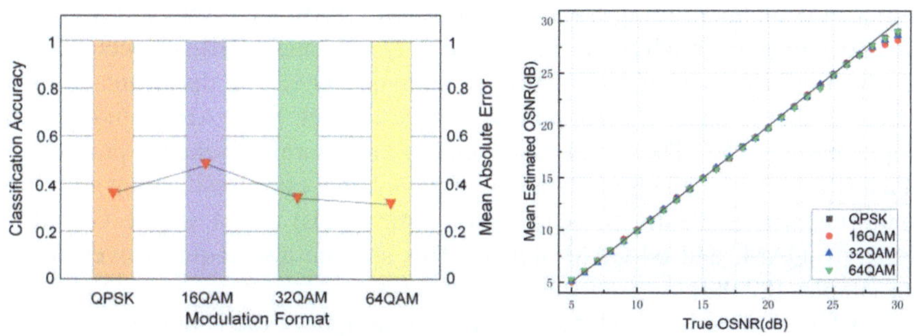

Fig. 5.27 (**a**) Classification accuracy and mean absolute error (**b**) Mean estimated OSNR and true OSNR for QPSK/16QAM/32QAM/64QAM using the proposed joint scheme

5.4 Adaptive Modulation Format Identification and OSNR Monitoring Using Transfer Learning with MTL-LSTM

With the advent of the cloud network era, data centers, cloud computing, the Internet of Things (IoT) and other new services are diversified, leading to the growing demand for network traffic [67]. The drawbacks of traditional optical networks are increasingly apparent, with low spectrum utilization and inflexible bandwidth allocation of dominant data services. To meet the continuously increasing demand for data services, the transmission capacity of EONs has steadily increased, and the network architecture is more dynamic, complex, and transparent, supporting mixed transmission of multiple modulation formats [68–72].

The MFI technology can automatically identify types based on the characteristics of the received signals, and provide the necessary information for DSP algorithms in the subsequent coherent receiver to achieve optimal equalization and demodulation performance [2]. OSNR is one of the important parameters for monitoring the physical state and signal quality of the optical network, reflecting the degree of noise interference in the transmission link and closely related to the bit error rate [6, 58, 60, 73–75].

Traditional MFI schemes can be roughly divided into likelihood-based (LB) and feature-based (FB) methods, where LB methods require a priori knowledge of channel parameters as a cost [76, 77], and FB methods usually use manual extraction of features to complete the classification [78]. Traditional OSNR monitoring schemes include out-of-band measurement and in-band measurement, but they are no longer suitable for dynamic networks [79]. The emergence of multitasking learning has further developed performance monitoring technology based on neural networks, achieving real-time joint monitoring of multiple optical network performance parameters [7, 25, 32, 80–83]. For example, Faisal Nadeem Khan has proposed the use of the deep neural network (DNN) in combination with AHs equalized by constant modulus algorithm (CMA) to simultaneously perform OSNR monitoring and MFI. The experimental results show that the mean estimation errors of OSNR monitoring for 112 Gbps PDM-QPSK, 112 Gbps PDM-16QAM, and 240 Gbps PDM-64QAM signals are 1.2, 0.4, and 1 dB, respectively. The classification accuracy of modulation formats is 100%. The disadvantage of this scheme is that it needs the support of a huge dataset, each AH includes 56,000 sample points [25]. The CNN-based end-to-end learning scheme proposed by Takahito Tanimura realizes modulation format/symbol rate classification and OSNR monitoring within 0.3 dB of mean square error (MSE) for 14 and 16 Gbaud DP-QPSK, 16-QAM, and 64-QAM signals. This scheme also requires over 500,000 data samples [82, 83].

Transfer learning is introduced to adjust the weight of the neural network based on prior knowledge instead of random initialization, which reduces the data samples and realizes rapid network reconstruction [34, 84–86].

In this section, an adaptive performance monitoring scheme for Elastic Optical Networks (EONs) using transfer learning with MTL-LSTM is proposed. This scheme completes MFI and OSNR monitoring by directly collecting four time-varying signals, without additional workload steps such as DSP algorithm compensation and data image conversion in advance. The SSMF length of the source task is 100km, four modulation formats of QPSK, 16QAM, 32QAM, and 64QAM are selected, OSNR is set to 17–25 dB, and data is collected to complete the training of the MTL-LSTM model. The accuracy rate of MFI is above 99.9%, and the monitoring error of OSNR is less than 0.1 dB. In the transfer learning stage, the transmission of the target task is set to 1000 and 200 km to simulate the different accumulations of nonlinear noise, dispersion, polarization mode dispersion, and other disturbances in the link condition. Fine-tune only the output layer and retain the network parameters of the input layer and LSTM layers of the source network model to achieve the training of the target network model with higher efficiency. Compared with the retraining method under the same conditions, the proposed scheme based on transfer learning in this paper can achieve a higher MFI accuracy and a lower OSNR MAE with a small amount of data. At the same time, the training parameters and time of the new model are reduced.

5.4.1 Proposed Scheme Based on Transfer Learning

In the field of dynamic optical networks, traditional performance monitoring technology faces challenges associated with additional workloads, including manual feature extraction, data type conversion, and DSP pre-compensation. Neural network-based algorithms necessitate large datasets as a foundation and require data reacquisition for training as link conditions change. To address the aforementioned issues, we propose an adaptive performance monitoring scheme using transfer learning with MTL-LSTM. As shown in Fig. 5.28, transfer learning involves the application of trained model parameters to a novel model. The weight parameters of the new network model are adjusted based on prior knowledge rather than random initialization, thereby resulting in higher initial performance, a faster improvement rate, and superior convergence effectiveness.

In the training phase of source task, the coherent receiver samples four time-varying electrical signals, namely XI, XQ, YI, and YQ, which capture the amplitude and phase information at different time points. These signals correspond to the polarization directions X and Y, and the in-phase (I) and quadrature (Q) components, respectively. To construct the dataset sample, the four data channels are combined into a [4×N] matrix sequence, where N represents the number of sampling points. Additionally, the label for each sample includes information such as the modulation

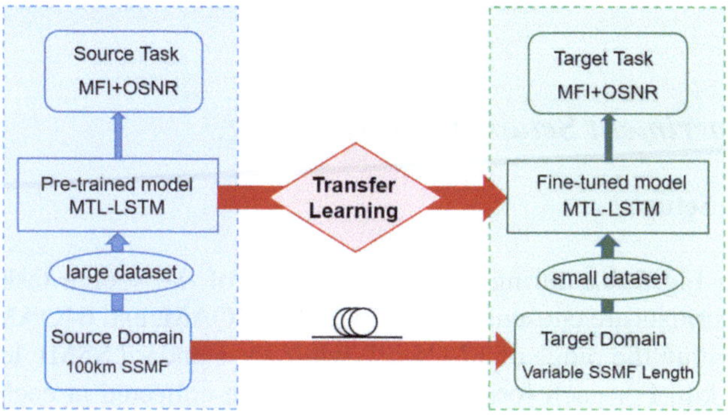

Fig. 5.28 Schematic diagram of transfer learning

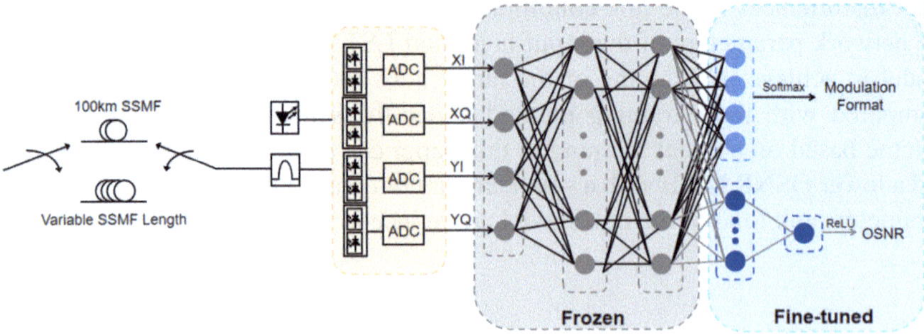

Fig. 5.29 The principle of adaptive performance monitoring for EONs using transfer learning with MTL-LSTM

format and OSNR, represented in the [modulation format, OSNR] format, enabling joint monitoring. As shown in Fig. 5.23, the neural network architecture consists of an input layer with 4 neurons, each corresponding to one of the four input data channels. The hidden layer utilizes LSTM cells, comprising 128 and 64 neurons with LeakyReLU activation function. To prevent overfitting, a dropout layer is incorporated into the model. The fully connected layer further maps the extracted features into lower dimensions. Two partial output layers are employed to output the predicted modulation format types and OSNRs, respectively. To achieve more accurate joint monitoring results, the entire training process employs the Adam optimization algorithm to minimize the loss function.

In the stage of transfer learning, we keep the network parameters of the input layer and the hidden layer frozen, meaning that their weights and biases remain unchanged throughout the training process. As shown in Fig. 5.29, the fine-tuning process only involves updating the network parameters of the output layer. This approach leads to a significant reduction in both training time and the number of samples required. In this paper, a 100 km SSMF link is selected as the source task, with the length of the optical fiber being modified to simulate various interferences. A small number of samples are collected for training the target task.

5.4.2 *Experiment Setup and Results*

Experiment Setup

As shown in Fig. 5.30, during the training phase of the source task model, four modulation formats are chosen: QPSK, 16QAM, 32QAM, and 64QAM. The OSNR is adjusted within the range of 17–25 dB, while the standard SSMF length remains fixed at 100 km. For each format, 959 samples are collected at each OSNR. The length parameter N is set to 512. The training and test sets are divided into an

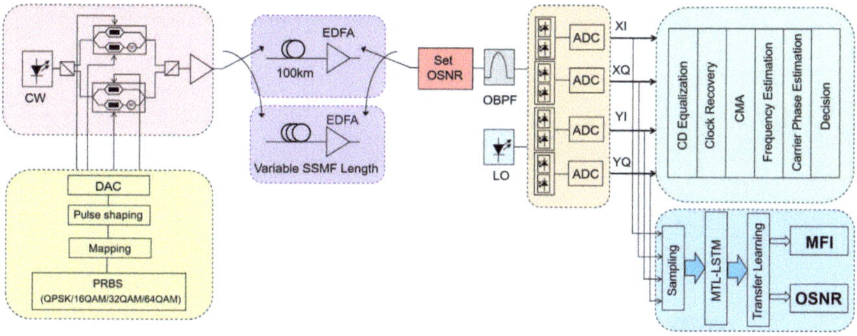

Fig. 5.30 Experiment system setup

80%:20% ratio. Then the fiber length is changed to 1000 and 200 km, and data samples are collected in the same way.

Results of Source Task

In the Fig. 5.31a shows that after more than 350 iterations, the loss gradually approaches 0, indicating that the source task model has achieved good results in both the training stage and the test stage. The MFI results in Fig. 5.31b, c show that recognition accuracy exceeding 99.9% for QPSK, 16QAM, and 32QAM. Meanwhile, Fig. 5.31d, e display that the predicted OSNR results are in substantial agreement with the true values. Figure 5.31f demonstrates that the MAE for the four modulation formats is 0.066, 0.042, 0.059, and 0.074 dB, respectively. Consequently, the experimental results validate the source task model's capability to accurately monitor various modulation formats and OSNR values.

Results of Target Task

In the target task phase, we adjust the fiber link length to 1000 km and continue to collect QPSK, 16QAM, 32QAM, and 64QAM signals under an OSNR range of 17–25 dB. We then compare the results of the transfer learning scheme with the retraining scheme across dataset sizes ranging from 5 to 60%. Figure 5.32 clearly illustrates that both MFI and OSNR monitoring results of transfer learning are superior to retraining, especially in the case of a small number of samples. Furthermore, when the prediction results are close, transfer learning requires roughly half the data samples compared to retraining. This experiment result shows the effectiveness of using transfer learning to fine-tune source network models.

To verify the applicability of the experimental conclusions, we change the fiber length to 200 km and keep the other conditions unchanged. Comparing the results of different dataset sizes in Fig. 5.33, it can still prove that the adaptive performance

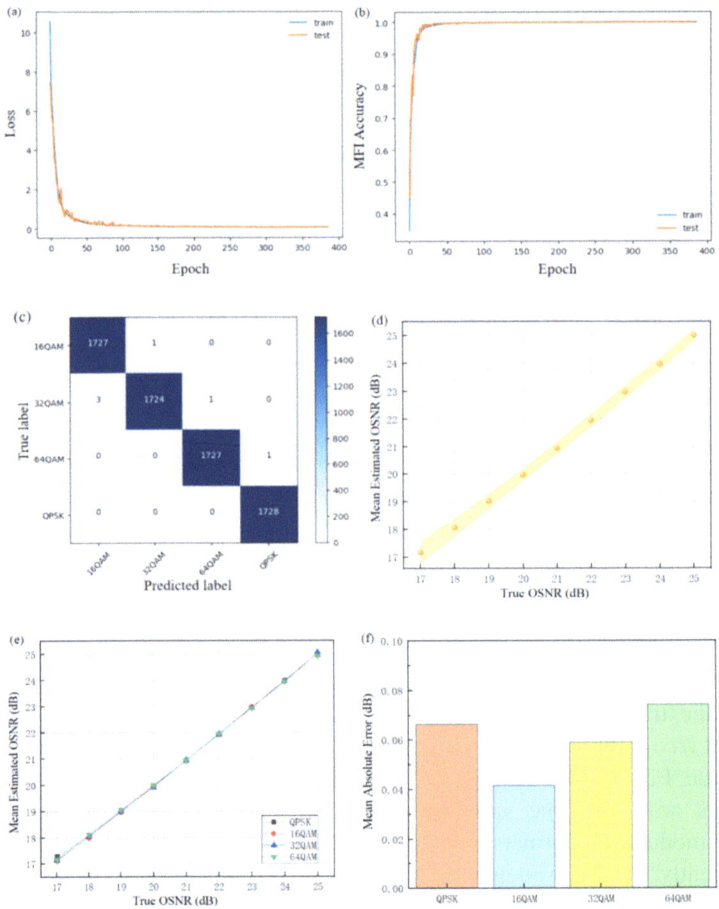

Fig. 5.31 The results of source task (100 km): (**a**) Training and test loss evolution over epochs, (**b**) MFI accuracy, (**c**) Confusion matrix of four modulation formats, (**d**) Overall error band, (**e**) True OSNR vs. Mean estimated OSNR, (**f**) MAE of OSNR prediction

monitoring scheme using transfer learning with MTL-LSTM proposed in this paper is better than the traditional retraining multi-task monitoring schemes.

In the training process of the network model, we set the batch size to 32, which represents the number of data samples required for each training step. By comparing the training time required for transfer learning and retraining under different dataset sizes and different lengths of SSMF in Fig. 5.34, it can be observed that the proposed scheme can effectively reduce the time for completing the network model parameter update.

Finally, we specifically compare the monitoring results of QPSK, 16QAM, 32QAM, and 64QAM. As shown in Fig. 5.35, under the condition of a small dataset size, the MFI and OSNR monitoring results of the proposed transfer learning-

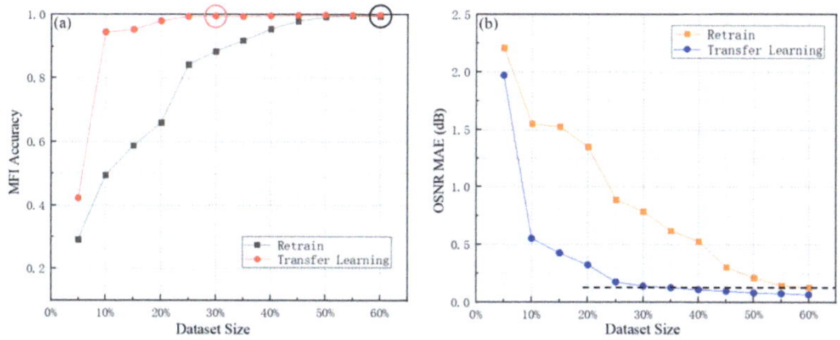

Fig. 5.32 The results of source task (1000 km): (**a**) MFI accuracy and (**b**) OSNR MAE for retraining and transfer learning under different dataset sizes

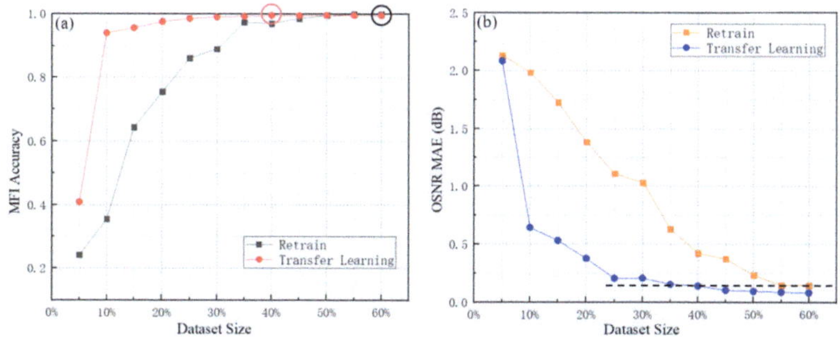

Fig. 5.33 The results of source task (200 km): (**a**) MFI accuracy and (**b**) OSNR MAE for retraining and transfer learning under different dataset sizes

Fig. 5.34 The training time required for each step of retrain and transfer learning for different dataset sizes and SSMF lengths

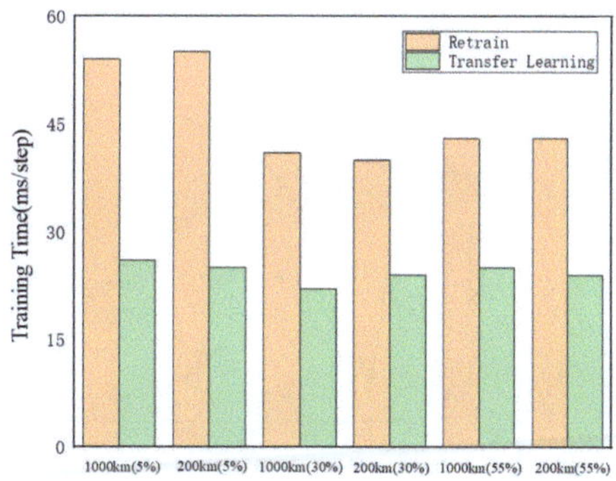

based scheme outperforms retraining. Additionally, it exhibits good adaptability to different modulation formats and different SSMF lengths, with modulation format recognition rates close to 100% and OSNR MAE less than 0.1 dB. Therefore, the proposed scheme in this paper enables accurate optical network performance monitoring with limited dataset samples under different conditions, while reducing the amount of data and time required for network model reconstruction.

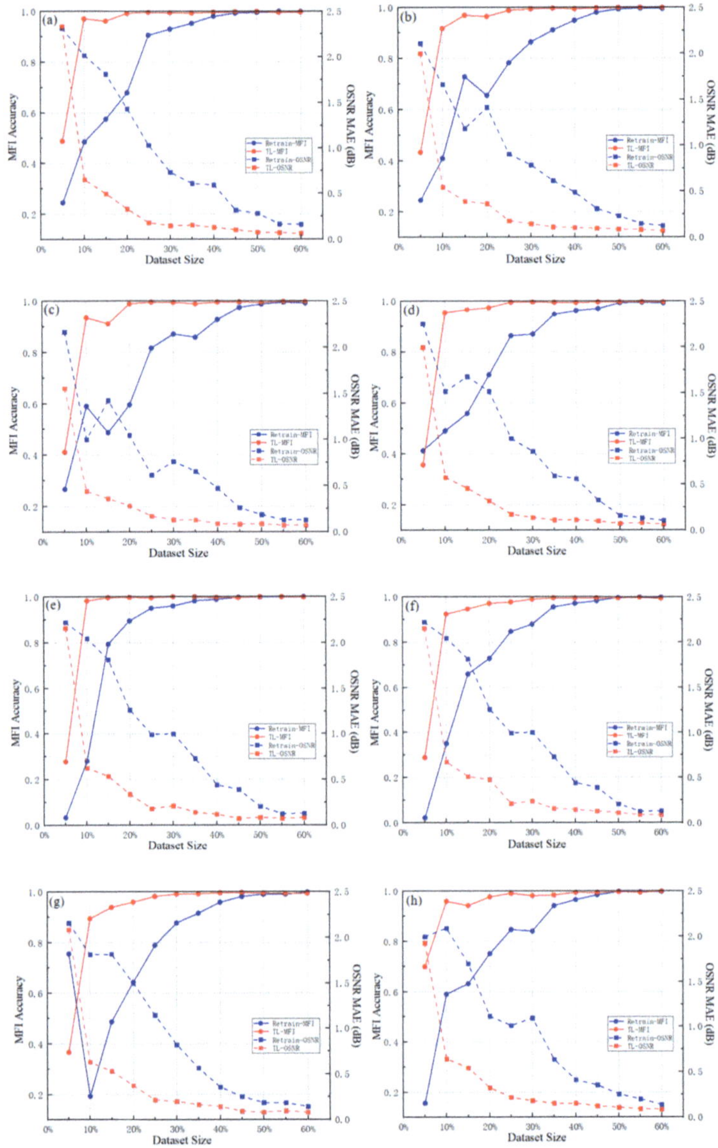

Fig. 5.35 Comparing results of retraining and transfer learning for MFI and OSNR monitoring of QPSK, 16QAM, 32QAM, and 64QAM at 1000 and 200 km SSMF

5.5 Summary

In this chapter, we discuss advanced modulation format identification methods in future optical communication systems. First, we propose a blind MFI method based on principal component analysis and singular value decomposition, which can be used without any prior information. The proposed MFI can accurately identify seven modulation formats in a large OSNR range of 5 to 35 dB. Then, a method based on signal constellation diagrams extraction and support vector machine is proposed to identify the modulation formats. Finally, a joint MFI and OSNR monitoring scheme based on LSTM is proposed by extracting features from the original IQ data samples. At the same time, transfer learning is introduced to optimize the joint monitoring scheme, reduce the dependence of modulation format identification and optical network parameter monitoring on large datasets, and improve the adaptability to channel conditions and modulation format changes.

References

1. X. Liu, H. Lun, M. Fu, Y. Fan, L. Yi, W. Hu, Q. Zhuge, AI-based modeling and monitoring techniques for future intelligent elastic optical networks. Appl. Sci. **10**(1), 363 (2020). https://www.mdpi.com/2076-3417/10/1/363
2. Z. Dong, F.N. Khan, Q. Sui, K. Zhong, C. Lu, A.P.T. Lau, Optical performance monitoring: a review of current and future technologies. J. Lightw. Technol. **34**(2), 525–543 (2016)
3. Y. Zhang, P. Zhou, Y. Liu, J. Wang, C. Li, Y. Lu, Fast adaptation of multi-task meta-learning for optical performance monitoring. Opt. Exp. **31**(14), 23183–23197 (2023)
4. J. Jiang, Q. Zhang, X. Xin, R. Gao, X. Wang, F. Tian, Q. Tian, B. Liu, Y. Wang, Blind modulation format identification based on principal component analysis and singular value decomposition. Electronics **11**(4), 612 (2022)
5. Z. Huang, Q. Zhang, X. Xin, H. Yao, R. Gao, J. Jiang, F. Tian, B. Liu, F. Wang, Q. Tian, et al., Modulation format identification based on signal constellation diagrams and support vector machine. Photonics **9**(12), MDPI, 927 (2022)
6. X. Chang, Z. Li, Q. Zhang, Y. Gao, Y. Wang, Q. Tian, F. Tian, X. Xin, Modulation format identification for coherent optical communication systems based on long short-term memory networks, in *2022 20th International Conference on Optical Communications and Networks (ICOCN)* (IEEE, Piscataway, 2022), pp. 1–3
7. X. Chang, Z. Li, Q. Zhang, C. Wang, Y. Wang, Q. Tian, F. Tian, X. Xin, Joint modulation format identification and OSNR monitoring based on LSTM, in *2022 Asia Communications and Photonics Conference (ACP)* (IEEE, Piscataway, 2022), pp. 639–641
8. F. Wang, H. Yao, Q. Zhang, J. Wang, R. Gao, D. Guo, M. Guizani, Dynamic distributed multi-path aided load balancing for optical data center networks. IEEE Trans. Netw. Service Manag. **19**(2), 991–1005 (2021)
9. M. Jinno, Elastic optical networking: roles and benefits in beyond 100-Gb/s era. J. Lightw. Technol. **35**(5), 1116–1124 (2016)
10. M. Xiang, Q. Zhuge, M. Qiu, X. Zhou, M. Tang, D. Liu, S. Fu, D.V. Plant, RF-pilot aided modulation format identification for hitless coherent transceiver. Opt. Exp. **25**(1), 463–471 (2017)
11. J. Lu, S. Fu, L. Deng, M. Tang, Z. Hu, D. Liu, C.K. Chan, Blind and fast modulation format identification by frequency-offset loading for hitless flexible transceiver, in *Optical Fiber Communication Conference* (Optica Publishing Group, Washington, 2018), pp. M2F–5
12. S. Fu, Z. Xu, J. Lu, H. Jiang, Q. Wu, Z. Hu, M. Tang, D. Liu, C.C.K. Chan, Modulation format identification enabled by the digital frequency-offset loading technique for hitless coherent transceiver. Opt. Exp. **26**(6), 7288–7296 (2018)
13. P. Isautier, K. Mehta, A.J. Stark, S.E. Ralph, Robust architecture for autonomous coherent optical receivers. J. Opt. Commun. Netw. **7**(9), 864–874 (2015)

14. L. Jiang, L. Yan, A. Yi, Y. Pan, M. Hao, W. Pan, B. Luo, An effective modulation format identification based on intensity profile features for digital coherent receivers. J. Lightw. Technol. **37**(19), 5067–5075 (2019)
15. Z. Zhao, A. Yang, P. Guo, A modulation format identification method based on information entropy analysis of received optical communication signal. IEEE Access **7**, 41492–41497 (2019)
16. J. Lu, Z. Tan, A.P.T. Lau, S. Fu, M. Tang, C. Lu, Modulation format identification assisted by sparse-fast-fourier-transform for hitless flexible coherent transceivers. Opt. Exp. **27**(5), 7072–7086 (2019)
17. G. Liu, R. Proietti, K. Zhang, H. Lu, S. Ben Yoo, Blind modulation format identification using nonlinear power transformation. Opt. Exp. **25**(25), 30895–30904 (2017)
18. R.A. Eltaieb, A.E. Farghal, H.d.H. Ahmed, W.S. Saif, A. Ragheb, S.A. Alshebeili, H.M. Shalaby, F.E. Abd El-Samie, Efficient classification of optical modulation formats based on singular value decomposition and radon transformation. J. Lightw. Technol. **38**(3), 619–631 (2020)
19. L. Jiang, L. Yan, A. Yi, Y. Pan, M. Hao, W. Pan, B. Luo, Blind optical modulation format identification assisted by signal intensity fluctuation for autonomous digital coherent receivers. Opt. Exp. **28**(1), 302–313 (2020)
20. A. Yi, L. Yan, L. Jiang, Y. Pan, B. Luo, W. Pan, Modulation format identification based on density distributions in stokes plane for digital coherent receivers, in *Asia Communications and Photonics Conference* (Optica Publishing Group, Washington, 2017), pp. Su2A–29
21. L. Jiang, L. Yan, A. Yi, Y. Pan, T. Bo, M. Hao, W. Pan, B. Luo, Blind density-peak-based modulation format identification for elastic optical networks. J. Lightw. Technol. **36**(14), 2850–2858 (2018)
22. H. Xu, L. Yang, X. Yu, Z. Zheng, C. Bai, W. Sun, X. Zhang, Blind and low-complexity modulation format identification scheme using principal component analysis of stokes parameters for elastic optical networks. Opt. Exp. **28**(14), 20249–20263 (2020)
23. R.A. Eltaieb, H.A. Abouelela, W.S. Saif, A. Ragheb, A.E. Farghal, H.E.d.H. Ahmed, S. Alshebeili, H.M. Shalaby, F.E. Abd El-Samie, Modulation format identification of optical signals: an approach based on singular value decomposition of stokes space projections. Appl. Opt. **59**(20), 5989–6004 (2020)
24. R. Zhao, W. Sun, H. Xu, C. Bai, X. Tang, Z. Wang, L. Yang, L. Cao, Y. Bi, X. Yu, et al., Blind modulation format identification based on improved pso clustering in a 2d stokes plane. Appl. Opt. **60**(31), 9933–9942 (2021)
25. F.N. Khan, K. Zhong, X. Zhou, W.H. Al-Arashi, C. Yu, C. Lu, A.P.T. Lau, Joint OSNR monitoring and modulation format identification in digital coherent receivers using deep neural networks. Opt. Exp. **25**(15), 17767–17776 (2017)
26. A. Yi, L. Yan, H. Liu, L. Jiang, Y. Pan, B. Luo, W. Pan, Modulation format identification and OSNR monitoring using density distributions in stokes axes for digital coherent receivers. Opt. Exp. **27**(4), 4471–4479 (2019)
27. W. Zhang, D. Zhu, Z. He, N. Zhang, X. Zhang, H. Zhang, Y. Li, Identifying modulation formats through 2d stokes planes with deep neural networks. Opt. Exp. **26**(18), 23507–23517 (2018)
28. D. Wang, M. Wang, M. Zhang, Z. Zhang, H. Yang, J. Li, J. Li, X. Chen, Cost-effective and data size–adaptive opm at intermediated node using convolutional neural network-based image processor. Opt. Exp. **27**(7), 9403–9419 (2019)
29. Y. Zhao, Z. Yu, Z. Wan, S. Hu, L. Shu, J. Zhang, K. Xu, Low complexity OSNR monitoring and modulation format identification based on binarized neural networks. J. Lightw. Technol. **38**(6), 1314–1322 (2020)
30. H.J. Cho, S. Varughese, D. Lippiatt, R. Desalvo, S. Tibuleac, S.E. Ralph, Optical performance monitoring using digital coherent receivers and convolutional neural networks. Opt. Exp. **28**(21), 32087–32104 (2020)
31. Z. Wan, Z. Yu, L. Shu, Y. Zhao, H. Zhang, K. Xu, Intelligent optical performance monitor using multi-task learning based artificial neural network. Opt. Exp. **27**(8), 11281–11291 (2019)

32. Y. Cheng, S. Fu, M. Tang, D. Liu, Multi-task deep neural network (MT-DNN) enabled optical performance monitoring from directly detected PDM-QAM signals. Opt. Exp. **27**(13), 19062–19074 (2019)
33. Z. Yu, Z. Wan, L. Shu, S. Hu, Y. Zhao, J. Zhang, K. Xu, Loss weight adaptive multi-task learning based optical performance monitor for multiple parameters estimation. Opt. Exp. **27**(25), 37041–37055 (2019)
34. Y. Cheng, W. Zhang, S. Fu, M. Tang, D. Liu, Transfer learning simplified multi-task deep neural network for PDM-64QAM optical performance monitoring. Opt. Exp. **28**(5), 7607–7617 (2020)
35. T. Mai, H. Yao, N. Zhang, L. Xu, M. Guizani, S. Guo, Cloud mining pool aided blockchain-enabled internet of things: an evolutionary game approach. IEEE Trans. Cloud Comput. **11**(1), 692–703 (2021)
36. Y. Gong, H. Yao, J. Wang, M. Li, S. Guo, Edge intelligence-driven joint offloading and resource allocation for future 6G industrial internet of things. IEEE Trans. Netw. Sci. Eng. **11**, 5644–5655 (2024)
37. Y. Zhu, H. Yao, T. Mai, W. He, N. Zhang, M. Guizani, Multiagent reinforcement-learning-aided service function chain deployment for internet of things. IEEE Internet Things J. **9**(17), 15674–15684 (2022)
38. Z. Xiao, J. Yang, T. Mao, C. Xu, R. Zhang, Z. Han, X.G. Xia, Leo satellite access network (LEO-SAN) towards 6G: challenges and approaches. IEEE Wireless Commun. 31, 2022.
39. Z. Xiao, Z. Han, A. Nallanathan, O.A. Dobre, B. Clerckx, J. Choi, C. He, W. Tong, Antenna array enabled space/air/ground communications and networking for 6G. IEEE J. Sel. Areas Commun. **40**(10), 2773–2804 (2022)
40. L. Zheng, S. Wang, L. Tian, F. He, Z. Liu, Q. Tian, Query-adaptive late fusion for image search and person re-identification, in *Proceedings of the IEEE Conference on Computer Vision and Pattern Recognition* (2015), pp. 1741–1750
41. M.C. Tan, F.N. Khan, W.H. Al-Arashi, Y. Zhou, A.P. Tao Lau, Simultaneous optical performance monitoring and modulation format/bit-rate identification using principal component analysis. J. Opt. Commun. Netw. **6**(5), 441–448 (2014)
42. R. Gao, Q. Zhang, X. Xin, Q. Tian, F. Tian, Chirped anti-resonant reflecting optical waveguide for the distributed sensing of pressure. Opt. Lett. **45**(3), 690–693 (2020)
43. X. Wang, Q. Zhang, X. Xin, R. Gao, K. Lv, Q. Tian, F. Tian, C. Wang, X. Pan, Y. Wang, et al., Robust and low-complexity principal component-based phase estimation algorithm for probabilistically shaped square-QAM systems. J. Lightw. Technol. **38**(22), 6153–6162 (2020)
44. X. Wang, Q. Zhang, X. Xin, R. Gao, Q. Tian, F. Tian, C. Wang, X. Pan, Y. Wang, L. Yang, Robust weighted k-means clustering algorithm for a probabilistic-shaped 64QAM coherent optical communication system. Opt. Exp. **27**(26), 37601–37613 (2019)
45. Z. Pan, C. Yu, A.E. Willner, Optical performance monitoring for the next generation optical communication networks. Opt. Fiber Technol. **16**(1), 20–45 (2010)
46. Z. Zhang, Y. Luo, C. Zhang, X. Liang, M. Cui, K. Qiu, Constellation shaping chaotic encryption scheme with controllable statistical distribution for OFDM-PON. J. Lightw. Technol. **40**(1), 14–23 (2022)
47. T. Wu, C. Zhang, C. Chen, H. Hou, H. Wei, S. Hu, K. Qiu, Security enhancement for OFDM-PON using brownian motion and chaos in cell. Opt. Exp. **26**(18), 22857–22865 (2018)
48. Q. Shi, Y. Karasawa, Robust maximum likelihood classification of quadrature amplitude modulation, in *2010 IEEE Wireless Communication and Networking Conference* (IEEE, 2010), pp. 1–5
49. J.L. Xu, W. Su, M. Zhou, Likelihood-ratio approaches to automatic modulation classification. IEEE Trans. Syst. Man Cybern. C (Appl. Rev.) **41**(4), 455–469 (2010)
50. W. Wei, J.M. Mendel, Maximum-likelihood classification for digital amplitude-phase modulations. IEEE Trans. Commun. **48**(2), 189–193 (2000)
51. D. Wang, M. Zhang, J. Li, Z. Li, J. Li, C. Song, X. Chen, Intelligent constellation diagram analyzer using convolutional neural network-based deep learning. Opt. Exp. **25**(15), 17150–17166 (2017)

52. S. Amiralizadeh, A. Yekani, L.A. Rusch, Discrete multi-tone transmission with optimized QAM constellations for short-reach optical communications. J. Lightw. Technol. **34**(15), 3515–3522 (2016)
53. S. Peng, H. Jiang, H. Wang, H. Alwageed, Y. Zhou, M.M. Sebdani, Y.-D. Yao, Modulation classification based on signal constellation diagrams and deep learning. IEEE Trans. Neural Netw. Learn. Syst. **30**(3), 718–727 (2018)
54. N. Dehghani, E. Kabir, A multi-focus image fusion method based on fractal dimension and guided filtering, in *2020 25th International Conference on Pattern Recognition (ICPR)* (IEEE, Piscataway, 2021), pp. 10697–10703.
55. C. Panigrahy, A. Seal, N.K. Mahato, Image texture surface analysis using an improved differential box counting based fractal dimension. Powder Technol. **364**, 276–299 (2020)
56. N. Sarkar, B.B. Chaudhuri, An efficient differential box-counting approach to compute fractal dimension of image. IEEE Trans. Syst. Man Cybern. **24**(1), 115–120 (1994)
57. S. Li, G. Gao, G. Hu, B. Gao, T. Gao, W. Wei, G. Wu, Aging feature extraction of oil-impregnated insulating paper using image texture analysis. IEEE Trans. Dielectrics Electrical Insul. **24**(3), 1636–1645 (2017)
58. D.K. Tizikara, J. Serugunda, A. Katumba, Machine learning-aided optical performance monitoring techniques: a review. Front. Commun. Netw. **2**, 756513 (2022)
59. D.K. Tizikara, J. Serugunda, A. Katumba, An overview of machine learning-aided optical performance monitoring techniques (2021, preprint). arXiv:2107.07338
60. W.S. Saif, M.A. Esmail, A.M. Ragheb, T.A. Alshawi, S.A. Alshebeili, Machine learning techniques for optical performance monitoring and modulation format identification: a survey. IEEE Commun. Surv. Tuts. **22**(4), 2839–2882 (2020)
61. L. Guesmi, H. Fathallah, M. Menif, Modulation format recognition using artificial neural networks for the next generation optical networks. Adv. Appl. Artif. Neural Netw. **11**, 12–28 (2018)
62. D. Wang, M. Zhang, Z. Li, J. Li, M. Fu, Y. Cui, X. Chen, Modulation format recognition and OSNR estimation using CNN-based deep learning. IEEE Photon. Technol. Lett. **29**(19), 1667–1670 (2017)
63. M. Wang, J. Liu, J. Zhang, D. Zhang, C. Guo, Modulation format identification based on phase statistics in stokes space. Opt. Commun. **480**, 126481 (2021)
64. W. Zhang, J. Ye, Z. Yue, Y. Wang, X. Zhang, X. Zhang, L. Xi, Identifying modulation formats using integrated clustering algorithm (2021, preprint). arXiv:2104.14475
65. P. Isautier, J. Pan, R. DeSalvo, S.E. Ralph, Stokes space-based modulation format recognition for autonomous optical receivers. J. Lightw. Technol. **33**(24), 5157–5163 (2015)
66. W.S. Saif, A.M. Ragheb, B. Nebendahl, T. Alshawi, M. Marey, S.A. Alshebeili, Performance investigation of modulation format identification in super-channel optical networks. IEEE Photon. J. **14**(2), 1–10 (2022)
67. S. Behera, G. Savva, K. Manousakis, G. Ellinas, Impairment-aware routing, modulation, spectrum, and core allocation with bit loading in spectrally–spatially flexible optical networks. J. Opt. Commun. Netw. **15**(6), 318–332 (2023)
68. M. Dallaglio, A. Giorgetti, N. Sambo, L. Velasco, P. Castoldi, Routing, spectrum, and transponder assignment in elastic optical networks. J. Lightw. Technol. **33**(22), 4648–4658 (2015)
69. S. Petale, S. Subramaniam, Machine learning aided optimization for balanced resource allocations in SDM-EONS. J. Opt. Commun. Netw. **15**(5), B11–B22 (2023)
70. P. Papanikolaou, K. Christodoulopoulos, E. Varvarigos, Optimization techniques for incremental planning of multilayer elastic optical networks. J. Opt. Commun. Netw. **10**(3), 183–194 (2018)
71. Y. Zhao, Q. Zhang, Z. Li, X. Xin, R. Gao, F. Chai, Y. Tao, Q. Tian, F. Tian, D. Chen, et al., Dynamic routing, modulation, and spectrum assignment based on fuzzy logic control in elastic optical networks. Appl. Opt. **61**(1), 223–230 (2021)

72. R.K. Jana, B.C. Chatterjee, A.P. Singh, A. Srivastava, B. Mukherjee, A. Lord, A. Mitra, Quality-aware resource provisioning for multiband elastic optical networks: a deep-learning-assisted approach. J. Opt. Commun. Netw. **14**(11), 882–893 (2022)
73. K. Kikuchi, Fundamentals of coherent optical fiber communications. J. Lightw. Technol. **34**(1), 157–179 (2015)
74. W. Zhao, Z. Yang, M. Xiang, M. Tang, Y. Qin, S. Fu, Accurate OSNR monitoring based on data-augmentation-assisted DNN with a small-scale dataset. Opt. Lett. **47**(1), 130–133 (2021)
75. Z. Dong, F.N. Khan, Q. Sui, K. Zhong, C. Lu, A.P.T. Lau, Optical performance monitoring: a review of current and future technologies. J. Lightw. Technol. **34**(2), 525–543 (2015)
76. O.A. Dobre, A. Abdi, Y. Bar-Ness, W. Su, Survey of automatic modulation classification techniques: classical approaches and new trends. IET Commun. **1**(2), 137–156 (2007)
77. Z. Zhu, A.K. Nandi, *Automatic Modulation Classification: Principles, Algorithms and Applications* (Wiley, Hoboken, 2015)
78. Y. Han, G. Wei, C. Song, L. Lai, Hierarchical digital modulation recognition based on higher-order cumulants, in *2012 Second International Conference on Instrumentation, Measurement, Computer, Communication and Control* (IEEE, 2012), pp. 1645–1648
79. S. Oda, J.Y. Yang, Y. Akasaka, K. Sone, Y. Aoki, M. Sekiya, J.C. Rasmussen, In-band OSNR monitor using an optical bandpass filter and optical power measurements for superchannel signals, in *39th European Conference and Exhibition on Optical Communication (ECOC 2013)* (IET, London, 2013), pp. 1–3
80. Q. Zhang, J. Chen, H. Zhou, J. Zhang, M. Liu, A simple artificial neural network based joint modulation format identification and OSNR monitoring algorithm for elastic optical networks, in *2018 Asia Communications and Photonics Conference (ACP)* (IEEE, Piscataway, 2018), pp. 1–3
81. C. Wang, S. Fu, H. Wu, M. Luo, X. Li, M. Tang, D. Liu, Joint OSNR and CD monitoring in digital coherent receiver using long short-term memory neural network. Opt. Exp. **27**(5), 6936–6945 (2019)
82. T. Tanimura, T. Hoshida, T. Kato, S. Watanabe, H. Morikawa, Data-analytics-based optical performance monitoring technique for optical transport networks, in *Optical Fiber Communication Conference* (Optica Publishing Group, Washington, 2018), pp. Tu3E–3
83. T. Tanimura, T. Hoshida, T. Kato, S. Watanabe, H. Morikawa, Convolutional neural network-based optical performance monitoring for optical transport networks. J. Opt. Commun. Netw. **11**(1), A52–A59 (2019)
84. L. Xia, J. Zhang, S. Hu, M. Zhu, Y. Song, K. Qiu, Transfer learning assisted deep neural network for OSNR estimation. Opt. Exp. **27**(14), 19398–19406 (2019)
85. D. K. Jha, J.K. Mishra, Transfer learning approach toward joint monitoring of bit rate and modulation format. Appl. Opt. **61**(13), 3695–3701 (2022)
86. J. Zhang, T. Xu, T. Jin, W. Jiang, S. Hu, X. Huang, B. Xu, Z. Yu, X. Yi, K. Qiu, Meta-learning assisted source domain optimization for transfer learning based optical fiber nonlinear equalization. J. Lightw. Technol. **41**(5), 1269–1277 (2022)

Chapter 6
Advanced Equalization Technology

With the repaid growth of high-definition video, virtual reality, augmented reality and other services, the demand for bandwidth and data traffic is constantly increasing, which is driving the development of coherent optical communication systems toward higher capacity and increased spectral efficiency (SE). To solve the signal impairments along with the long-haul fiber link and high capacity, advanced Equalization technologies have received much attention in recent years. In this chapter, we proposed the optimized principal component-based phase estimation (OPCPE) algorithm for PS square-QAM systems to solve the incompatibility between PCPE and PS at high SNR [1]. Then, the LR-DBP algorithm based on logistic regression digital backpropagation which is proposed to compensate the effects of dispersion and fiber nonlinearities in long haul optical transmission systems is introduced [2]. Meanwhile, we present a novel CPR algorithm for PS QAM systems that is based on the unscented Kalman filter [3].

6.1 Robust and Low-Complexity Principal Component-Based Phase Estimation Algorithm

Among various emerging technologies, probabilistic shaping (PS) has swept across the entire optical transmission research community [4–8]. Compared to regular square-QAM formats, the PS technique makes the constellation symbols occur with different probability, thereby decreasing the gap between the capacity of communication systems and the Shannon limit [9–13]. However, the impact of PS on the chain of digital signal processing (DSP) algorithms of a coherent receiver still needs further investigation [14–16]. For instance, Dris et al. have pointed out that, conventional polarization demultiplexing implementation which works for uniform-shaped formats, is sub-optimal when applied to formats using PS [14]. Yan et al. have investigated the applications of traditional frequency offset estimation (FOE)

X. Xin, *Advanced Optical Fiber Transmission Systems*, Wireless Networks,
https://doi.org/10.1007/978-3-032-09516-9_6

algorithms in PS-MQAM constellations, and observed significant performance degradations when the strength increases [15]. Furthermore, Mello et al. have investigated the interplay of PS and the blind phase search (BPS) algorithm in [16], and found that the PS impact on BPS can affect the overall system performance, specially at low SNRs, which suggests the need for alternative carrier phase recovery (CPR) algorithms to be deployed in probabilistically-shaped transmissions.

Recently, Diniz et al. have proposed a low-complexity principal component-based phase estimation (PCPE) algorithm for square-QAM modulation formats [17]. Since the PCPE method outperforms BPS algorithm at low SNR values when applied to uniform shaped square-QAM formats, it is a potential candidate for phase recovery in PS square-QAM systems. As will be clearly shown in the rest of the paper, PCPE is capable of achieving better generalized mutual information (GMI) results in a PS-64QAM system at low SNRs. However, for high SNR, PCPE shows a serious penalty compared to BPS especially in strong shaping case. Different from facing a uniform constellation, the hybrid two-stage implementation combining PCPE and BPS can no longer effectively reduce the performance penalty, which leads to the optimization techniques we proposed based on mirroring and partially scaling the received constellation (see Sect. 6.1.2).

In this section, a robust and low-complexity phase recovery method for PS square-QAM systems is introduced. By introducing two optimization techniques, namely the constellation mirroring method and the partial scaling method, the proposed optimized principal component-based phase estimation (OPCPE) algorithm can solve the incompatibility between PCPE and PS at high SNR, while keeping the better GMI performance compared to BPS at low SNR. It is worth mentioning that the proposed algorithm retains the advantages of low cycle slip rate (CSR) and low computational complexity of PCPE. To further improve the accuracy of the proposed algorithm, a hybrid implementation combining the proposed algorithm and BPS is also investigated. Compared to the well-known two-stage BPS (2S-BPS), the proposed simple and hybrid methods can reduce the required SNR to reach the 0.8 normalized generalized mutual information (NGMI) threshold by about 0.74 and 0.98 dB respectively for a PS-64QAM system with 0.03 shaping parameter when the combined linewidth is 500 kHz. Compared with BPS, the proposed algorithm can better ensure the shaping gain brought by PS technology, which is a feasible phase recovery alternative for practical probabilistically-shaped transmissions.

6.1.1 System Model

A. Probabilistic Shaping (PS)

When applying the PS scheme to square QAM modulation formats, the in-phase and quadrature components of the transmitted constellation can be considered as two independent pulse-amplitude-modulation (PAM) signals, whose signal levels are distributed with non-equal probabilities following the Maxwell-Boltzmann (MB) distribution [6]. The MB distribution can be expressed as:

$$P_X(x_i) = \frac{\exp(-\nu |x_i|^2)}{\sum_{k=1}^{M} \exp(-\nu |x_k|^2)} \tag{6.1}$$

where x_i denotes the i-th complex M-QAM constellation point and ν denotes the shaping parameter. Thanks to the PS technique, symbols with larger amplitudes are transmitted with lower probabilities, which can improve the system performance and obtain a shaping gain in certain SNR conditions. The entropy of PS-MQAM can be calculated by:

$$H = -\sum_{i=1}^{M} P_X(x_i)\boldsymbol{log}_2 P_X(x_i) \tag{6.2}$$

B. Principal Component-Based Phase Estimation Algorithm
Principal component analysis (PCA) is a well-known algorithm that converts a set of likely correlated observations into a set of linearly decorrelated principal components (PC) [18]. Although the PCs directly extracted from a QAM constellation are independent of phase rotation, Diniz et al. pointed out that, for the received constellation after squaring process, the angle of the first PC is proportional to the original constellation rotation [17]. Therefore, the phase estimation for square-QAM signals can be transformed into the extraction of the first PC from the corresponding squared constellation.

Since the PCPE algorithm is based on block-wise parallel processing architecture, the squared constellation signal as input is first divided into multiple blocks of length N. Then the real and imaginary parts of each block are separated to form a $2 \times N$ real matrix. For instance, the kth matrix can be expressed as:

$$\mathrm{A}_k = \begin{bmatrix} \mathrm{Re}(x_k^2) \\ \mathrm{Im}(x_k^2) \end{bmatrix} \tag{6.3}$$

where$x_k^2 = \left[x_k^2[1]\ x_k^2[2] \ldots x_k^2[N]\right]^{r}$denotes the kth squared constellation block, while Re(.) and Im(.) represent the real and imaginary parts respectively. Then the covariance matrix for each block is calculated as:

$$C_k = A_k A_k^T \tag{6.4}$$

After that, the first PC v_k of each block can be updated by:

$$cv_k = C_k v_{k-1} \tag{6.5}$$

$$v_k = \frac{v_k}{\sqrt{v_k^2[1] + v_k^2[2]}} \tag{6.6}$$

It is worth noticing that the PC as an initial value is assumed to be $\nu_0 = \left[1\ 0\right]^T$.In order to avoid the error introduced by the mismatch between the initial value and the actual phase rotation, the first iteration will be repeated 3 times to ensure correct convergence.

The estimated phase noise for each block is given by:

$$\hat{\phi}_k = \frac{1}{2}\arctan\left\{\frac{\nu_k[2]}{\nu_k[1]}\right\} - \frac{\pi}{4} \tag{6.7}$$

Finally, a phase unwrapping process is performed to eliminate phase ambiguities, and the estimated phase to be compensated is:

$$\hat{\phi}_k = \hat{\phi}_k + \left\lfloor \frac{1}{2} + \frac{\hat{\phi}_{k-1} - \hat{\phi}_k}{\pi/2} \right\rfloor \frac{\pi}{2} \tag{6.8}$$

Moreover, the PCPE algorithm can also be combined with BPS to achieve better accuracy. For the PCPE and BPS hybrid scheme (PCPE-BPS), after performing the PCPE for the first stage, the second stage performs a BPS procedure with test phases as:

$$\phi_b = \eta\pi\left(\frac{2b-1}{4B_2} - \frac{1}{4}\right), \quad b \in \{1, 2, \cdots, B_2\} \tag{6.9}$$

where η is the phase aperture while B_2 is the number of test phases of the BPS stage.

C. Interplay of PS and PCPE

Since the PCPE algorithm can achieve satisfactory performance in uniformly distributed Square-QAM systems, it is natural to assume that the algorithm can obtain similar results in PS Square-QAM systems. However, our numerical simulation results show that, whether it is the PCPE or PCPE-BPS algorithm, the performance will experience a severe degradation as the shaping factor increases.

To evaluate the interplay of PS and PCPE, Monte Carlo simulations were carried out to calculate the GMI between the transmitted and received symbols. Square M-QAM symbols following MB distribution were generated with 1 sample per symbol. In the simulated system, phase noise consisted of two parts, one was a constant phase shift of $\phi_{const} = \pi/6$, and the other was random laser phase noise modeled as a Wiener process as [19]

$$\phi(n) = \phi(n-1) + w(n) \tag{6.10}$$

where $\phi(n)$ was the random phase noise of the n-th symbol, while {w(n)} was a sequence of Gaussian-distributed random variables with mean zero and variance

$$\sigma^2 = 2\pi\Delta f\tau \tag{6.11}$$

where $\Delta f = 200\,\text{kHz}$ was the combined linewidth of transmitter and receiver and $\tau = (32\,\text{GBd})^{-1}$ was the symbol duration. Additive white Gaussian noise (AWGN) was then added to the complex signals, the SNR values considered only the power of the signals and AWGN. 2S-BPS, PCPE, and PCPE-BPS were employed in the phase estimator respectively to compensate the phase noise. The number of test phases in both stages of 2S-BPS was 11, while the PCPE-BPS was performed with $B_2 = 11$ and $\eta = 1/11$, which ensured the same search accuracy of the two algorithms in the second stage. The range of SNR used in the simulation was 0 dB to 25 dB. For each SNR, the corresponding GMI result is the average value obtained by 100 realization with 256 blocks of 64 symbols each. According to [16, 20], the range of shaping factor for which PS should be implemented is 0 to 0.05 for 64-QAM. In this paper, we selected 4 different shaping factors from 0.02 to 0.05 at the same interval for evaluation. In addition, cycle slips were ideally compensated to simplify the GMI computation [17]. To be more specific, at the transmitting end of the simulation, the actual phase rotation per block is recorded. At the receiving end, for each block, the phase estimation result is added to 1 4 times of $\pi/2$, and the four results are compared with the recorded actual phase rotation. The result with the smallest error is used for final phase recovery, thereby eliminating the impact of cycle slips.

As shown in Fig. 6.1, the performance results of 2S-BPS are closely related to the SNR value, when the SNR is low, there is a clear gap between its result and the ideal

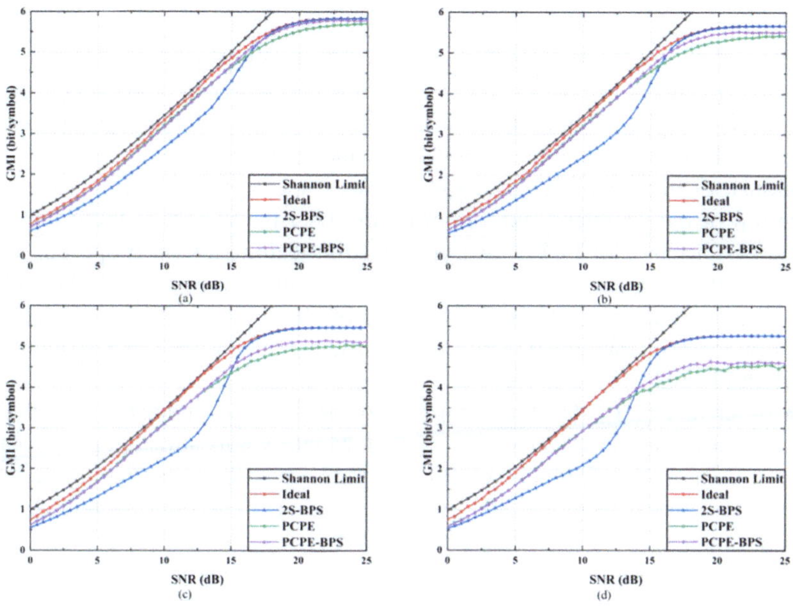

Fig. 6.1 GMI results versus SNR with different phase recovery algorithms in PS 64-QAM systems, the values of shaping parameters are (**a**) 0.02, (**b**) 0.03, (**c**) 0.04 and (**d**) 0.05

GMI. Take inset (b) as an example, for PS 64-QAM with shaping parameter equal to 0.03, the GMI performance of 2S-BPS is actually worse than uniform 64-QAM when SNR is lower than 15 dB, which indicates that the performance degradation caused by BPS exceeds the shaping gains expected by PS. This phenomenon proves the importance of deploying alternative phase recovery algorithms in PS systems. As a contrary, PCPE outperforms 2S-BPS significantly in the low SNR range, especially when the shaping parameter is large.

When the SNR is high enough, even for strongly shaped signals, the performance of 2S-BPS is close to ideal. On the other hand, the problem of incompatibility between PCPE and PS will become more obvious as the shaping factor increases. When the shaping parameter is 0.02, the hybrid algorithm PCPE-BPS can compensate for the performance impairment caused by PCPE to a certain extent, yet for shaping parameters larger than 0.02, even if the two-stage scheme is used, the GMI performance still cannot correctly converge to the entropy value of the PS signal, which causes a serious performance penalty compared to 2S-BPS.

The above phenomena indicate that PCPE can be used as an alternative phase recovery scheme to ensure the shaping gains expected by PS, but a feasible optimization method is urgently needed to improve the robustness of PCPE in PS systems.

6.1.2 Optimized Principal Component-Based Phase Estimation

After carefully analyzing the characteristics of multiple sets of data, we found that the squared constellation diagrams whose phase noise cannot be correctly compensated by PCPE often have at least one of the two following features. One is the significant mismatch between the two sides of the axis of symmetry, the other is the lack of high-power constellation points, especially the highest power point at the top of the symmetry axis. In order to offset the impact of these two issues on the performance of PCPE, two optimization techniques are proposed, namely the constellation mirroring method and the partial scaling method. We refer to the novel phase recovery algorithm enabled by these two optimization techniques as optimized principal component-based phase estimation, its block diagram is as shown in Fig. 6.2.

In order to clearly state the reasons and ideas of the optimization techniques we used in Sects. 6.3.1 and 6.3.2, another Monte Carlo simulation was carried out. We generated 10^8 blocks of PS 64-QAM symbols with shaping parameter of 0.05, and block length of 64. The generated symbols only contained a fixed phase rotation of $\pi/6$, and were free of AWGN and random phase noise. PCPE was used to perform phase recovery for these symbols. During the PCPE process, each block performed only one iteration, and the initial PC in PCPE was set as $\nu_0 = \left[\cos(\pi/6)\ \sin(\pi/6)\right]^T$.

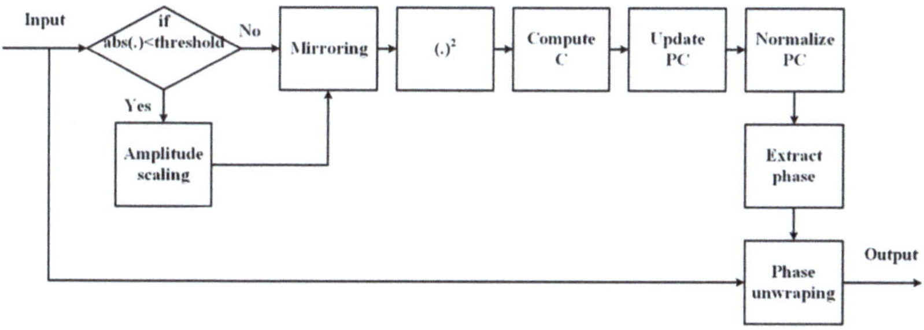

Fig. 6.2 Block diagram of the proposed optimized principal component-based phase estimation algorithm

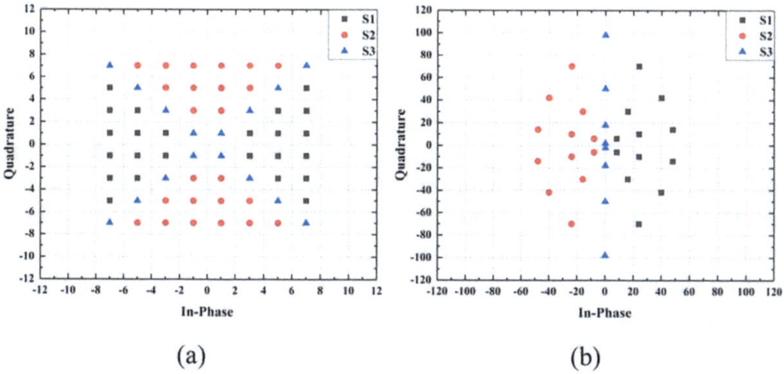

Fig. 6.3 Grouping for 64QAM constellation: (**a**) before squaring; (**b**) after squaring

During each Monte Carlo realization, two characteristic values of each block, the power mismatch ratio (PMR) and number of outer constellation points, were recorded respectively, along with the residual phase errors of PCPE.

A. Constellation Mirroring

The PMR represents the degree of mismatch of the constellation diagram. To be more specific, we separate the symbols of a QAM constellation into 3 groups, S_1, S_2 and S_3, as shown in Fig. 6.3a, representing the three different cases where the absolute values of the real part of the coordinate are greater than, less than or equal to the absolute value of the imaginary part. After squaring the received constellation, S_3 will become the axis of symmetry of $S_1$and S_2, as shown in Fig. 6.3b.

To evaluate the degree of mismatch of the squared constellation, PMR is defined as:

$$PMR = \frac{\sum_{x_i \in S_1} |x_i|^2}{\sum_{x_j \in S_2} |x_j|^2} \tag{6.12}$$

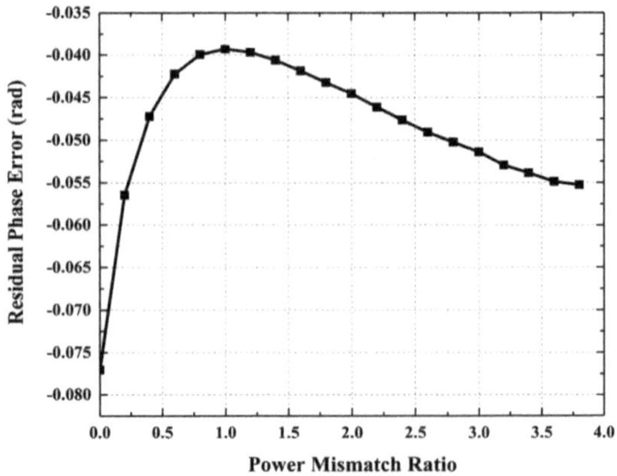

Fig. 6.4 Measured residual phase error versus PMR in a PS-64QAM system with PCPE phase recovery and v = 0.05

We classified the PMR values obtained by the Monte Carlo simulation at intervals of 0.2, and plotted the relationship between the PMR value and the corresponding average residual phase error as shown in Fig. 6.4. Simulation results show that, when the PMR value is 1, the performance of PCPE algorithm reaches the optimal. And when the PMR value deviates from 1, no matter whether it is increased or decreased, the performance will be degraded accordingly.

Based on the result shown in Fig. 6.4, the first optimization required for PCPE in PS systems is to make the value of PMR approach 1 as much as possible. Although the value of PMR is determined by the transmitting end, as long as a simple mirroring operation is performed at the receiving end, the PMR value of the input fed into PCPE can be made to 1. For received signal block x_k, the mirrored signal can be expressed as:

$$\hat{x}_k = x_k \exp(i\frac{\pi}{2})| \tag{6.13}$$

Thanks to the 90° symmetry of square-QAM constellation, the constellation point that originally belonged to S_1 will become part of S_z after 90° rotation without changing its power, and vice versa. Combine the mirrored signal with the original $x_k$and input it into the PCPE process, the optimized A_k obtained by constellation mirroring method can be expressed as:

$$\mathbf{A}_k = \begin{bmatrix} \mathrm{Re}(x_k^2) & \mathrm{Re}(\hat{x}_k^2) \\ \mathrm{Im}(x_k^2) & \mathrm{Im}(\hat{x}_k^2) \end{bmatrix} \tag{6.14}$$

B. Partial Scaling

After determining the relationship between the performance of PCPE and PMR, we further investigated the impact of the number of outer constellation points on PCPE performance. It should be noted that during this simulation, the previously mentioned mirroring optimization has been deployed. During our simulation for PS 64-QAM, the outer constellation points refer to symbols whose amplitude greater

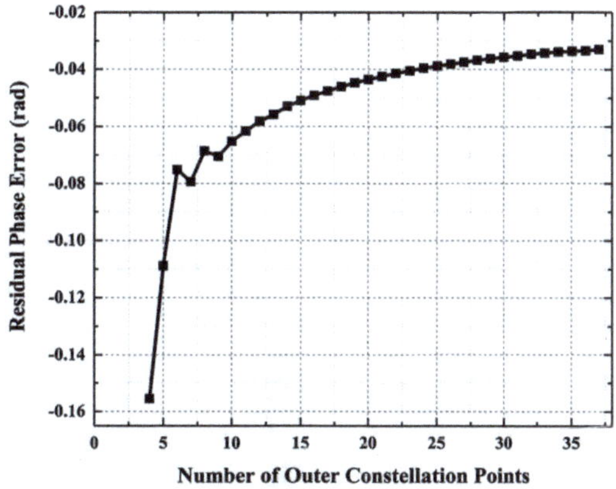

Fig. 6.5 Measured residual phase error versus number of outer constellation points in a PS-64QAM system with PCPE phase recovery and $v = 0.05$

than $3\sqrt{2}$. As shown in Fig. 6.5, during 108 realizations, the minimum number of outer constellation points is 4, and the residual phase error is the most severe in this case. For cases where the number of outer constellation points is less than 7, the performance of PCPE increases rapidly as the number increases. Although the curve shows some fluctuations when the number is 7 and 9, the overall trend of the curve shows that as the number gradually increases, the residual error gradually decreases.

Since the purpose of PS is to reduce the appearance probability of outer constellation points, it is unrealistic to increase the number of outer points at the transmitting end. Therefore, we propose an optimization method based on partial scaling. At the receiving end, the inner points will be separated and scaled according to the amplitude. For instance, as shown in Fig. 6.6, for PS 64-QAM signals, the inner constellation points with amplitude less than $3\sqrt{2}$ are selected and their amplitudes are 7/3 times magnified.

6.1.3 Simulation Results and Discussions

In this section, we benchmark the proposed OPCPE and the hybrid scheme of OPCPE and OBPS (OPCPE-OBPS) against the 2S-OBPS through three aspects: GMI performance, cycle slip tolerance and computational complexity. For the first two indicators, the results are obtained through Monte Carlo simulation.

In the simulation, two different cases with combined linewidths of 200 and 500 kHz were implemented separately to thoroughly investigate the performance comparison between the proposed algorithm and 2S-OBPS. The optimal block length is 128 when the linewidth is 200 kHz, and 64 in the other case.

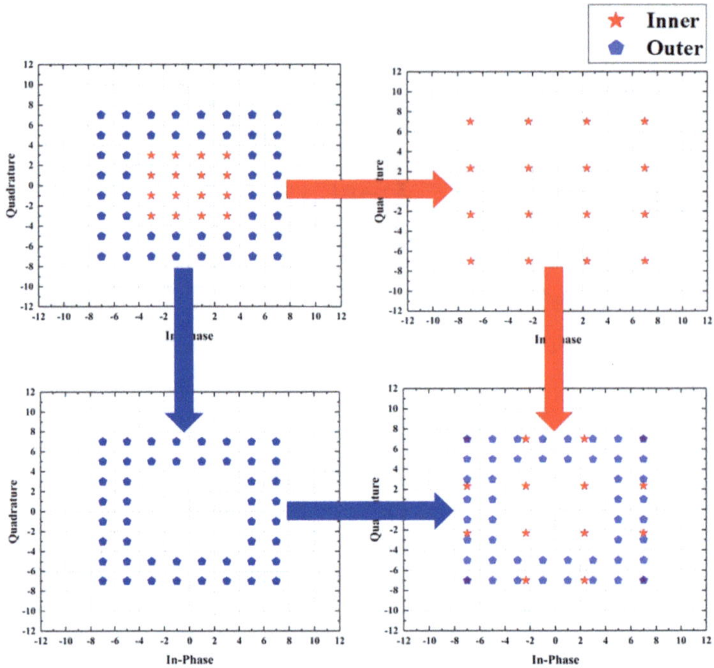

Fig. 6.6 Schematic of partial scaling for PS 64-QAM at the receiving end

A. Results of GMI Performance

Firstly, the simulation results of GMI performance with a combined linewidth of 200 kHz are as shown in Fig. 6.7. Simulation results prove that the OPCPE algorithm can effectively solve the problem of incompatibility between PCPE and PS at high SNR, while retaining the GMI performance advantage of PCPE over BPS at low SNR. Taking the 0.8 NGMI threshold as a reference, under the four different shaping parameters investigated, the SNR required gains that OPCPE can achieve compared to 2S-OBPS are around 0.19, 0.37, 0.41, and 0.26 dB, respectively, while the gains that OPCPE-OBPS can achieve are approximately 0.30, 0.54, 0.66, and 0.53 dB.

When the combined linewidth is 500 kHz, the results are as shown in Fig. 6.8. Taking the 0.8 NGMI threshold as a reference, under the four different shaping parameters investigated, the SNR required gains that OPCPE can achieve compared to 2S-OBPS are around 0.64, 0.74, 0.55, and 0.17 dB, respectively, while the gains that OPCPE-OBPS can achieve are approximately 0.80, 0.98, 0.90, and 0.51 dB.

In conclusion, the proposed OPCPE algorithm can robustly solve the incompatibility between PCPE and PS at high SNR for all the shaping parameters and linewidths investigated. It should be noted that, the standard 2S-OBPS algorithm outperforms all other schemes when the SNR is high enough. This is because the OPCPE algorithm is not based on test phases. By deploying the hybrid OPCPE-

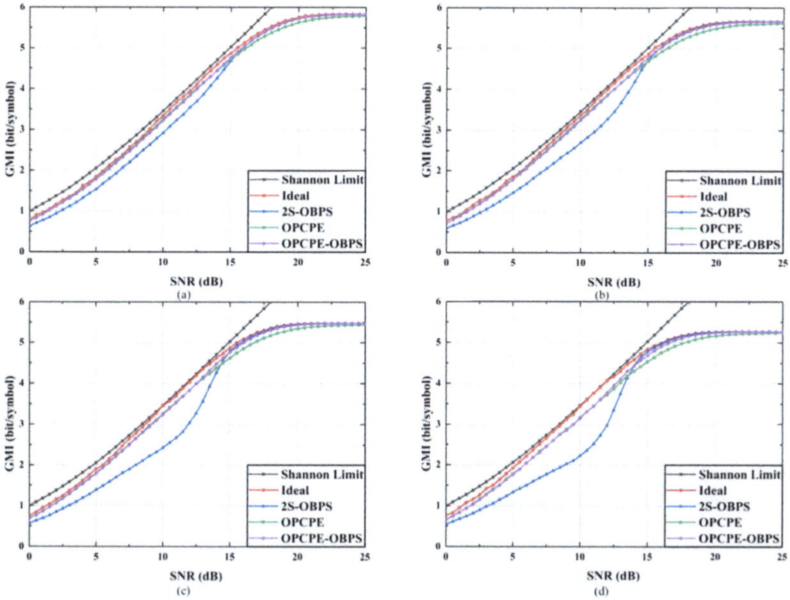

Fig. 6.7 GMI results versus SNR with different phase recovery algorithms in PS 64-QAM systems when the combined linewidth is 200 kHz, the values of shaping parameters are (**a**) 0.02, (**b**) 0.03, (**c**) 0.04 and (**d**) 0.05

OBPS algorithm, the performance difference between OPCPE and 2S-OBPS at high SNR can be reduced. On the other hand, under low SNR conditions, the proposed algorithms retain the performance advantage of PCPE over BPS. When considering the 0.8 NGMI threshold, the OPCPE-OBPS algorithm can achieve a lower SNR penalty than 2S-OBPS under all conditions studied, which makes it a more suitable phase recovery algorithm for PS systems than the traditional BPS algorithm.

B. Results of Cycle Slip Tolerance

Secondly, we investigated the cycle slip tolerance of the proposed algorithm benchmarking the 2S-OBPS. For OPCPE-OBPS, the phase wrapping is done in the OPCPE stage, which means its cycle slip tolerance is the same as OPCPE. Therefore, for this part of simulation evaluation, only OPCPE and 2S-OBPS are considered. As shown in Fig. 6.9, under the two linewidths studied, both the CSRs of OPCPE and BPS increase with the increase of shaping parameter. Throughout the measured SNR range, the CSRs of OPCPE (red lines) are significantly lower than BPS (black lines), which proves the better cycle slip tolerance of the proposed algorithm.

C. Computational Complexity Analysis

Finally, the computational complexities of the three above mentioned CPR algorithms are investigated. In [17], the complexity comparison between 2S-BPS

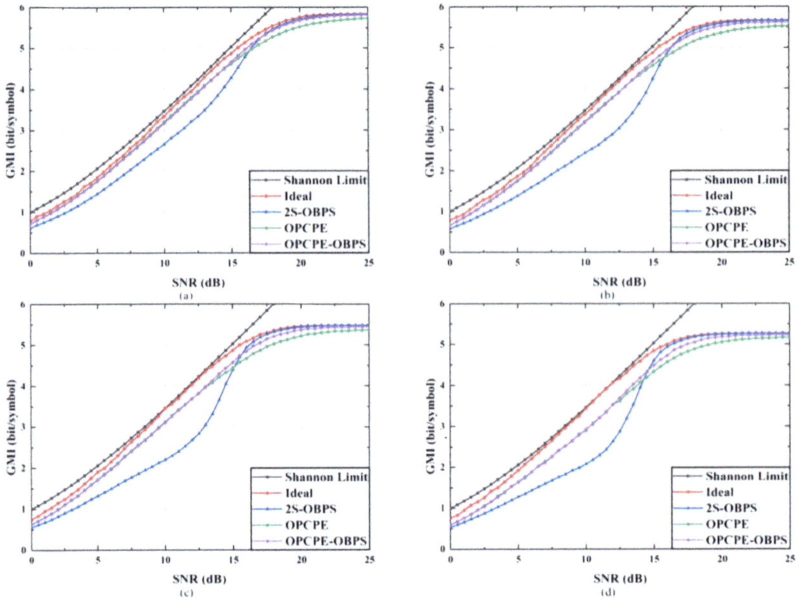

Fig. 6.8 GMI results versus SNR with different phase recovery algorithms in PS 64-QAM systems when the combined linewidth is 500 kHz, the values of shaping parameters are (**a**) 0.02, (**b**) 0.03, (**c**) 0.04 and (**d**) 0.05

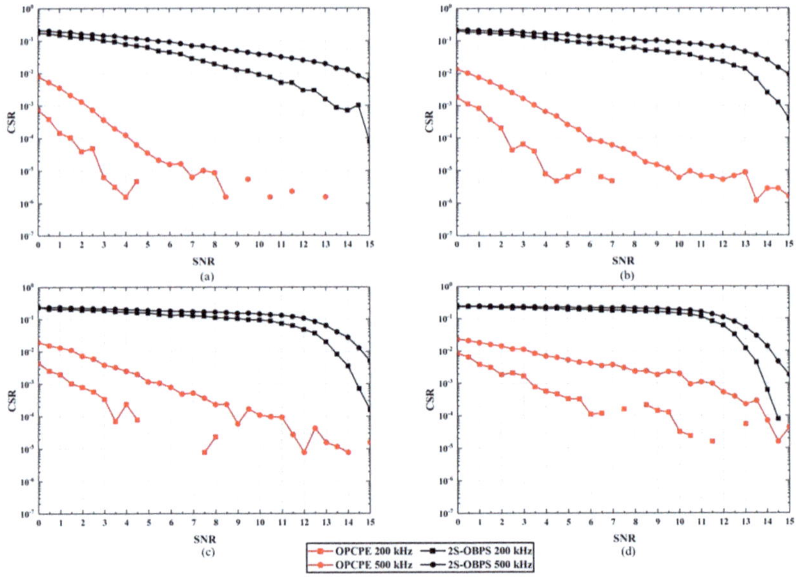

Fig. 6.9 CSR results versus SNR with different phase recovery algorithms in PS 64-QAM systems, the values of shaping parameters are (**a**) 0.02, (**b**) 0.03, (**c**) 0.04 and (**d**) 0.05

Table 6.1 Computational complexity for CPR algorithms

Operation	2S-BPS	OPCPE	OPCPE-BPS
Addition	5NB_T+2N+3	13N+3	5NB_2+15N+3
Multiplication	6NB_T+4N	22N+8+2N_L	6NB_2+26N+8+2N_L
Square root	0	1	1
LUT	B_2+2	3	3
Decision	NB_T	N	N+N
Comparison	B_T-2	0	B_2-1

Table 6.2 Calculated computational complexity for CPR algorithms

	Operation	2S-BPS	OPCPE	OPCPE-BPS
2	Addition	3587 (100%)	419 (11.68%)	2243 (62.53%)
	Multiplication	4352 (100%)	776 (17.83%)	3016 (69.30%)
	Decision	704 (100%)	32 (4.55%)	384 (54.55%)
4	Addition	7171 (100%)	835 (11.64%)	4483 (62.52%)
	Multiplication	8704 (100%)	1544 (17.74%)	6024 (69.21%)
	Decision	1408 (100%)	64 (4.55%)	768 (54.55%)
28	Addition	14339 (100%)	1667 (11.63%)	8963 (62.51%)
	Multiplication	17408 (100%)	3080 (17.69%)	12040 (69.16%)
	Decision	2816 (100%)	128 (4.55%)	1536 (54.55%)

and PCPE is divided into six parts, namely real additions (or subtractions), real multiplications (or divisions), computation of root square, access to look up tables (LUT) for trigonometric computations, decisions and comparisons. Based on the results given in [17], the complexity of OPCPE can be easily obtained by calculating the complexity increasement caused by symbol mirroring and scaling, which can be found in the appendix. The number of each type of operation is shown in Table 6.1, where N represents the length of symbol sequence per block, B_T indicates the total number of test phases in both stages of 2S-BPS, while B_2 denotes the number of test phases in the second stage, and N_L is the number of symbols in the current block whose amplitude is lower than the threshold.

For several realistic scenarios, the calculated values for N-related operations for each algorithm are shown in Table 6.2. The block lengths evaluated are 32, 64 and 128, respectively, and the value of N_L is its upper bound, that is, $N_L = N$. In addition, for the BPS process, $B_1 = B_2 = 11$.

6.1.4 Conclusions

The interplay of PS and phase recovery algorithm is a major obstacle for the transmission performance of optical communication systems. Previous studies have shown that, for uniform square-QAM constellations, PCPE can achieve better performance than BPS in the low SNR range. In this paper, the performance comparison between PCPE and BPS in PS systems is investigated by simulation. The results show that although PCPE can still achieve better GMI performance in the low SNR range, it fails to converge correctly at high SNR. In order to improve the robustness of PCPE, an optimized PCPE algorithm, namely the OPCPE algorithm, is proposed. The proposed algorithm not only solves the problem of compatibility between PCPE and PS, but also retains the advantages of low CSR and low computational complexity. Compared with BPS, the proposed algorithm can better ensure the shaping gain brought by PS technology, which is a feasible phase recovery alternative for PS system.

6.2 A Low Complexity Nonlinear Compensation Algorithm Based on Logistics Regression

With the development of communication services, such as industrial internet of things (IOT), cloud computing and high-definition video, long-distance backbone optical transmission systems have been promoted to satisfy the continuously increasing capacity and bandwidth demands [21–23]. To solve the signal impairments along with the long-haul fiber link, digital signal processing (DSP) technology has received much attention in recent years [24]. To combat the fiber Kerr nonlinearity in long-distance optical transmission systems, nonlinear compensation (NLC) techniques become an indispensable component of DSP [25, 26]. There are various techniques reported for reducing the large computational complexity associated with DBP, namely through split-step Fourier methods (SSFM) [27–29], Volterra series nonlinear equalizers (VSNE) [30–32] and machine-learning algorithms [25, 33–35]. The principle of digital backpropagation (DBP) is realized by numerically approximating the nonlinear Schrödinger equation (NLSE) solution with SSFM. The NLSE can precisely describe the nonlinear effect in a single-mode fiber. However, the computation complexity of the conventional DBP is excessively high, which reduces the efficiency of DSP. In [24], FY. Zhang et al. experimentally demonstrated a 34.94G baud dual-polarization 16QAM single-channel transmission with a blind adaptive DBP algorithm and achieved 0.2–1.4 dB improvement. Furthermore, the DBP algorithm based on the SSFM and a memory polynomial (MP) model is proposed and the complexity is reduced by 50.7% compared to the standard DBP-SSFM [27]. In [28], JW Zhang al. experimentally demonstrated a PDM-16QAM system over 205 km SSMF by applying joint LUT pre-distortion and post DBP nonlinearity compensation. The Volterra series transfer functions have been proposed to mitigate fiber nonlinearities, which are modeled and compensated in the time domain or frequency domain by using a set of nth order nonlinear transfer functions [30]. V. Vgenopoulou et al. demonstrated a novel Volterra-based nonlinear equalizer and revealed a 0.8 dB improvement after transmitting 6×32 Gbaud PM-16QAM signals over 1040 km [31]. The weighted VSNE has been proposed, which reduce the computer complexity by 60% [32]. However, the Volterra series have not been widely applied because the performance will be worse when the computational complexity is reduced. Recently, the machine-learning algorithms have become research hotspot for compensating various impairments in the optical transmission system. In [25], Bertold Ian Bitachon et al demonstrated a (deep learning)-DBP(DL-DBP) in a 32 Gbaud 16QAM transmission across 1200 km and achieved a 1.9 dB gain with DBP. Among various models and algorithms, the Logistic regression algorithm plays an important role in the optical communication system. Chuang YC et al used the logistic regression classification for mitigating high noise-ratio advisement light panel in visible light communications [33]. The logistic regression algorithm can predict the category and get the probability of prediction through Sigmoid mapping meanwhile. In this section, the LR-DBP algorithm based on logistic regression digital backpropagation which is proposed to compensate the effects of dispersion and fiber nonlinearities in long haul optical transmission systems is introduced. The LR-DBP can evaluate nonlinear effects by using machine learning to adjust the step structure of DBP. The proposed algorithm is experimentally demonstrated in an 80 Gbit/s 16QAM signal transmitted across 1000 km single-mode fiber. The BER of LR-DBP can satisfy the HD-FEC threshold when the launched power is in the range of − 5 to 2 dBm, which has a 0.2 dB improvement over ADBP, 0.6 dB improvement over DBP and a 1.8 dB improvement over CDE. The proposed algorithm could obtain with 19.2% computational complexity decrease and 19.6% CPU running time decrease by calculating the average numbers of real multiplications and CPU running times.

6.2.1 Structure and Principal of LR-DBP

The structure of LR-DBP is shown in Fig. 6.10. The logical regression decision (LRD) is processed after dispersion compensation in each step. We divide each step

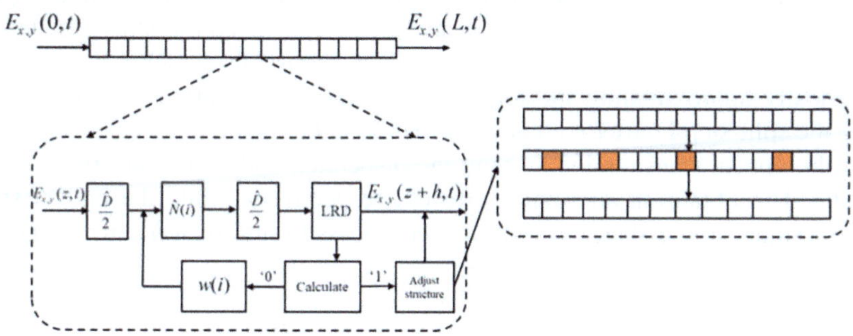

Fig. 6.10 The structure of LR-DBP (LRD: logical regression decision)

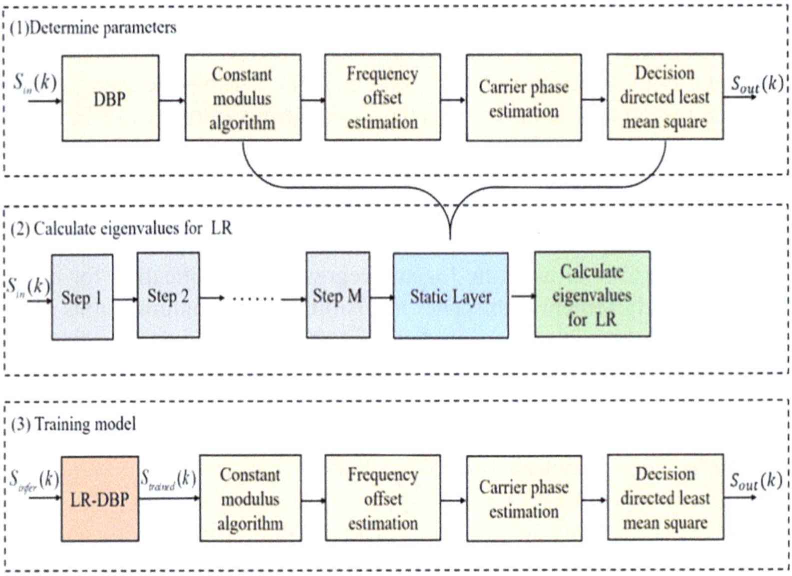

Fig. 6.11 The principal of LR-DBP (1) Determine static layer parameters (2) Calculate eigenvalues (3) Training model

into two categories and the LRD can correct the nonlinear calculation error of each step and update the structure of DBP.

Figure 6.11 shows the proposed training method which is divided into three stages. In the first stage, we estimate the parameters of the static layers, which are not updated during the training process [25]. The static layers include the constant multi-modulus algorithm (CMMA), frequency offset estimation (FOE), carrier phase estimation and decision-directed least mean square (DD-LMS). The static layers preliminarily predict the long-haul transmission environment, allowing LR to estimate the eigenvalues at the bit level.

In the second stage, the eigenvalues and the parameters are calculated for logistic regression. the principle of LR-DBP is analyzed step by step for adjusting the steps of DBP and the training parameters in logistic regression, w. DBP is realized by numerically approximating the nonlinear Schrödinger equation (NLSE) solution using the split-step Fourier method (SSFM) [36–39].

In the optical transmission system, the dispersion and nonlinear effects can be calculated to obtain an approximate result, respectively. while the optical signal passes a small short distance h. The transfer process from z to $z+h$ can be expressed as [25]:

$$E_{x,y}(z+h,t) = \exp\left(h\frac{\hat{D}}{2}\right)\exp(h\hat{N})\exp\left(h\frac{\hat{D}}{2}\right)E_{x,y}(z,t) \tag{6.15}$$

Where $E_{x,y}$ presents the complex field envelope in the x/y states of polarization. z and t denote the spatial and temporal coordinates. α, β_2 and γ are defined as the attenuation coefficient, dispersion parameter and nonlinear coefficient respectively. we use the linear and nonlinear operators $\hat{D}$ and $\hat{N}$ instead to simplify the equation, which can be simplified expressed as:

$$\hat{D} = \frac{\alpha}{2} E_{x,y} + j\frac{\beta_2}{2}\frac{\partial^2 E_{x,y}}{\partial t^2}| \tag{6.16}$$

$$\hat{N} = -j\frac{8\gamma}{9}\left(\left|E_x\right|^2 + \left|E_y\right|^2\right) \tag{6.17}$$

The conventional DBP method is accomplished by Eq. (6.15) into the linear and nonlinear operators $\hat{D}$ and $\hat{N}$. j represents the imaginary unit. However, the current traditional algorithms can't effectively estimate the nonlinearity because the calculation of the nonlinear operator $\hat{N}$ is not accurate enough. Therefore, we describe the nonlinear operator $\hat{N}$ in detail which can be written as:

$$\hat{N} = -j\frac{8\gamma}{9}\left(|E_x|^2 + |E_y|^2\right) + \frac{j}{\omega_0}\frac{1}{E_{x,y}}\frac{\partial}{\partial T}[(|E_x|^2 + |E_y|^2)E_{x,y}] \tag{6.18}$$

Where ω_0 presents the center frequency of light wave. Therefore, the estimation of the nonlinear operator in each step determines the accuracy and complexity of the entire nonlinear compensation algorithm. In general, DBP use approximate integrals $\int_i^{i+1} \hat{N}(z')dz'$ to estimate nonlinear operators $\hat{N}$:

$$\int_i^{i+1} \hat{N}(i)di \approx \frac{h}{2}[\hat{N}(i) + \hat{N}(i+1)] \tag{6.19}$$

i represents the imaginary unit and using $\int_i^{i+1} \hat{N}(i)di$ can improve the accuracy of SSFM compared to $\hat{N}(i)$. However, $\hat{N}(i+1)$ is unknown in the calculation process, which will be realized by iterating multiple times and increasing the complexity. In this process, the weight is updated through the loss function to achieve the approximation of the nonlinear operator $\hat{N}(i)$. The bit decision of the logistic regression is used to optimize the DBP step structure.

The cross-entropy error measures the difference between the predicted value of the model and the target value. We approximate the minimum loss of nonlinear operators by training the parameters in advance, which can be expressed as [33]:

$$\begin{aligned}
&\frac{\partial}{\partial w(i)}\hat{N}_L(i) = 2(\hat{N}(i) - \hat{N}_w(i)) \times \frac{\partial}{\partial w(i)}(\hat{N}(i) - \hat{N}_w(i)) \\
&w(i) \leftarrow w(i) - S\frac{\partial}{\partial w(i)}\hat{N}_L(i) \\
&= w(i) + S(\hat{N}(i) - \hat{N}_w(i)) \times \hat{N}_w(i)(1 - \hat{N}_w(i)) \times \frac{\partial}{\partial w(i)} w \bullet i
\end{aligned} \tag{6.20}$$

w represents the output coefficient and $\hat{N}_w(i)$ represents the nonlinear operator of logistic regression. S represents the learning rate. The output of the logistic regression $\hat{N}_w(i)$ could approximate to the actual nonlinear factor by updating the coefficient $w(i)$. We calculate the eigenvalues for LR which consists of smoothed mean square error(MSE), the adaptive step length (ASL) and the power for measuring w and $\hat{N}_w(i)$ in each step accurately [24, 25]. The factor vector can be computed through a series of iterations, which can be written as [24]:

$$\mu_\psi(i+1) = \mu_\psi(i) - \rho\frac{\sigma^2(i) - \sigma^2(i-1)}{\mu_\psi{}^2(i) - \mu_\psi{}^2(i-1)} \tag{6.21}$$

Where $\mu_\psi(i)$ denotes the ASL, ρ represents the small positive constant used to control the adaptive processing. The normalized noise variance of the symbols at the output of the static layers is represented by $\sigma^2(i)$.

The MSE parameter is used to estimate the overall transmission condition, while the factor vector represents the dispersion and nonlinearity compensation effect between DBP steps. Moreover, the power correlation factor is applied to differentiate the predictions for LR-DBP in each step. Meanwhile, employing the static layers will simplify to calculate the MSE between the LR-DBP output symbol and the transmitted symbols.

6.2.2 Experiment Results and Discussions

The experimental setup for 80 Gbit/s 16QAM coherent optical communication system is shown in Fig. 6.12. The continuous optical wave is generated from an external cavity laser (ECL) with less than 100 kHz linewidth and 15 dBm output power. Then the light source at 1550 nm is modulated by the I/Q modulator to achieve electrical/optical conversion. The I/Q modulator has a 3 dB bandwidth of 33 GHz and an insertion loss of 6 dB. The two parallel Mach Zehnder modulators (MZMs) in the I/Q modulator are both biased at the null point and driven at the full swing, while the phase difference between the upper and lower branches of the I/Q modulator was fixed at $\pi/2$. We used an arbitrary waveform generator with 25 GSa/s and 3 dB bandwidth of 14 GHz to convert the two independent digital

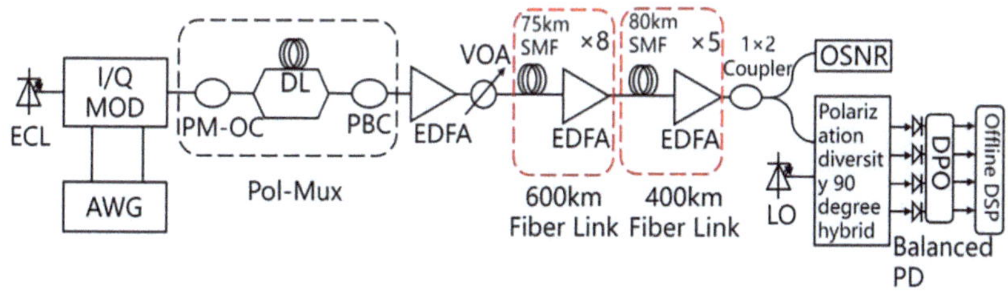

Fig. 6.12 Experiment Setup for the 16QAM coherent optical communication system (ECL: External cavity laser, I / Q MOD: I / Q modulator, Pol-MUX: polarization multiplexer, OC: optical coupler; DL: delay line; PBC: polarized beam combiner; PBS: polarizing beam splitter; EDFA: erbium doped fiber amplifier; LO: local oscillator)

symbol sequences to analog radio frequency (RF) signals. The polarization multiplexing of the signal is realized by the polarization-multiplexer, which comprises a polarization-maintaining optical coupler (PM-OC), optical delay line (DL) to provide delay and a polarization beam combiner (PBC) to recombine the signal. we use an erbium-doped fiber amplifier (EDFA) with a 5.5 dB noise figure and variable optical attenuator (VOA) to adjust the launched signal power into the fiber link.

The transmission link consists of 8 unrepeated spans of 75 km and 5 unrepeated spans of 80 km G.652 single-mode fiber (SMF) with an EDFA for amplification without any inline dispersion. We utilize an ECL with a linewidth less than 100 kHz as an optical local oscillator (LO). To implement the optical polarization diversity and phase diversity of the received signal and the optical LO, we used an integrated coherent receiver (ICR). Two 90° optical hybrids, four 32 GHz balanced photodetectors, and a four-channel digital phosphor oscilloscope with 100 GSa/s and 20 GHz bandwidth were used to receive the signal. We then processed the offline digitized signal using MATLAB. The offline DSP is performed at the receiver which includes low pass filter, LR-DBP, cascaded multi-modulus Algorithm (CMMA) equalization, frequency offset estimation (FOE), carrier phase estimation (CPE), decision-directed least mean square (DD-LMS) algorithm and the BER calculation.

Firstly, we analyze the characteristic data required by the logistic regression. The convergence of MSE is analyzed before the beginning of the training stage for verify the effectiveness of the smoothed MSE as a training feature. Figure 6.13 shows the MSE as a function of the number of iterations for LR-DBP. The MSE will eventually reach $2 \times 10 - 2$ with the number of iterations increasing. It is worth emphasizing that we normalized the data before analyzing the MSE characteristic.

Figure 6.14 shows the curves of BER versus launch power for three different cases including compensation of dispersion effect (CDE), traditional DBP, ADBP (advanced DBP) and LR-DBP [22]. The BER performance of CDE can satisfy the SD-FEC threshold when the launch power is smaller than 2.6 dB. The BER gradually decreases as launched power increases for all four compensation algorithms at first. At this time, the ASE noise is the dominant factor affecting the BER performance under this circumstance. The fiber Kerr nonlinearity determines the system

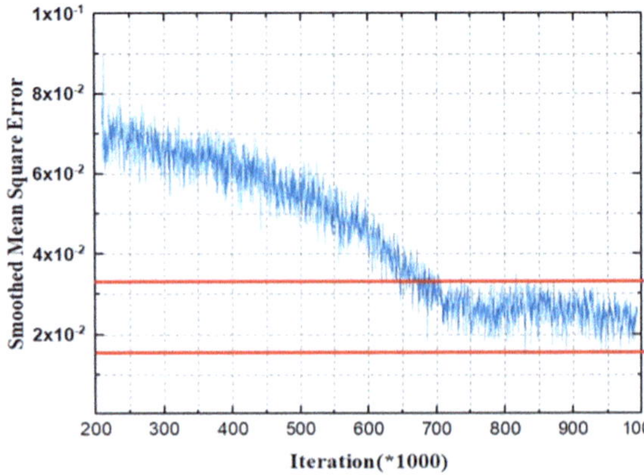

Fig. 6.13 Smoothed mean square error (MSE) as a function of iterations for LR-DBP

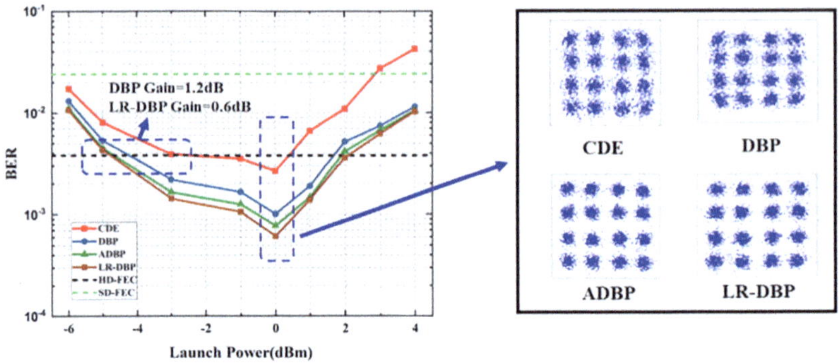

Fig. 6.14 Measured BER curves versus launched power in the 80Gb/s 16 QAM coherent optical communication system

performance when launched power increases, which lead to nonlinear distortion and the BER performance declines. LR-DBP has a better BER performance than DBP and ADBP within the entire measurement launch power range. The BER of LR-DBP can satisfy the HD-FEC threshold when the launched power is in the range of – 5 to 2 dBm, t, which has a 0.2 dB improvement over ADBP, 0.6 dB improvement over DBP and a 1.8 dB improvement over CDE.

The training stage of LR-DBP was further investigated on the basis of measuring the performance of OSNR and BER. We divided the data into the data set and the training set when extracting the features of the LR algorithm. The data is divided into data sets and training sets. The ratio of training set has a certain impact on BER performance. The total number of data sets is relatively small when partitioning the data sets, because feature extraction is based on the number of DBP steps. During the training process, the division of the test set and training set will have an impact on the final BER performance. Figure 6.15 shows the influence of different training levels on BER performance. We test five sets of data and found that 80% is the optimal threshold for partitioning. When the training set is less than 80%, the lack

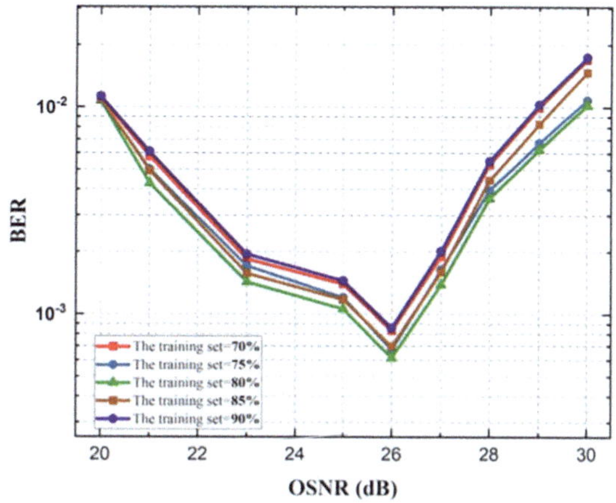

Fig. 6.15 Measured BER curves versus OSNR in the 80Gb/s 16QAM coherent optical communication system

of training leads to the underfitting of logistic regression and the BER performance declines. When the training set is higher than 80%, the training set overfitting and the BER performance will also be affected.

In all kinds of nonlinear compensation algorithms, the complexity is an important factor that restricts the feasibility of algorithms and will affect the efficiency of the whole DSP. Therefore, it is necessary to discuss the algorithm complexity of LR-DBP.

The overlap preservation technique is employed when a longer sample sequence is processed by the SSFM algorithm. We use the method of comparing the complexity of DBP in [10] to evaluate the advantage of LR-DBP with 16 spans/step for 16QAM. Table 6.3 presents the complexity of LR-DBP in comparison with ADBP and DBP-SSFM in terms of NRMS, GR and CPU Running Times(s) when launched power is 0 dBm. NRMS is defined as number of real multiplications respectively. In Table 6.3, we have adopted the reduction gain (GR) as a figure of merit (GR = 1−O/Oref, where O and Oref are the number of operations required by LR-DBP or ADBP and DBP). The CPU calculation time were performed on the computer with AMD Ryzen 5 3600 6-core processor at 3.6 GHz, 16 GB Random Access Memory and Windows 10 Professional operating system. Moreover, the number of real multiplications and real additions per processing sample required for a step in the SSFM algorithm are shown in [24].

The average NRMS (real multiplications) as a function of launched power are shown in Fig. 6.16a. It can be seen that LR-DBP can not only improve the system performance but also reduce the overall computational complexity compared to

Table 6.3 THE improvement in computational complexity using various DBP algorithms

Method	DBP	ADBP	LR-DBP
Average NRMS	1674.88	1497.53	1369.10
GR (%)	–	10.59	18.26
CPU running times(s)	13.31	12.34	11.00

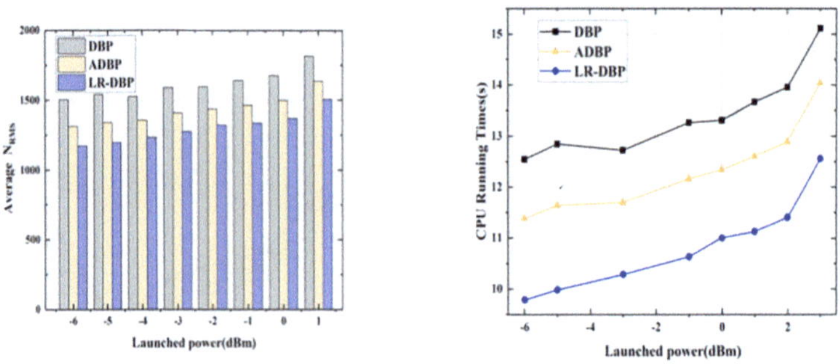

Fig. 6.16 (**a**) Average NRMS as a function of launched power. (**b**) CPU running time as a function of launched power

DBP and ADBP. It is worth noting that the long-distance transmission will suffer more nonlinear ductile damage when launched power increases. To counteract the nonlinear effect, more iterations and computational complexity will be introduced. The complexity of LR-DBP is reduced by 22.0 and 10.6% compared to DBP and ADBP when the launched power is -6 dBm. The complexity falls by 18.26 and 8.6% when launched power is 0 dBm. The average reduction in complexity is 19.2%. We compared the CPU running times of DBP and LR-DBP. It can be seen in Fig. 6.16b that the CPU running time has slight differences with NRMS, but the overall trend is the same. The average decrease of CPU running time is 19.6%.

6.2.3 *Conclusions*

In conclusion, we proposed a novel algorithm based on logistic regression digital backpropagation (LR-DBP) in order to mitigate the effect of dispersion and nonlinear effect in long haul optical transmission systems. LR-DBP can evaluate nonlinear effect via machine learning by employing a cross-entropy error factor that adjusts the step structure of DBP. The proposed algorithm is experimentally demonstrated in an 80 Gbit/s 16QAM signal transmission over a 1000 km single-mode fiber. The BER of LR-DBP can satisfy the HD-FEC threshold when the launched power is in the range of -5 to 2 dBm, t, which has a 0.2 dB improvement over ADBP, 0.6 dB improvement over DBP and a 1.8 dB improvement over CDE. The proposed algorithm could obtain with 19.2% computational complexity decrease and 19.6% CPU running time decrease by calculating the average numbers of real multiplications and CPU running times. We believe that our novel algorithm design is suitable candidate for practical long-haul coherent optical communication system.

6.3 Unscented Kalman Filter with Joint Decision Scheme for Phase Estimation in Probabilistically Shaped QAM Systems

The growing need for communication services that offer both high speed and high capacity has posed challenges for traditional coherent optical communication systems. These challenges primarily revolve around system capacity and spectral efficiency (SE) [40–43]. Probabilistic shaping (PS) has been widely employed in coherent optical communication to improve system capacity and reduce the gap with the Shannon limit.

However, the introduction of PS results in additional performance degradation and will impair these decision-directed digital signal processing (DSP) algorithms. For instance, Sasai et al. have investigated phase noise tolerances of high-order PS signals and QAM signals in [44] and suggested that we should take the laser linewidth and carrier phase recovery (CPR) algorithms into account when choosing a signal shaping form. In [45], Barbosa et al. have evaluated CPR algorithms for PS transmission and indicated that the blind phase search (BPS) stage is highly affected by PS at moderate and low optical signal-to-noise ratio (OSNR). In [1], Mello et al. have researched the interplay of PS and BPS algorithms and recommended the necessity for alternate CPR methods to be deployed in PS transmissions. Wang et al. have developed an optimized principal component-based phase estimation algorithm which outperforms the BPS algorithm at low signal-to-noise ratio (SNR) for PS QAM systems [46]. As an effective compensation method for linear impairments, the Kalman filter has received interest for applications in coherent optical communication systems [47–49]. Because of the fast-tracking capability and high accuracy for parameter estimation, several Kalman-filter-based algorithms have been suggested and implemented such as carrier recovery [50–52] and polarization state tracking [53, 54]. As an alternative to the BPS algorithm, the CPR algorithm based on the Kalman filter can also function as a blind method for carrier recovery. However, the Kalman filter's enhanced performance is accompanied by increased computational complexity and additional delay [48]. Moreover, the typical Kalman filter-based CPR algorithms use decision-directed operations, such as ML estimates, to update the measurement state, which is negatively affected by PS. As a potential alternative to these existing CPR algorithms, the Kalman filter in PS transmissions requires additional study.

In this section, we present a novel CPR algorithm for PS QAM systems that is based on the unscented Kalman filter. The proposed technology combines the UKF-based algorithm with a novel joint decision scheme to mitigate the effects of PS and enhance PS transmission performance. The UKF-based structure ensures the estimation precision of the proposed algorithm. By partitioning the constellation symbols sensibly and employing both the MAP and ML decisions, the joint decision scheme ensures that the proposed technique is immune to the influence of PS in PS transmissions. Compared to conventional carrier phase recovery algorithms, the proposed technology maintains PS resilience and achieves a lower optical signal-to-noise ratio (OSNR) penalty when the normalized generalized mutual information (NGMI) threshold is taken into account. Monte Carlo simulations and experiments are used to assess the efficacy of the proposed algorithm. In comparison to conventional BPS, the proposed technology can reduce the SNR required to meet the 0.8 NGMI threshold for PS 16QAM systems. The proposed UKF-based algorithm can be applied to PS systems more effectively than the conventional CPR algorithm, resulting in an improvement in NGMI performance.

6.3.1 Operation Principle of the Proposed Phase Estimation Algorithm

The Kalman filter provides rapid tracking capabilities and high precision for parameter estimation, making it a useful instrument for compensating linear impairments. Consequently, numerous Kalman-filter-based CPR techniques have been suggested and confirmed to be effective. In order to use the Kalman filter approach for carrier phase estimation, it is important to define the model of the received signal before to CPR. If the signal has undergone DSP processing, including IQ imbalance compensation, chromatic dispersion (CD) compensation, constant modulation algorithm (CMA), and frequency offset estimation (FOE), the primary impairment is laser phase noise (LPN). The received signal can be defined as

$$r(n) = s(n)exp^{j\phi(n)} + \eta(n) \tag{6.22}$$

where $r(n)$ and $s(n)$ represents the n-th received and transmitted signal. $\phi(n)$ represents the phase noise which is modeled as $\varphi(\text{n}) = \varphi(\text{n} - 1) + \omega(\text{n})$. $\omega(n)$ is independent and identically distributed Gaussian random variables with zero mean and variance $\sigma^2 = 2\pi\Delta f\tau$. Δf is the combined line width of transmitter and receiver lasers and τ is the symbols period. $\eta(n)$ represents the collective amplified spontaneous emission (ASE) noise that is assumed to be additive Gaussian with zero mean.

In the implementations of linear Kalman filter (LKF) and extended Kalman filter (EKF) proposed in [55, 56], the received sampled signal is processed sample-by-sample, and the state parameter vector to be estimated by the Kalman filter is $x_k = [\theta_k]$. The equation for the state update is $\theta_{k+1} = A\theta_k + \eta_p$, where A represents the state transition matrix, which is defined as 1 in the case for a Wiener process. The variable η_p represents the process noise between two successive samples. The observation model proposed in [55] is then given by $\tilde{\theta_k} = \theta_k + \eta_\theta$, where $\tilde{\theta_k}$ represents the actual value for the phase noise and η_θ is the observation noise. Both η_p and η_θ are assumed to be white Gaussian with zero mean and covariances denoted by Q and R, respectively.

In order to accurately estimate the phase noise for each sample and minimize the observation noise, the Kalman filtering algorithms use the prediction-update frame.

According to the state update equation indicated above, a prediction $\hat{x}_k^-$ of the state parameter vector x_k is made at time instance $k-1$. We get a prediction of the measurement y_k^- at time instance k using $\hat{x}_k^-$. The actual measurement $\tilde{y}_k$ at time instance k is then made and the final estimate of the Kalman filter at time instance k is given as

$$\hat{x}_k = \hat{x}_k^- + \mathrm{K}_k(\widetilde{y_k} - y_k^-) \tag{6.23}$$

where K_k represents the Kalman gain which is updated according to the measurement matrix, process, and measurement error covariance in each time instance. By minimizing the innovation vector ($\widetilde{y_k} - y_k^-$), the prediction ($\hat{x}_k - \hat{x}_k^-$) approaches zero and the filter algorithm can lock to the immediatc change in state parameters.

The unscented Kalman filter, as opposed to LKF and EKF, employs the unscented transform to obtain a set of 'sigma points' instead [48]. The probability distribution function and statistical moments of the state parameters x_k can be captured by those points. Although UKF requires more computational effort than LKF and EKF, it outperforms LKF and EKF in terms of system performance and can estimate non-linear systems with greater precision. Thus, the UKF is utilized to develop the proposed algorithm. In [1], Mello investigates the interaction between PS and BPS algorithm and concludes that BPS is influenced by PS due to correlated incorrect decisions made in the phase recovery scheme. PS influences the decisions made by the BPS algorithm, resulting in a performance decrease in excess of the anticipated shaping gain. Moreover, the accuracy of the estimate is influenced by the decisions taken about measurement acquisition during the observation stage. It suggests that the effects of PS also have an influence on a UKF-based CPR algorithm. Applying UKF in the PS transmission thus requires taking the erroneous decision into account.

Maximum a posteriori probability (MAP) detection has recently been suggested [57], and numerous works implementing MAP detection in the CPR scheme have been reported recently [58]. For PS M-QAM modulation formats, the received symbols have different prior probabilities. For the PS constellation, the maximum likelihood detection that follows the rule of choosing the constellation symbols that are closest in terms of Euclidean distance is suboptimal. Figure 6.17a illustrates how the decision boundaries for MAP detection differ from the conventional boundaries for ML decisions. In comparison with the ML decision, the MAP decision considers the prior probability of the different transmitted symbols to maximize the a posteriori probability of an accurate determination. The MAP decision rule can be expressed as [57]

$$\min_i |y_k - x_D|^2 + 2N_0 ln(P(x_D)) \tag{6.24}$$

where y_k represents the k-th received symbol, x_D represents the surrounding constellation symbols in terms of Euclidean distance, N_0 is the one-dimension noise variance. $P(x_D)$ represents the prior probabilities which are generally known at the transmitter. The first term in Eq. (6.24) is the squared absolute Euclidean

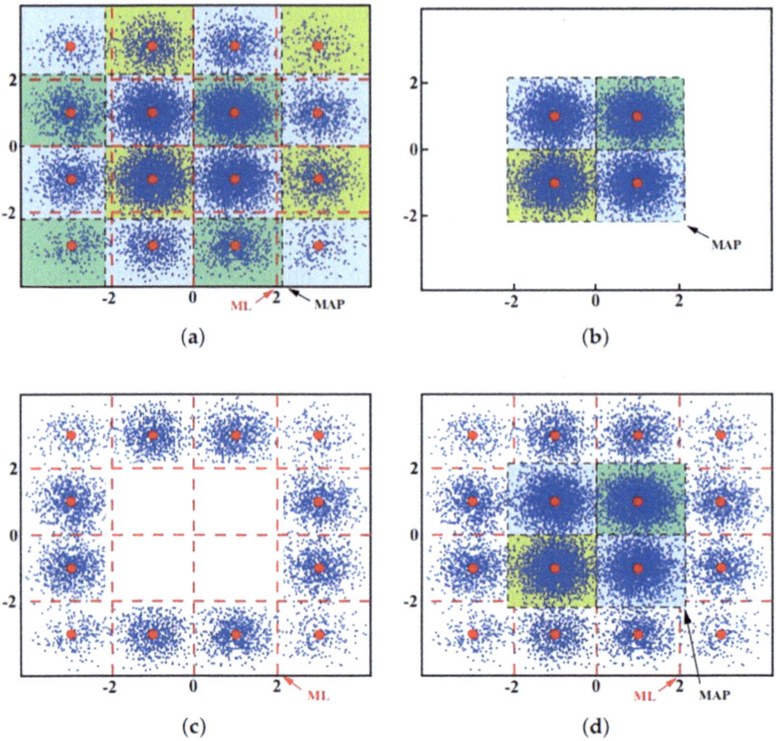

Fig. 6.17 Schematic of boundary setting for the proposed joint decision scheme: (**a**) Boundary comparison of MAP and ML. (**b**) Boundaries for inner constellation points using MAP detection. (**c**) Boundaries for outer constellation points using ML detection. (**d**) Boundaries for joint decision scheme. The colored areas are the decision regions of MAP, the dashed lines are the decision boundaries of ML, and the red points are the ideal constellation points

distance which is the same metric used in the decisions of the traditional BPS. For constellation symbols under the PS scheme, the second term in Eq. (6.24) is related to the noise variance and prior probabilities, which changes the decision boundary. The symbols with a higher probability occupy larger areas, which decreases the likelihood of making a mistaken decision regarding these symbols and, as a result, reduces the overall probability of decision error.

However, additional investigation and simulation indicate that implementing MAP detection into the UKF-based CPR algorithm does not result in the expected performance improvement. In this observation model, the observed signal's amplitude noise is accounted for by a complex observation noise, allowing the UKF to monitor both amplitude and phase variations. Despite the fact that a complex observation model improves estimation precision, constellation symbols with larger amplitudes have a greater impact on the innovation vector.

The decision regions for symbols with lower amplitudes and higher probabilities are expanded by MAP detection, whereas the regions for symbols with higher

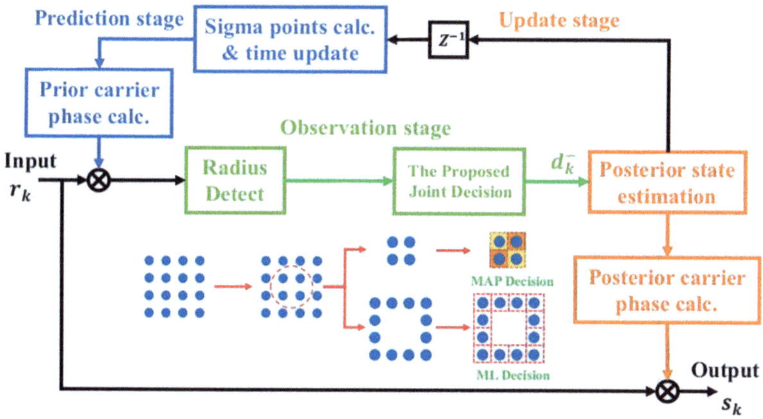

Fig. 6.18 Block representation of the proposed CPR algorithm with joint decision scheme

amplitudes in the ideal constellation map are reduced. We infer that additional decision errors of these symbols with larger amplitudes occur after MAP detection which has an impact on the UKF update.

We, therefore, propose the jointed decision method to prevent the above-mentioned additional decision errors in order to apply MAP detection to the UKF in order to resolve the PS-induced correlated incorrect decisions. The proposed joint decision scheme is shown in Fig. 6.17.

We set different decision boundaries for the inner and outer constellation points in the proposed joint detection method. Figure 6.17b illustrates how to detect inner constellation points using MAP's detection regions, maintaining MAP's advantage of lowering incorrect decisions in these regions. The boundaries for the outer constellation point, shown in Fig. 6.17c, follow the regular pattern of the ML detection. By doing so, the relative narrowing of the outer constellation decision region could be prevented, the number of erroneous decisions made for the outer constellation symbols could be decreased, and both inner and outer symbols could be recognized using the proper boundaries. The overall decision boundary set is shown in Fig. 6.17d.

Based on the joint detection scheme, we propose a UKF-based CPR algorithm, referred to as UKF-MAP/ML. The operation frame of UKF-MAP/ML is shown in Fig. 6.18.

The majority of the block diagram depicts the method by which the UKF framework iteratively estimates the phase noise via the input signal. The process can be divided into three stages: prediction, observation, and update. The input signal s(k) is divided into two channels for phase estimate and impairment compensation respectively. At the prediction stage, the prior estimates of the current state vector $\hat{x}_k$ and the error covariance P_k are obtained according to the posterior estimates $\hat{x}_{k-1}$ in the previous instant [48].

$$\begin{aligned} \hat{x}_k^- &= A\hat{x}_{k-1} \\ P_k^- &= AP_kA^T + Q \end{aligned} \tag{6.25}$$

where $(.)^T$ represents the conjugate transpose operation. The sigma points set is then calculated using the unscented transform and the prior estimations, given as [24]:

$$X_k^- = \begin{cases} \hat{x}_k^- & i = 0, \\ \hat{x}_k^- + \left(\sqrt{(n+\lambda)P_k^-}\right)^{i-1} & i = 1, \ldots, n, \\ \hat{x}_k^- - \left(\sqrt{(n+\lambda)P_k^-}\right)^{i-n-1} & i = n+1, \ldots, 2n. \end{cases} \tag{6.26}$$

where $\left(\sqrt{(n+\lambda)P_k^-}\right)^{i-1}$ represents the (i-1)th column of the lower triangular matrix of the matrix $(n+\lambda)P_k^-$ after the Cholesky decomposition and n represents the state dimension. The parameter λ is given as $\lambda = \alpha^2(n+k) - n$. The unscented transformation introduces four external parameters: $\alpha, \beta, \kappa, \lambda$. The values of these external parameters determine the distribution of the sigma points.

At the observation stage, the received signal s(k) is rotated by the sigma points. The smaller flowchart below in Fig. 6.18 depicts the observation stage and the proposed joint decision scheme in more detail. The rotated signal s_k^- is then fed into the radial decision module. According to the threshold, this module calculates the amplitude of the rotated symbols and distinguishes between symbols that correspond to inner constellation points and outer ones. The distinguished signal is then sent to the joint decision module to get the observation of the received signal. Those symbols after the joint decision module d_k^- are used to update the measurements:

$$\hat{Z}_k^- = h(X_k^-) = Y_k - d_k^- X_k^- \tag{6.27}$$

where Y_k represents the actual measurement of the received signal. The measurements are utilized to calculate the posterior estimations [59]:

$$\begin{aligned} \hat{z}_k^- &= \textstyle\sum_{i=0}^{i=2n} W_m^{(i)} \hat{Z}_{i,k}^- \\ P_{z_k,z_k} &= \textstyle\sum_{i=0}^{i=2n} W_c^{(i)} [\hat{Z}_{i,k}^- - \hat{z}_k^-][\hat{Z}_{i,k}^- - \hat{z}_k^-]^T \\ P_{z_k,x_k} &= \textstyle\sum_{i=0}^{i=2n} W_c^{(i)} [X_{i,k}^- - \hat{x}_k^-][\hat{Z}_{i,k}^- - \hat{z}_k^-]^T \end{aligned} \tag{6.28}$$

where $W_m^{(i)}$ and $W_c^{(i)}$ represents the the weight of the sigma points when calculating the approximate mean and approximate covariance, respectively. The weights are

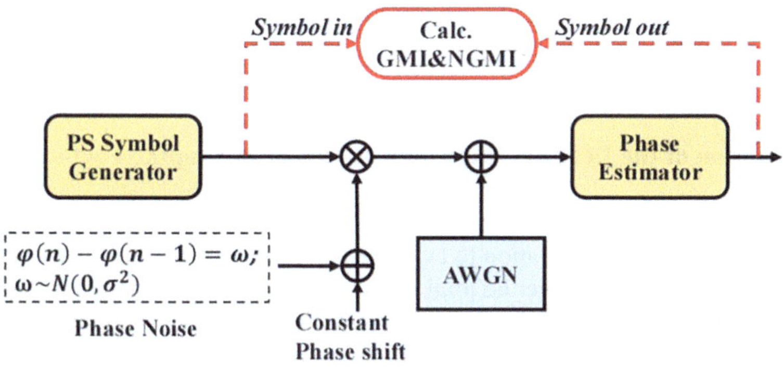

Fig. 6.19 Simulation setup. AWGN: additive white Gaussian noise; GMI: generalized mutual information; NGMI: normalized generalized mutual information

given as [59]:

$$\begin{aligned} K &= P_{z_k,x_k} / P_{z_k,z_k} \\ \hat{x}_k &= \hat{x}_k^- + K(z_k - \hat{z}_k^-),\ z_k = 0 \\ P_k &= P_k^- - K P_{z_k,z_k} K^T \end{aligned} \tag{6.29}$$

The symbols after phase recovery $s_k = r_k \hat{x}_k^*$ are obtained from the output port. The posterior estimate is also fed back to the processing of the next instant.

6.3.2 Simulation Results and Discussions

Figure 6.19 depicts the simulation setup. To evaluate the performance of the proposed algorithm, Monte Carlo simulations were carried out to calculate the GMI and NGMI between the transmitted and received symbols, which are represented as dashed red lines in Fig. 6.19. Square 16QAM symbols following Maxwell-Boltzmann (MB) distribution were generated with 1 sample per symbol.

The channel model considered only LPN and AWGN. The linear phase noise consisted of a constant phase shift of $\pi/6$ and random laser phase noise simulated by the model described in Sect. 6.2.1. The constant phase shift is used to test whether the algorithm can compensate for the phase rotation accurately. The random laser phase is used to test whether the algorithm can monitor the change of laser phase noise in the actual system. AWGN is then added to the symbols rotated by the laser phase noise. The output of the phase estimator is used to calculate the GMI and NGMI. The symbol rate was set to 32 GBaud and the combined linewidth between the transmitter laser and the local oscillator (LO) was set to 200 kHz. The SNR values considered only the signal power and AWGN. The GMI and NGMI results were obtained from the mean of 200 realizations with 16,384 symbols each to avoid the random error of the simulation results.

Optimization of the Decision Threshold at Variables Shaping Factors and Fixed SNR

We used Monte Carlo simulation to evaluate the GMI performance of the proposed algorithm in order to validate the joint decision scheme's underlying principle. The SNR was fixed to the condition that the NGMI obtained 0.8, which works as the threshold for 27.5% soft-decision forward error correction (SD-FEC) [46].

We evaluated various thresholds for each of the four shaping variables for PS 16-QAM, which ranged from 0.1 to 0.2. As the channel model in this simulation only accounted for linear phase noise and AWGN, the signal scale after the channel is equivalent to the ideal 16QAM constellation points. The proposed joint decision scheme requires determining the symbol amplitude and differentiating between symbols based on the threshold. In order to include the amplitude range of the ideal 16QAM constellation points, the threshold was specified to be between 0 and 5 with intervals of 0.5.

Figure 6.20 illustrates that the GMI performance of the proposed algorithm is related to the threshold value at various shaping factors. By comparing the performance of the GMI at various thresholds of the same shaping factor, it

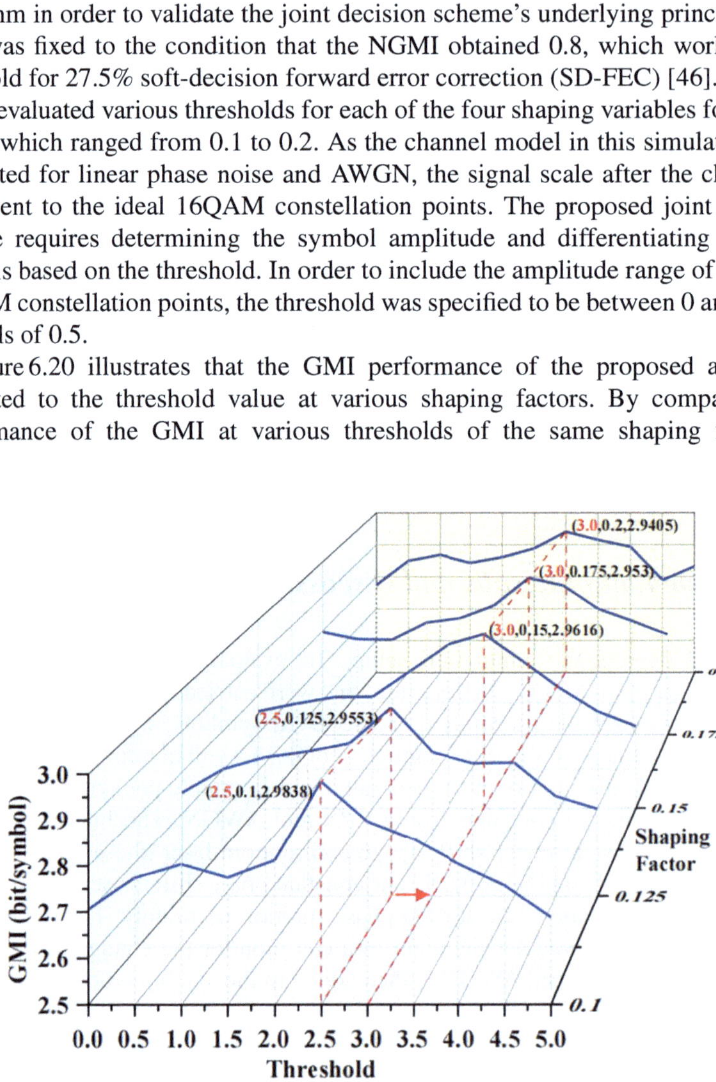

Fig. 6.20 GMI performance results versus thresholds under different shaping factors when the SNR was set to 9.5 dB

is possible to analyze the relationship between the GMI and the threshold and determine the threshold value at which the GMI reaches its optimal value. The efficacy of the GMI improved as the threshold value rose from 0 to 2.5, reaching its optimum at 2.5. In addition, the GMI efficacy declined when the threshold value increased from 2.5 to 5. When the threshold was too low, the joint detection scheme was unable to take full advantage of the performance enhancement provided by the MAP. Nonetheless, if the threshold exceeded the optimal level, MAP would introduce additional decision errors and degrade performance. This phenomenon demonstrated that the performance of the joint decision method improved when the threshold value was appropriately selected. As the shaping factor increased, so did the threshold at which the GMI reached its optimum value, as shown in Fig. 6.20.

Due to the fact that the proposed scheme is equivalent to MAP if the threshold is set too high, we can verify the inference that MAP would introduce additional decision errors and degrade performance if the threshold was set above the optimal level. In the results with a shaping factor of 0.1, the GMI performance with a threshold of 5 was worse than the GMI performance with a threshold of 0. When the threshold was set to 0, the ML decided all symbols, whereas the MAP decided all received symbols when the threshold was set to 5. Consequently, the results indicated that the performance loss caused by the additional decision error was greater than the performance improvement anticipated by the MAP, thereby establishing the necessity of the proposed joint decision scheme.

By comparing the GMI performance at varying thresholds of various shaping factors, we can verify the effectiveness of the proposed technology under different parameters. When the shaping factor was 0.1 or 0.125, the GMI performance achieved its peak when the threshold was set to 2.5. And when the shaping factor was 0.15, 0.175, or 0.2, the GMI performance reached its peak when the threshold was set to 3. The threshold value at which the GMI reaches its optimal value varies with the shaping factor. This phenomenon occurred because, as the shaping factor increased, the probabilities of exterior constellation symbols decreased further. Given that the KF update was less impacted by the detection error of outside symbols, a high threshold was considered suitable. The phenomenon is consistent with the characteristics of the constellation point distribution varying with the shaping factor, further validating the scheme's rationality.

Comparison with Different CPR Algorithms at Variable Shaping Factors

In the preceding section, the performance enhancement of the proposed algorithm was validated. In this section, we compared the performance of the proposed UKF-based CPR algorithm to that of the existing LKF- and EKF-based CPR algorithms, as well as BPS. We examined the performance of the GMI and NGMI for PS 16-QAM, using SNR intervals of 0.5 and various shaping factors, with SNR ranging from 5 to 12.5. The threshold was set appropriately for each shaping factor in order to improve performance. The block length of BPS was 64 and there were 11 test

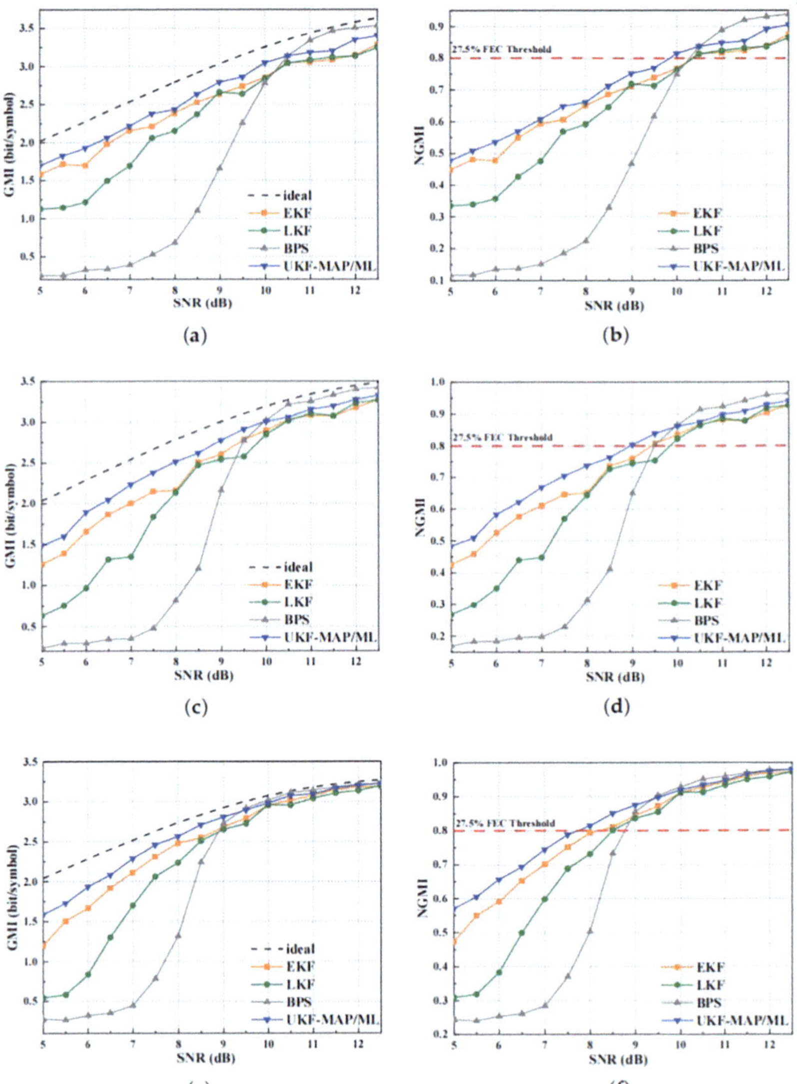

Fig. 6.21 GMI and NGMI performance results with different CPR algorithms in PS 16QAM systems when the shaping factors are: (**a, b**) 0.1, (**c, d**) 0.15, and (**e, f**) 0.2. UKF-MAP/ML, the proposed UKF-based CPR algorithm with a joint detection scheme. The red dashed line represents the 27.5% FEC threshold

phases. Under varying SNRs, the covariance Q for KF-based algorithms was 1×10^{-5}, and the covariance R was adjusted.

The performance outcomes were closely correlated with the SNR value, as illustrated in Fig. 6.21. In terms of GMI performance, there was a clear distinction between the BPS and the other algorithms when the SNR was low. Compared to the BPS, these KF-based algorithms demonstrated superior tolerance to the effects of PS in the low SNR region. And performance results demonstrated that the proposed system outperformed the EKF and LKF-based CPR algorithm across the entire SNR region. The NGMI results are displayed in Fig. 6.21. Using the 0.8 NGMI threshold as a benchmark, the proposed method could obtain SNR gains of approximately 0.51, 0.50, and 0.52 dB when compared to the EKF. The required SNR gains were approximately 0.42, 0.52, and 1.11 dB relative to the BPS.

It should be noted that the BPS algorithm performed better than the other methods when the SNR was high enough. This was due to the fact that KF-based techniques, which are independent of test phases, estimate phase noise by updating the state vector, and the estimation's precision is highly dependent on the initial state vector and the default covariance. Nonetheless, the proposed algorithm maintained superior robustness in the low SNR zone.

Taking into consideration the 0.8 NGMI threshold, the proposed algorithm could accomplish a smaller OSNR penalty than the BPS under the investigated conditions. Increasing the shaping factor decreased the performance disparity between the proposed algorithm and BPS when the SNR was high. It demonstrated that if the shaping factor was high enough, the performance of the proposed algorithm would be comparable to BPS. Consequently, the proposed algorithm is more appropriate for PS systems than the conventional CPR algorithms.

6.3.3 *Experimental Results and Discussions*

The experimental verification of the proposed UKF-based CPR algorithm was carried out in a 32 GBaud single polarization coherent optical system, as shown in Fig. 6.22.

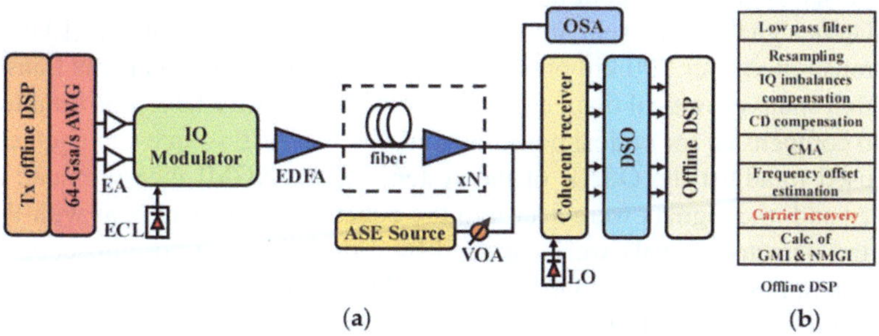

Fig. 6.22 (**a**) Experiment setup. AWG, arbitrary waveform generator; EA, electrical amplifier; ECL, external cavity laser; EDFA; erbium-doped fiber amplifier; VOA, variable optical attenuator; ASE source, amplified spontaneous emission source; LO, local oscillator; OSA, optical spectrum analyzer; DSO, digital storage oscilloscope. (**b**) Offline digital signal processing. CD compensation, chromatic dispersion compensation; CMA, constant modulus algorithm

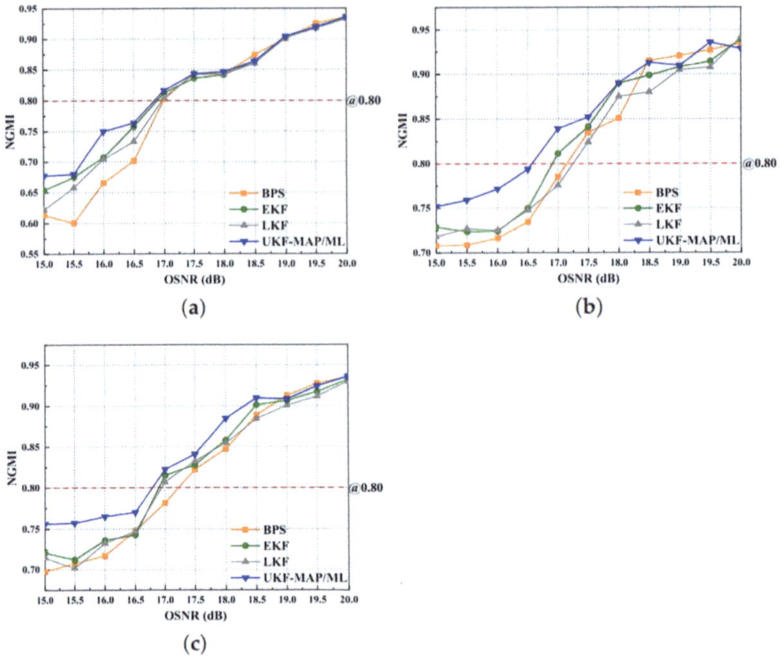

Fig. 6.23 NGMI versus the OSNR for 32 GBaud PS 16QAM when the shaping factors are: (**a**) 0.1, (**b**) 0.15, and (**c**) 0.2

At the transmitter end, an offline DSP applied in MATLAB was used to generate PS 16QAM signals following MB distribution. The shaping factor was set as 0.1, 0.15, and 0.2. Digital-to-analog conversion was realized by an arbitrary waveform generator (AWG) with a sampling rate of 64 GSa/s. The analog signals were then amplified by two electrical amplifiers and fed into the IQ modulator. An external cavity laser generated a continuous light wave with a 100 kHz line width. The modulated signals were sent into an Erbium-doped fiber amplifier (EDFA) and then transmitted over 500 km optical fiber link. The optical signal was transmitted through 5 spans of optical fiber, each of length 100 km. An EDFA was set after each span to compensate for optical power attenuation. An ASE source and another VOA were used to adjust the OSNR of the system for the NGMI measurement. At the receiver end, the signals were sent into the coherent receiver to beat with the LO. Then the detected signals were captured by a digital storage oscilloscope (DSO) with a sampling rate of 100 GSa/s. Offline DSP, which was shown in Fig. 6.22b, was used to recover the transmitted data from the received signals.

Figure 6.23 shows the NGMI performance at different shaping factors in the experimental demonstration. The block length of BPS was 64 and there were 11 test phases. Under varying SNRs, the covariance Q for KF-based algorithms was 1×10^{-5} and the covariance R was adjusted. The performance of the NGMI improved as the OSNR increased for all CPR algorithms. Due to the effect of PS, there was a performance disparity between BPS and the other algorithms when the OSNR was low. Using the 0.8 NGMI threshold as a benchmark, the proposed method could obtain SNR gains of approximately 0.10, 0.60, and 0.20 dB when compared to EKF. In comparison to BPS, the requisite SNR enhancements were approximately 0.40, 1.10, and 0.90 dB, respectively. As the OSNR increased, the performance disparity between BPS and the other algorithms decreased. Because the impact of PS on the CPR algorithms decreased as the OSNR increased, and because the performance disparity between KF and BPS was not readily apparent when the OSNR was sufficiently high. Although the performance difference between several algorithms was not readily apparent when the OSNR was high, the proposed method was able to obtain a lower OSNR penalty than the conventional BPS and KF-based algorithm when compared to the 0.8 NGMI threshold. The results of the experiment validate the claim that the proposed UKF-based CPR algorithm with joint detection is more suitable for the PS system.

6.3.4 Conclusions

Conventional CPR algorithms are impacted by PS when implemented in PS systems. In this paper, Monte Carlo simulation is utilized to compare the performance of the conventional BPS- and KF-based CPR algorithms during PS transmission. Due to additional decision errors, implementing MAP detection, which has been shown to be optimal for PS systems, directly into the UKF-based CPR algorithm does not result in the anticipated performance enhancement. Therefore, a UKF-based CPR algorithm with a joint detection scheme is proposed in order to prevent making erroneous decisions, and it is verified to be effective for PS systems by both simulation and experiment. By correctly partitioning the constellation symbols and maximizing the use of both MAP and ML detection, the performance of the proposed method can be enhanced. The proposed algorithm is more suitable for PS systems than the conventional BPS- and KF-based algorithms.

6.4 Summary

In this chapter, we discuss advanced Equalization technologies in coherent optical communication systems to solve the signal impairments along with the long-haul fiber link and high capacity. We first proposed the optimized principal component-based phase estimation (OPCPE) algorithm for PS square-QAM systems to solve the incompatibility between PCPE and PS at high SNR. Then, we introduced the LR-DBP algorithm based on logistic regression digital back propagation which is proposed to compensate the effects of dispersion and fiber nonlinearities in long haul optical transmission systems. In addition, we present a novel CPR algorithm for PS

QAM systems which combines the UKF-based algorithm with a novel joint decision scheme to mitigate the effects of PS and enhance PS transmission performance.

References

1. X. Wang, Q. Zhang, X. Xin, R. Gao, K. Lv, Q. Tian, F. Tian, C. Wang, X. Pan, Y. Wang, et al., Robust and low-complexity principal component-based phase estimation algorithm for probabilistically shaped square-QAM systems. J. Lightw. Technol. **38**(22), 6153–6162 (2020)
2. C. Wang, F. Tian, Q. Zhang, R. Gao, Y. Wang, Q. Tian, X. Wang, X. Pan, X. Xin, Experimental demonstration of low complexity nonlinear compensation algorithm based on logistic regression. Microw. Opt. Technol. Lett. **64**(7), 1154–1161 (2022)
3. Y. Gao, Z. Li, D. Guo, Z. Dong, L. Zhu, H. Chang, S. Zhou, Y. Wang, Q. Tian, F. Tian, et al., Unscented kalman filter with joint decision scheme for phase estimation in probabilistically shaped QAM systems. Electronics **12**(19), 4075 (2023)
4. H.C. Chien, et al., Approaching terabits per carrier metro-regional transmission using beyond-100GBd coherent optics with probabilistically shaped DP-64QAM modulation. J. Lightw. Technol. **37**(8), 1751–1755 (2019)
5. B. Liu, et al., Adaptive dynamic wavelength and bandwidth allocation algorithm based on error-back-propagation neural network prediction. Opt. Commun. **437**, 276–284 (2018)
6. X. Li, et al., 1-Tb/s millimeter-wave signal wireless delivery at D-band. J. Lightw. Technol. **37**(1), 196–204 (2019)
7. K. Wang, J. Zhang, L. Zhao, X. Li, J. Yu, Mitigation of pattern-dependent effect in SOA at O-band by using DSP. J. Lightw. Technol. **38**(3), 590–597 (2020)
8. Y. Zhu, et al., Spectrally-efficient single-carrier 400G transmission enabled by probabilistic shaping, in *Optical Fiber Communication Conference (OFC)* (2017), paper M3C.1
9. R. Maher, et al., Constellation shaped 66 GBd DP-1024QAM transceiver with 400 km transmission over standard SMF, in *Proceedings of the European Conference on Optical Communication (ECOC)* (2017), paper Th.PDP.B.2
10. F. Buchali, et al., Rate adaptation and reach increase by probabilistically shaped 64-QAM: an experimental demonstration. J. Lightw. Technol. **34**(7), 1599–1609 (2016)
11. B. Liu, et al., Probabilistic shaping for ROF system with heterodyne coherent detection. APL Photon. **2**(5), 056104 (2017)
12. G. Böcherer, F. Steiner, P. Schulte, Bandwidth efficient and rate-matched low-density parity-check coded modulation. IEEE Trans. Commun. **63**(12), 4651–4665 (2015)
13. J. Zhang, et al., 280 Gb/s IM/DD PS-PAM-8 transmission over 10 km SSMF at O-band for optical interconnects, in *Optical Fiber Communication Conference (OFC)* (2020), paper M4F.1
14. S. Dris, S. Alreesh, A. Richter, Blind polarization demultiplexing and equalization of probabilistically shaped QAM, in *Optical Fiber Communication Conference (OFC)* (2019), paper W1D.2
15. Q. Yan, L. Liu, X. Hong, Blind carrier frequency offset estimation in coherent optical communication systems with probabilistically shaped M-QAM. J. Lightw. Technol. **37**(23), 5856–5866 (2019)
16. D.A.A. Mello, F.A. Barbosa, J.D. Reis, Interplay of probabilistic shaping and the blind phase search algorithm. J. Lightw. Technol. **36**(22), 5096–5105 (2018)
17. J.C.M. Diniz, et al., Low-complexity carrier phase recovery based on principal component analysis for square-QAM modulation formats. Opt. Exp. **27**(11), 15617–15626 (2019)
18. H. Abdi, L.J. Williams, Principal component analysis. Wiley Interdisc. Rev. Comput. Stat. **2**(4), 433–459 (2010)
19. T. Sasai, A. Matsushita, M. Nakamura, S. Okamoto, F. Hamaoka, Y. Kisaka, Laser phase noise tolerance of uniform and probabilistically shaped QAM signals for high spectral efficiency systems. J. Lightw. Technol. **38**(2), 439–446 (2020)

20. F.A. Barbosa, S.M. Rossi, D.A.A. Mello, Phase and frequency recovery algorithms for probabilistically shaped transmission. J. Lightw. Technol. **38**(7), 1827–1835 (2020)
21. P. Lu, L. Zhang, X. Liu, J. Yao, Z. Zhu, Highly-efficient data migration and backup for big data applications in elastic optical inter-data-center networks. IEEE Netw. **29**(5), 36–42 (2015)
22. Z. Zhu, W. Lu, L. Zhang, N. Ansari, Dynamic service provisioning in elastic optical networks with hybrid single-/multi-path routing. J. Lightw. Technol. **31**(1), 15–22 (2013)
23. X. Wang, Q. Zhang, X. Xin, R. Gao, Q. Tian, F. Tian, C. Wang, X. Pan, Y. Wang, L. Yang, Robust weighted K-means clustering algorithm for a probabilistic-shaped 64QAM coherent optical communication system. Opt. Exp. **27**, 37601–37613 (2019)
24. F. Zhang, Q. Zhuge, M. Qiu, X. Zhou, M.Y.S. Sowailem, T.M. Hoang, M. Xiang, D.V. Plant, Blind adaptive digital backpropagation for fiber nonlinearity compensation. J. Lightw. Technol. **36**, 1746–1756 (2018)
25. B.I. Bitachon, A. Ghazisaeidi, M. Eppenberger, B. Baeuerle, M. Ayata, J. Leuthold, Deep learning based digital backpropagation demonstrating SNR gain at low complexity in a 1200 km transmission link. Opt. Exp. **28**, 29318–29334 (2020)
26. J. Zhou, Y. Wang, Y. Zhang, Blind back-propagation method for fiber nonlinearity compensation with low computational complexity and high performance. Opt. Exp. **28**, 11424–11438 (2020)
27. J. Gonçalves, C.S. Martins, F.P. Guiomar, T.R. Cunha, J.C. Pedro, A.N. Pinto, P.M. Lavrador, Nonlinear compensation with DBP aided by a memory polynomial. Opt. Exp. **24**, 30309–30316 (2016)
28. J. Zhang, J. Yu, H. Chien, 1.6Tb/s (4x400G) unrepeatered transmission over 205-km SSMF using 65-GBaud PDM-16QAM with joint LUT pre-distortion and post DBP nonlinearity compensation, in *Optical Fiber Communication Conference* (Optical Society of America, Washington, 2017), p. Th2A.51
29. J. Zhang, X. Li, Z. Dong, Digital nonlinear compensation based on the modified logarithmic step size. J. Lightw. Technol. **31**(21), 3546–3555 (2013)
30. C.S. Martins, L. Bertignono, A. Nespola, A. Carena, F.P. Guiomar, A.N. Pinto, Low-complexity time-domain dbp based on random step-size and partitioned quantization. J. Lightw. Technol. **36**(13), 2888–2895 (2018)
31. V. Vgenopoulou, N.P. Diamantopoulos, I. Roudas, S. Sygletos, MIMO nonlinear equalizer based on inverse volterra series transfer function for coherent SDM systems, in *Optical Fiber Communication Conference (OFC) 2019* (Optical Society of America, Washington, 2019), p. Th2A.50
32. F.P. Guiomar, A.N. Pinto, Simplified volterra series nonlinear equalizer for polarization-multiplexed coherent optical systems. J. Lightw. Technol. **31**(22), 3879–3891 (2013)
33. Y. Chuang, C. Chow, Y. Liu, C. Yeh, X. Liao, K. Lin, Y. Chen, Using logistic regression classification for mitigating high noise-ratio advisement light-panel in rolling-shutter based visible light communications. Opt. Exp. **27**(21), 29924–29929 (2019)
34. C. Häger, H.D. Pfister, R.M. Bütler, G. Liga, A. Alvarado, Model-based machine learning for joint digital backpropagation and PMD compensation, in *Optical Fiber Communication Conference (OFC) 2020* (Optical Society of America, Washington, 2020), p. W3D.3
35. C. Häger, H.D. Pfister, Nonlinear interference mitigation via deep neural networks, in *Optical Fiber Communication Conference* (Optical Society of America, Washington, 2018), p. W3A.4
36. J. Zhang, J. Yu, H.C. Chien, High symbol rate signal generation and detection with linear and nonlinear signal processing. J. Lightw. Technol. **36**(3), 408–415 (2018)
37. J. Zhang, J. Yu, H.C. Chien, Advanced linear and nonlinear compensations for 16QAM SC-400G unrepeatered transmission system. Opt. Commun. **409**, 34–38 (2018)
38. J.H.C. Júnior, T. Sutili, S.M. Rossi, R.C. Figueiredo, D.A.A. Mello, Fast adaptive digital back-propagation algorithm for unrepeatered optical systems, in *Optical Fiber Communication Conference (OFC) 2020* (Optical Society of America, Washington, 2020), p. T4I.2
39. S. Marco, M. Domenico, F. Enrico, Enhanced split-step Fourier method for digital backpropagation, in *European Conference on Optical Communication (ECOC)* (2014)

40. F. Zhang, et al., XPM model-based digital backpropagation for subcarrier-multiplexing systems. J. Lightw. Technol. **33**(24), 5140–5150 (2015)
41. S. Marco, M. Domenico, F. Enrico, Single-step digital backpropagation for nonlinearity mitigation, in *European Conference on Optical Communication (ECOC)* (2015)
42. F. Chai, Q. Zhang, H. Yao, X. Xin, R. Gao, M. Guizani, Joint multi-task offloading and resource allocation for mobile edge computing systems in satellite IoT. IEEE Trans. Veh. Technol. **72**, 7783–7795 (2023)
43. F. Wang, H. Yao, Q. Zhang, J. Wang, R. Gao, D. Guo, M. Guizani, Dynamic distributed multi-path aided load balancing for optical data center networks. IEEE Trans. Netw. Serv. Manag. **19**, 991–1005 (2022)
44. S. Zhou, X. Liu, R. Gao, Z. Jiang, H. Zhang, X. Xin, Adaptive Bayesian neural networks nonlinear equalizer in a 300-Gbit/s PAM8 transmission for IM/DD OAM mode division multiplexing. Opt. Lett. **48**(2), 464–467 (2023)
45. L. Zhu, H. Yao, H. Chang, Q. Tian, Q. Zhang, X. Xin, F. Yu, Adaptive optics for orbital angular momentum-based internet of underwater things applications. IEEE Internet Things J. **9**(23), 24281–24299 (2022)
46. X. Xu, X. Tang, M. Guo, J. Qi, S. Liu, Y. Lu, Y. Qiao, Wide-linewidth coherent systems with parallelized extended Kalman filter for data center inter-connection. Opt. Commun. **458**, 124740 (2020)
47. J. Jignesh, B. Corcoran, A. Lowery, Parallelized unscented kalman filters for carrier recovery in coherent optical communication. Opt. Lett. **41**(14), 3253 (2016)
48. K. Shibahara, T. Mizuno, Y. Miyamoto, MIMO carrier phase recovery for carrier-asynchronous SDM-MIMO reception based on the extended Kalman filter. Opt. Exp. **29**(11), 17111 (2021)
49. H. Zhong, W. Ye, Z. Zuo, D. Huang, Y. Guo, Kalman filter-enabled parameter estimation for simultaneous quantum key distribution and classical communication scheme over a satellite-mediated link. Opt. Exp. **30**(4), 5981 (2022)
50. W. Sanyal, S. Sharma, P. Krishnamurthy, Multi-rate Kalman filter for carrier phase recovery in 200 Gbps PDM coherent optical receivers, in *2021 National Conference on Communications (NCC), Kanpur, India* (2021), pp. 1–4
51. S. Sharma, P. Krishnamurthy, Reduced sampling rate Kalman filters for carrier phase and frequency offset tracking in 200 Gbps 16 QAM coherent communication system. Sci. Rep. **11**, 1991 (2021)
52. Q. Zhang, Y. Yang, K. Zhong, J. Liu, X. Wu, Y. Yao, Joint polarization tracking and channel equalization based on radius-directed linear Kalman filter. Opt. Commun. **407**, 142–147 (2018)
53. J. Cho, P. Winzer, Probabilistic constellation shaping for optical fiber communications. J. Lightw. Technol. **37**(6), 1590–1607 (2019)
54. H. Xu, X. Zhang, X. Tang, C. Bai, L. Xi, W. Zhang, H. Zheng, Joint scheme of dynamic polarization demultiplexing and PMD compensation up to second order for flexible receivers. IEEE Photon. J. **9**(3), 1–15 (2017)
55. L. Pakala, B. Schmauss, Extended Kalman filtering for joint mitigation of phase and amplitude noise in coherent QAM systems. Opt. Exp. **24**(6), 6391 (2016)
56. T. Inoue, S. Namiki, Carrier recovery for M-QAM signals based on a block estimation process with Kalman filter. Opt. Exp. **22**(12), 15376 (2014)
57. S. Hu, W. Zhang, X. Yi, Z. Li, F. Li, X. Huang, M. Zhu, J. Zhang, K. Qiu, MAP detection of probabilistically shaped constellations in optical fiber transmissions, in *Optical Fiber Communication Conference (OFC) 2019, San Diego, CA, USA* (2019), p. W1D.3
58. Z. Chen, S. Fu, M. Tang, Z. Zhang, Y. Qin, Maximum probability directed blind phase search for PS-QAM with variable shaping factors. Opt. Exp. **30**(1), 550 (2022)
59. X. Xie, Y. Pi, Phase noise filtering and phase unwrapping method based on unscented Kalman filter. J. Syst. Eng. Electron. **22**(3), 365–372 (2011)

Chapter 7
Nonlinear Compensation Technology Based on Artificial Intelligence

In the ever-evolving optical communication system, the pursuit of higher data rates and improved signal integrity in the presence of various channel damages has led to important research on nonlinear compensation techniques. This chapter explores the intersection of advanced nonlinear compensation strategies and artificial intelligence (AI), and proposes a new method to improve the performance of coherent optical communication systems.

We begin this exploration in Sect. 7.1, where we introduce a robust weighted K-means clustering algorithm tailored for PS-64QAM coherent optical communication systems. This section explains the principle of probabilistic shaping and its integration with the weighted K-means algorithm, then elaborates on the experimental setup and gives a comprehensive discussion of the results obtained [1]. Continuing to Sect. 7.2, we will again explore decision techniques based on Gaussian hybrid clustering in the context of coherent optical communication systems. This section provides theoretical analysis and operational principles of the technique and complements insights into experimental Settings and results [2]. Finally, Sect. 7.3 highlights a data-defined naive Bayes (DNB) decision scheme for nonlinear mitigation. This section discusses the principle behind the DNB-based decision method, describes the experimental setup in detail, and analyzes the experimental results [3].

In this chapter, we highlight the key role of artificial intelligence in enhancing nonlinear compensation techniques, showing how AI-driven algorithms can significantly improve the accuracy and efficiency of optical communication systems. The experimental results not only validate the theoretical model, but also open up a new way for practical application.

X. Xin, *Advanced Optical Fiber Transmission Systems*, Wireless Networks,
https://doi.org/10.1007/978-3-032-09516-9_7

7.1 Robust Weighted K-Means Clustering Algorithm for PS-64QAM Coherent Optical Communication System

With the rapid growth of high-definition video, 4K TV, mobile Internet and other services, global network traffic is growing rapidly, to satisfy the continuous increasing capacity and bandwidth demands, the development of long-distance backbone optical transmission systems has been promoted [4–7]. The past 10 years have witnessed a growing interest amongst both academics and service-suppliers in the high-speed optical transceiver with digital signal processing based on coherent detection [8].

In order to cope with the forever growing broadband traffic, applying high-order quadrature amplitude modulation (QAM) formats in optical communication systems can be a feasible solution to increase spectral efficiencies (SE) and data rates without occupying extra spectrum [9, 10]. Unfortunately, higher-level modulation formats require higher optical signal to-noise ratio (OSNR), which would result in a limited transmission distance.

The use of new-type optical fibers can reduce span losses and nonlinear effects, thus improving signal performance and increasing single span length. For instance, a new ultra-large effective area fiber (ULAF) with Raman amplification can be adopted to effectively extend the transmission distance compared with the standard single mode fiber (SSMF) with Erbium-doped fiber amplifier (EDFA) [7]. However, the large-scale application of this scheme in practical network has not been widely applied, due to the high cost of re-installing fibers and replacement of existing amplifiers.

In order to fully exploit the potential of the existing fiber networks, probabilistic shaping (PS) has swept across the entire optical transmission research community [11–15]. It has been demonstrated to yield OSNR sensitivity gain and transmission distance improvement. In [12], H. Chien et al. demonstrated a probabilistically-shaped dual polarization (DP) 64QAM scheme at 105 GBaud, achieving per wavelength 1.26-Tbps line rate and 905.8-Gbps net rate after 800-km single-channel and 400-km WDM (SE: 8.05 bits/s/Hz) transmissions, respectively. In [13], J. Yu et al. experimentally demonstrate PS 64QAM outperforms hybrid-32/64QAM by 37.5

Another research focus in the optical transmission research society is the application of machine-learning algorithms. Among various models and algorithms, one of the most popular choices is k-means clustering algorithm, which plays an important role in nonlinearity mitigation, modulation format identification, phase estimation and data detection [16–20]. In [16], J. Zhang et al. proposed two k-means-clustering-based algorithms to efficiently mitigate the fiber nonlinearity for 64QAM signal. In [21], L. Sun et al. proposed a likelihood ratio estimation method based on K-means clustering, which was experimentally demonstrated for soft decision of PAM4 to mitigate the level nonlinearity and level dependent noise. Moreover, a joint algorithm based on look-up table (LUT) and k-means was proposed in [22] to compensate the nonlinearity distortion caused by saturation effect of optical/electrical devices in a photonics-aided microwave communication system at K-band.

To sum up, probabilistic shaped high-order modulation formats and machine-learning based clustering algorithms are recognized as cost-effective methods to bring significant performance gains with extended distance or increased capacity. It is natural to assume that the combination of them can further improve the system tolerance of high optical power. Nevertheless, the joint algorithm works only under certain OSNR range conditions. That is, when OSNR is low, there will be certain deviations between the high-power points centroids obtained by K-Means and the ideal positions, while when the nonlinear effect is nonnegligible, signal points belonging to the same cluster would be incorrectly divided into two parts because the characteristic values of the signal points at the constellation diagram's corners will be ignored by the algorithm due to low probability of occurrence (discussed in detail in Sect. 2.2). To the best of our knowledge, few studies in the literature deal with the robustness of K-Means when it comes to non-equally distributed signal constellations. This is the most important motivation for this paper.

A novel Weighted K-Means scheme for PS 64QAM signal is proposed in order to locate the optimal decision points more accurately and enhance the robustness of clustering algorithm. By using a weighting factor following the reciprocal of Maxwell-Boltzmann distribution, the proposed algorithm can combine the advantages of PS and K-Means robustly while reducing the overall computational complexity of the clustering process. Experimental verification of the proposed clustering technique was demonstrated in a 120-Gb/s probabilistic shaped 64QAM coherent optical communication system. The results show that the proposed algorithm has outperformed K-Means with respect to Bit Error Rate (BER), clustering robustness and iteration times in both back-to-back and 375 km transmission scenarios. For back-to-back situation, the proposed algorithm is capable of achieving about 0.6 and 1.8 dB OSNR gain over K-Means clustered signals and unclustered signals. For the case of transmission, the proposed clustering procedure can robustly locate the optimal decision points with launched signal power ranging from −5 to 5 dBm, while the working range for K-Means procedure is only −4 to 2 dBm. In addition, the proposed weighted algorithm takes less iteration times than K-Means to converge, especially when the signal impairments caused by fiber Kerr nonlinearity is severe.

7.1.1 Principal of Probabilistic Shaping and Weighted K-Means Algorithm

In this section, we first present the principal of PS. Then, we introduce the principle of Weighted K-Means.

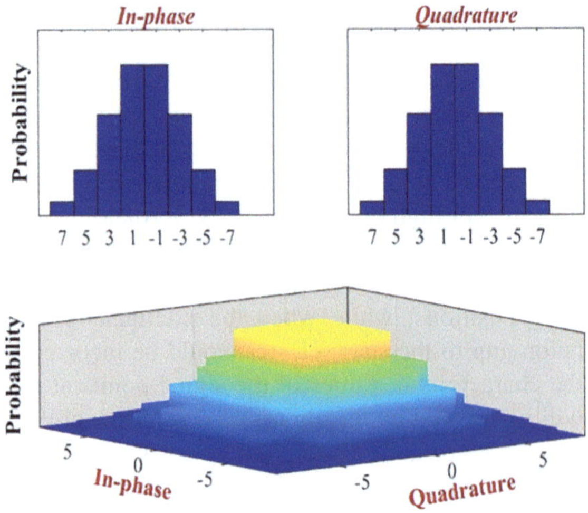

Fig. 7.1 A schematic illustration for the PS of the 64QAM signal

Principal of PS

In the probabilistic shaping scheme, each dimension of the transmitted higher-level QAM signal constellations is distributed with non-equal probabilities following the Maxwell-Boltzmann (MB) distribution [23–26]. The MB distribution can be expressed as:

$$P_X(x_i) = \frac{\exp(-\nu(\mathrm{Re}(x_i)^2 + \mathrm{Im}(x_i)^2))}{\sum_{k=1}^{M} \exp(-\nu(\mathrm{Re}(x_k)^2 + \mathrm{Im}(x_k)^2))} \tag{7.1}$$

Where x_i denotes the ith complex M-QAM constellation point and v denotes the shaping parameter. With the help of PS, the inner constellation points with lower energy are transmitted with a higher probability than the outer constellation points with higher energy. Therefore, the signal constellation after probabilistic shaping has a lower average power than the original one, which leads to a shaping gain. Figure 7.1 gives a schematic illustration for the probabilistic constellation shaping for 64QAM. For example, when $v = 0.05$, the PAM level distribution for PS 64QAM is [0.022, 0.073, 0.163, 0.242, 0.242, 0.163, 0.073, 0.022].

Given the probability mass function (PMF) $P_X(x_i)$, the source entropy of PS 64QAM signal can be calculated by:

$$H = -\sum_{i=1}^{64} P_X(x_i) \log_2 P_X(x_i) \tag{7.2}$$

Principal of Weighted K-Means

K-Means clustering, or Lloyd's algorithm [27], is an iterative, data-partitioning algorithm that assigns N observations to exactly one of k clusters defined by centroids, where k is chosen before the algorithm starts. Although K-Means is an effective way to refine the decision points and improve BER performance for uniform QAM signals, when being applied to probabilistic shaped QAM signals, the mentioned algorithm suffers from some limitations mainly concerning the value range of OSNR. The constellation diagrams shown in Fig. 7.2 are obtained by numerical simulation for 25 GBaud PS 64QAM after 480 km (the span length is 80 km, and the number of spans is 6) transmission, the corresponding launched signal powers for the three subfigures are −8, −3d, and 5 dBm, and the OSNRs are 19, 24, and 32 dB respectively. When the OSNR is low, the received constellation diagram is severely corrupted due to Amplified Spontaneous Emission (ASE) noises, which leads to certain amounts of misjudgment between nearby signal points. At the corner of the signal constellation, due to the probability shaping technique, the number of occurrences of the inner signal points is times over the outer points. Therefore, as shown in Fig. 7.2a, the modulus of the clustering centroids (yellow triangles) at the corner of the constellation are significantly lower than ideal value (red squares). Figure 7.2b shows that when the OSNR is slightly higher yet the nonlinear effects are not severe, K-Means algorithm can correctly identify the centroids of the received signal. However, when the OSNR is high and the signal constellation is suffering from strong nonlinear effects, the Nonlinear

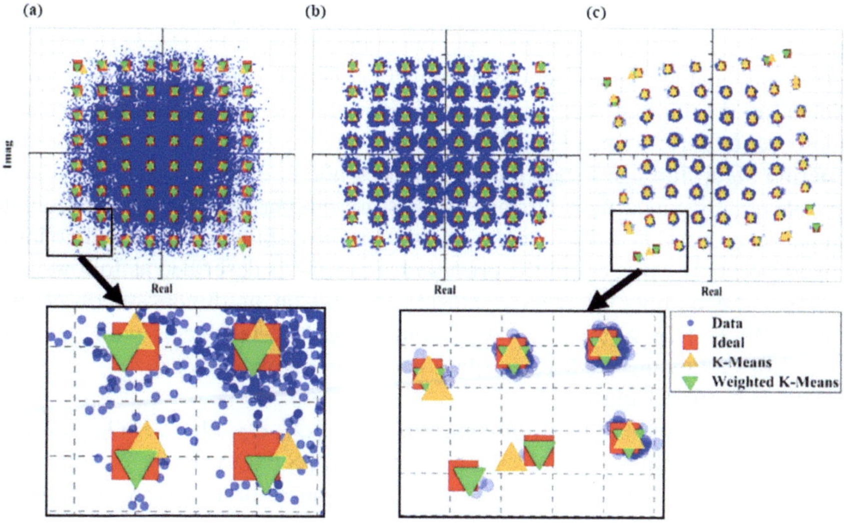

Fig. 7.2 Illustration of deploying clustering algorithms to PS 64QAM, when: (**a**) the OSNR is 19 dB; (**b**) the OSNR is 24 dB yet the nonlinear effects are not severe; (**c**) the OSNR is 32 dB and the signal constellation is suffering from strong nonlinear effects

Phase Noise (NLPN) caused by Kerr effect would lead to the wrong convergence of the K-Means clustering procedure. As shown in Fig. 7.2c, certain cluster centroids at the four corners of the constellation would be mispositioned to other clusters, since these high-power signals points only occur with low probability, and therefore their characteristic values are overlooked by the K-Means procedure. This misplacement would result in serious misjudgment during the decision and demodulation process, since signal points belonging to the same cluster would be incorrectly divided into two parts, hence a severe BER performance degradation.

As analyzed above, further research is still required with respect to the robustness of applying K-Means algorithm to probabilistic shaped QAM signals. This article presents a Weighted K-Means algorithm that aims to solve the robustness problem through a simple yet effective linear weighting approach. Since the K-Means procedure belongs to unsupervised learning algorithms, the clustering process does not have the information to exclude the incorrectly grouped samples out of the current cluster. Therefore, the most intuitive and effective way to avoid incorrect convergence is to distinguish the received samples according to their amplitudes. In our proposed algorithm, we give different weights to samples grouped within the same cluster according to their distances to the origin, which emphasizes the characteristic values of the desired samples originated from the clusters at the constellation diagram's corners, rather than those samples that are incorrectly grouped. To be more specific, we introduce a MB distributed weighting factor α_i into the centroids determining step of K-Means procedure, which can be expressed as:

$$\alpha_i = \frac{\sum_{k=1}^{M} \exp(-\nu(\mathrm{Re}(X_i)^2 + \mathrm{Im}(X_i)^2))}{\exp(-\nu(\mathrm{Re}(X_i)^2 + \mathrm{Im}(X_i)^2))} \tag{7.3}$$

The mathematical ideas behind introducing such a weighting factor is similar to importance sampling (IS), where certain samples are given more importance to achieve better performance [28]. With help of this weighting factor, the optimized clustering algorithm can locate the centroids of higher energy constellation points more accurately, since their attribute values are emphasized during the weighted summation process, in order to make sure the low probability clusters will no longer be overlooked. The process of the proposed algorithm is described as follows.

Input: The number of desired clusters k, and an m dimension observation database $X = \{X_1, X_2, X_3, \ldots, X_N\}$ containing N data objects with non-equal probabilities following the MB distribution of shaping parameter υ.

Output: A set of k clusters.

Steps 1 : Choose initial cluster centroids $O = \{O_1, O_2, O_3, \ldots, O_k\}$, which in this case are the ideal signal constellation points of PS 64QAM.

Steps 2 : Compute point-to-cluster-centroid distances of all observations to each centroid, which can be described as:

$$D(X_i, O_j) = \sqrt{\sum_{q=1}^{m} \left(X_{iq} - O_{jq}\right)^2} \quad 1 \le i \le N, 1 \le j \le K \tag{7.4}$$

Where X_i represents the ith observation point's coordinate, O_j denotes the jth cluster centroid's coordinate, while X_{iq} is the ith observation point's attribute value of the qth dimension, and O_{jq} is the jth cluster centroid's attribute value of the qth dimension. For coherent optical communication systems, $m = 2$.

Steps 3 : Batch update—Assign each observation to k clusters S_k with the closest centroid according to the previously calculated $D(X_i, O_j)$.

Steps 4 : Centroids determining—Compute the weighted average of the observations in each cluster to obtain k new centroid locations according to Eq. (7.5).

$$O_j = \frac{\sum_{X_i \in S_j} \alpha_i X_i}{\left|S_j\right| \times \sum_{X_i \in S_j} \alpha_i} \quad 1 \le i \le \left|S_j\right|, 1 \le j \le k \tag{7.5}$$

Where S_j represents the jth cluster, and $|S_j|$ indicates the total number of data objects that are assigned to S_j.

Steps 5 : Repeat steps 2 through 4 until cluster assignments do not change, or the maximum number of iterations is reached.

The simulation results of Weighted K-Means (green triangles) are also shown in Fig. 7.2, which proves the effectiveness of our proposed algorithm in exactly the same scenarios.

7.1.2 Experiment Setup

Figure 7.3 shows the experimental setup for the generation and delivery of 120-Gb/s 64QAM coherent optical communication system. For both uniform 64QAM and PS 64QAM, the in-phase and quadrature components of the transmitted signal can be considered as two Pulse Amplitude Modulation (PAM) signals. The random formation of pseudo-random binary sequences (PRBS) was realized in a MATLAB program. Especially for PS 64QAM, a constant composition distribution matcher (CCDM) [29, 30] was applied to generate PAM-8 output symbols with Maxwell-Boltzmann distribution. The parameters of CCDM consist of arbitrary target distribution $P_X(x_i)$ and the optimal type number n_c. In this work, we chose $P_X(x_i) = [0.022, 0.073, 0.163, 0.242, 0.242, 0.163, 0.073, 0.022]$ and $n_c = 16384$ to generate the desired PAM-8 output symbols with an empirical distribution of

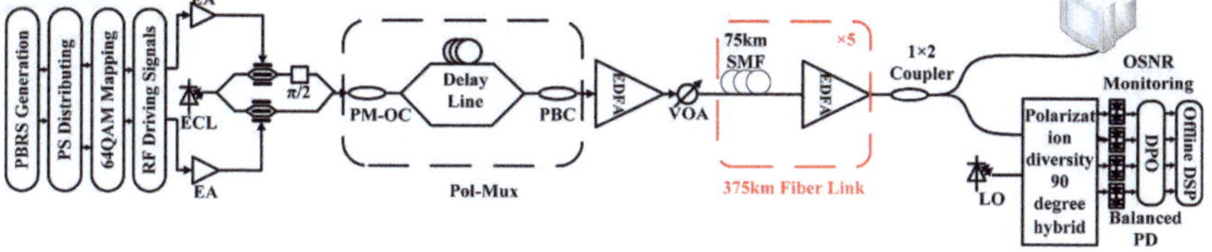

Fig. 7.3 Experiment setup for the 64QAM coherent optical communication system

$P_{\bar{X}}(x_i) = [360, 1196, 2671, 3965, 3965, 2671, 1196, 360]/16384$ with $\upsilon = 0.05$. After that, we used an arbitrary waveform generator (Tektronix AWG70002) with 25 GSa/s and 3 dB bandwidth of 14 GHz to convert the two independent digital symbol sequences to analog radio frequency (RF) signals. Before being fed into the I/Q modulator (Fujitsu FTM7961EX) as driving signals, two linear broadband amplifiers (SHF S807) were applied to boost the peak-to-peak voltage from 0.25 to 2 Vpp. An external cavity laser (ECL) with less than 100 kHz linewidth and 15 dBm output power was used to generate continuous optical wave. The light source at 1550 nm was then modulated by the I/Q modulator to achieve electrical/optical conversion. The two parallel Mach-Zehnder modulators (MZMs) in the I/Q modulator were both biased at the null point and driven at the full swing, while the phase difference between the upper and lower branches of the I/Q modulator was fixed at $\pi/2$. A polarization-division-multiplexing (PDM) module, which consists of a polarization beam splitter (PBS), one polarization beam combiner (PBC) and an optical delay line (DL), was utilized to realize polarization multiplexing. The modulated optical signal was amplified by an EDFA with 5.5 dB noise figure. The following variable optical attenuator (VOA) was applied to adjust the launched signal power into the fiber link.

The transmission link consists of five unrepeated spans of G.652 fiber with length of 75 km and the total length was 375 km. Each span was followed by an EDFA with 18 dB gain to compensate the attenuation. An ECL with a linewidth less than 100 kHz was utilized as an optical local oscillator (LO). We used an integrated coherent receiver (ICR) to implement the optical polarization diversity and phase diversity of the received signal and the optical LO. The integrated optical hybrid includes two PBSs and two 90-degree optical hybrids. The bandwidth of each balanced photodetector is 32 GHz. After that, a four-channel digital phosphor oscilloscope (Tektronix DPO72004C) with 100 GSa/s and 20 GHz bandwidth is used to implement analog-to-digital conversion, which was followed by offline DSP applied to four-channel sampled data sequence. In addition, an optical spectrum analyzer (OSA) was applied to monitor the OSNR. The flow chart of the offline DSP is as shown in Fig. 7.4. The processing consists of low-pass filter, amplitude normalization, frequency domain chromatic dispersion (CD) compensation, clock recovery and resampling based on Gardner algorithm, Gram-Schmidt orthogonalizing process (GSOP), constant modulus algorithm (CMA) equalization, frequency offset estimation (FOE), carrier phase estimation (CPE) based on blind phase search

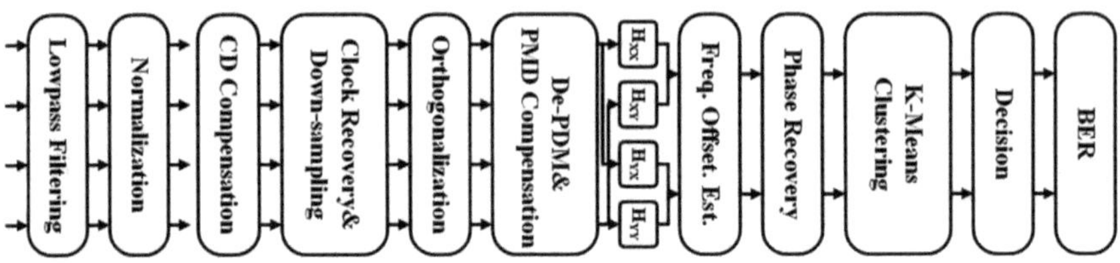

Fig. 7.4 Flow chart of the offline digital signal processing

(BPS), synchronization, K-Means clustering, decision, 64QAM demodulation and BER counting. During the polarization de-multiplexing process, in order to optimize the convergence performance, we applied a 2-tap CMA algorithm for three times when different convergence coefficients are descending as $2 \times 10-3$, $8 \times 10-4$, and $2 \times 10-5$ respectively. The inputs are the same, while the results of the four adaptive filters were inherited by the next iteration. Based on multiple tests, the above coefficients proved to have the best performance. During the clustering process, since the outputs of the K-Means or Weighted K-Means procedure are used as the decision points for QAM demodulation, the number of centroids was strictly constrained to the constellation size, that is, for both 64QAM and PS 64QAM signals, the number of desired clusters k equals to 64. Moreover, the number N of data we chose for the clustering procedure was 16,384, which is large enough for both algorithms to converge to their optimal results respectively. It is worth noting that the maximum number of iterations we chose was 1000 to prevent the clustering process from failing to converge.

Results and Discussions

Figure 7.5 shows the optical spectra of the optical 64QAM signals generated by I/Q modulator. When comparing the system performances between uniform 64QAM signals and PS 64QAM signals, it is only fair to compare the BER when the net rates of both modulation formats are the same. In this work, the baud rates we chose were 10 and 11.5 GBaud respectively. This is because that the source entropy of PS 64QAM when $\upsilon = 0.05$ is 5.273 bits/symbol according to Eq. (7.2), while the entropy of uniform 64QAM is 6 bits/symbol. Since 11.5 GBaud $\times$ 5.273 bits/symbol $\approx$ 10 GBaud $\times$ 6 bits/symbol, the net rate of PS 64QAM was slightly larger than uniform 64QAM, which proves the fairness of the following

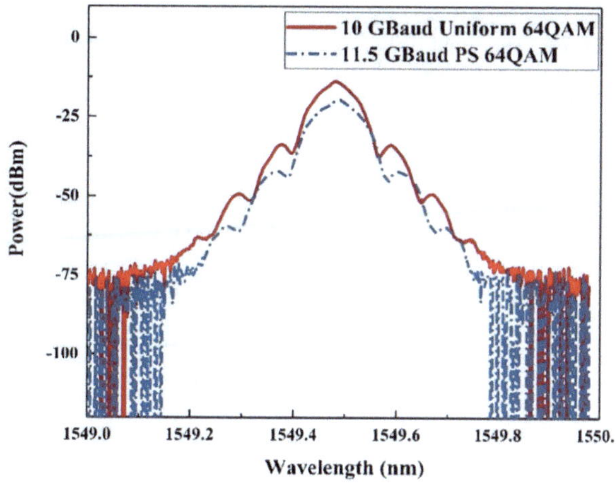

Fig. 7.5 Optical spectra of 10 GBaud Uniform 64QAM and 11.5 GBaud PS 64QAM

results comparison. Moreover, the optical spectra of both modulation formats were captured when the output power of ECL was fixed at 15 dBm and the peak-to-peak voltages of RF driving signals were fixed at 2 Vpp, which demonstrates how the PS technique can significantly reduce the transmitted power at the cost of a containable loss of net bit rate to achieve an overall system performance improvement.

Generally, the probabilistic shaping is combined with forward error correction (FEC) to approach Shannon limit in practical transceivers. Due to the significant error-correcting capability of FEC coding combined with PS, the post-FEC BER can be reduced to 1 × 10-12 or even zero. Such low BER values cannot be reliably estimated by off-line experiments [31]. Since our research focuses on the symbol-level decision optimizing, and is compatible with forward error correction, we use the same method as [14, 32, 33] to evaluate the performances of our system without FEC encoding and decoding. For optical communication systems, the most commonly standards for FEC thresholds are 2 × 10-2 for soft-decision FEC (SD-FEC) [34, 35] and 3.8 × 10-3 for hard-decision FEC (HD-FEC) [36–38]. Therefore, we applied these two values to measure the performance in our experimental verification. Firstly, we investigated the BER performance in a back-to-back (BTB) scenario. A variable optical attenuator and an Erbium-doped fiber amplifier was applied to change the OSNR for BER measurement. The curves of BER versus OSNR are measured in five different cases as shown in Fig. 7.6.

The BER performances of both the uniform 64QAM and PS 64QAM can satisfy the SD-FEC threshold with OSNR higher than 25 dB, which proves the feasibility of the system. For both modulation formats, the BER decreases as the OSNR gradually increases. It can be clearly seen that, by introducing the PS technique, a much better OSNR sensitivity can be obtained. When the OSNR is 30 dB, the BER of PS 64QAM can reach below the HD-FEC threshold, while the best BER performance of uniform 64QAM is 1.1 × 10-2.

Utilizing the K-Means algorithm can enhance the system performance for both modulation formats, yet the improvement for PS 64QAM is obviously less than

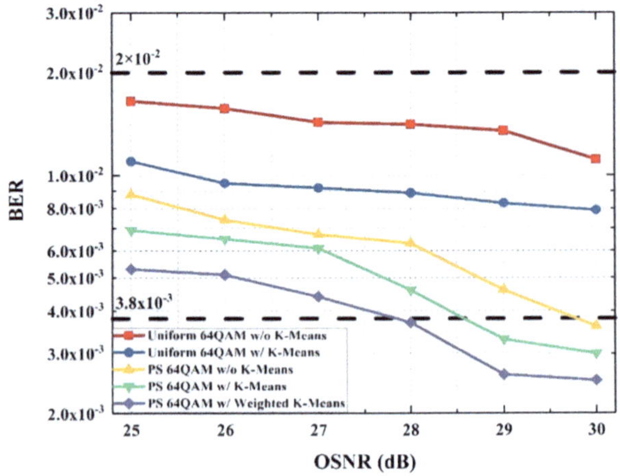

Fig. 7.6 Measured BER curves versus OSNR in the 120-Gb/s 64 QAM coherent optical communication system for back-to-back case

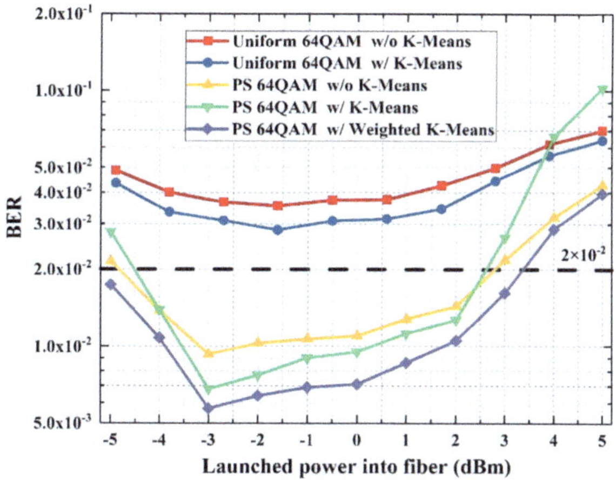

Fig. 7.7 Measured BER curves versus launched signal power into 375 km SMF (dBm) in the 120-Gb/s 64 QAM coherent optical communication system

uniform 64 QAM. Taking the HD-FEC threshold as a reference, the K-Means clustering technique can obtain around 1.2 dB OSNR gain for PS 64QAM.

By employing the proposed Weighted K-Means algorithm, the clustering centroids can be found more accurately, since the characteristic values of high-power signal points will be emphasized due to the weighting factor. Therefore, an additional 0.6 dB performance gain can be achieved, which proves the superiority of the proposed algorithm.

Secondly, the robustness of the proposed Weighted K-Means algorithm is verified in a single wave transmission system with a total link distance of 375 km. The launched signal power into the SMF, abbreviated as P_{in}, was varied by a VOA to measure BER curves.

For both modulation formats, the BER first gradually decreases as P_{in} increases, as shown in Fig. 7.7. That is because the OSNR of the received signal is positive correlated with the launched power into fiber, whose value is rather low in this range. Therefore, the ASE noise is the dominant factor affecting the BER performance under this circumstance. On the other hand, once the launched power exceeds a certain value (−1.6 dBm for uniform 64QAM, −3 dBm for PS 64QAM), the system performance is determined by the fiber Kerr nonlinearity, which results in nonlinear distortion especially for outer symbol constellation points. Within the entire measurement OSNR range, PS 64QAM has a better BER performance compared to uniform 64QAM. Especially when the launched power is in the range of −4.5 to 2.5 dBm, its BER can satisfy the SD-FEC threshold.

The K-Means algorithm can bring a clustering gain for uniform 64QAM robustly, yet the improvement is not as effective as PS technique. It is worth noting that, when being employed to PS 64QAM, the K-Means procedure can only correctly converge and improve system performance when the launched power is in the range of −4 to 2 dBm. Instead, when the value is outside the abovementioned range, the incorrect convergence will severely worsen the BER performance, noticeably for signals suffering from serious nonlinear phase noises. To be more specific, when P_{in}

is around −4 dBm, the green curve (PS 64QAM w/ K-Means) intersects the yellow curve (PS 64QAM w/o K-Means), which indicates that when the launched power is lower than approximately −4 dBm, the traditional K-Means can no longer robustly converge, the incorrectly clustering results would worsen the system performance instead of improving it. Similarly, when the launched power is higher than 2.4 dBm, the BER performance of the green curve deteriorates drastically, when P_{in} reaches 4.0 dBm, the clustering result is so unreasonable that it offsets the shaping gain achieved by PS technique.

For comparison, our proposed Weighted K-Means procedure outperformed the traditional K-Means algorithm with respect to robustness and BER performance throughout the entire measured power range from −5 to 5 dBm. It is worth emphasizing that, even when the launched signal power is up to 5 dBm, and the signal impairments caused by fiber Kerr nonlinearity is severe, the Weighted K-Means algorithm can still robustly locate the optimal decision points and therefore enhance the BER performance. In short, the robustness of the proposed clustering procedure is tremendously improved. Moreover, taking the SD-FEC threshold as a reference, the tolerance of the system to nonlinear effects caused by high launched signal power is improved by 0.5 dB compared to classic PS 64QAM system.

The constellation diagram of the recovered PS 64QAM with the launched power at 5 dBm is as shown in Fig. 7.8. The insert (a) gives the clustering result of K-Means procedure. As can be seen from the part indicated by the red arrow, certain cluster centroids at the four corners of the constellation were mispositioned to inner clusters, thus, signal points belonging to the same cluster were incorrectly divided into two parts. Since the cluster centroids were used as decision points in the BER calculation, the incorrect convergence caused a major performance degradation. As a comparison, with the help of the weighting factor, the above problem has been solved in exactly the same scenario, which proves the robustness of our proposed

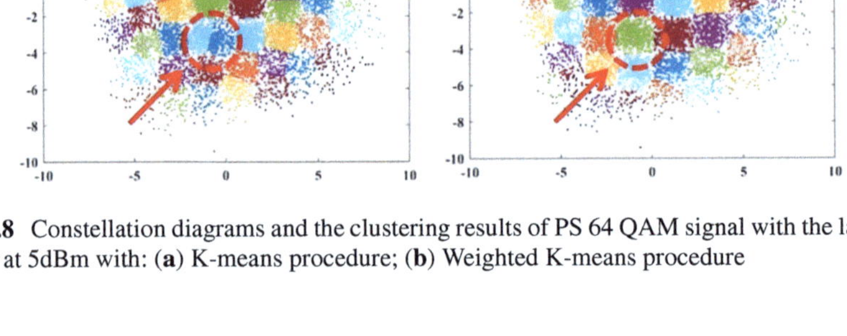

Fig. 7.8 Constellation diagrams and the clustering results of PS 64 QAM signal with the launched power at 5dBm with: (**a**) K-means procedure; (**b**) Weighted K-means procedure

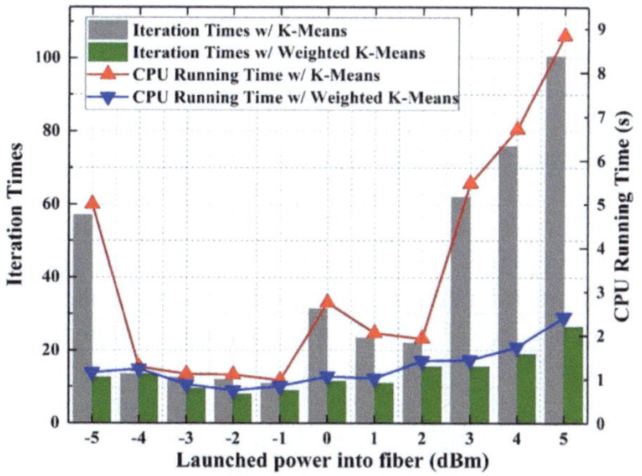

Fig. 7.9 Iteration times and CPU running time versus launched signal power into 375 km SMF P_{in} (dBm) in the 120-Gb/s 64 QAM coherent optical communication system

Weighted K-Means algorithm. The clustering result of the proposed Weighted K-Means procedure was correct and reliable as shown in insert (b).

Finally, the iteration times and CPU running times of both algorithms were collected as shown in Fig. 7.9. The CPU running time tests were performed on an Intel Personal Computer with process Core i7-7700k CPU at 4.2 GHz, 16 GB Random Access Memory and Windows 10 Professional operating system. As can be seen from Fig. 7.9, the proposed Weighted K-Means algorithm can not only improve system performance, but also reduce the overall computational complexity. To be more specific, although the average CPU runtime per iteration is slightly increased from 88.44 to 92.43 ms due to the extra computation of the weighting operation, yet the weighting factor accelerates the searching of correct centroids and avoids incorrect convergence. Throughout the working range of launched power for K-Means from −4 to 2 dBm, the iteration times and CPU running times of Weighted K-Means are not higher than the traditional K-Means. More importantly, when the traditional K-Means procedure fails to correctly converge, multiple invalid iterations make the computational complexity skyrockets, while the proposed algorithm can still correctly locate the clustering centroids within a reasonable number of iterations.

7.2 Mixture-of-Gaussian Clustering-Based Decision Technique for Coherent Optical Communication System

The coherent optical communication system with the combination of high-order modulation formats, coherent detection, and digital signal processing (DSP) has significantly impacted the field of optical communication [39]. The coherent optical communication system can simultaneously achieve high spectral efficiency and long-haul transmission. However, the performances of a high-speed long-haul coherent optical communication system are significantly deteriorated by nonlinearities caused by devices and transmission link. Therefore, system nonlinearity mitigation has been considered as a key technology to increase the optical communication system capacity and performance [40]. It is significant to further research on the advanced DSP algorithms for system nonlinearity compensation. To mitigate nonlinearity, many DSP algorithms have been proposed, such as Volterra-series transfer-function-based algorithms [41–43], digital back-propagation (DBP) [44–46], support vector machines (SVMs) [47, 48], support vector regression (SVR) [49], neural networks [50, 51], Parzen window (PW) classifier [40], k-nearest neighbor (KNN) [52, 53], expectation maximization (EM) algorithm [54], and k-means [16]. An adaptive finite impulse response (FIR) linear equalizer with a Volterra-series scheme can effectively compensate for both linear and nonlinear effects [55]. DBP utilizes the backpropagation algorithm in the digital domain to solve the inverse nonlinear Schrödinger equation of the fiber link based on the split-step Fourier method and calculate the transmitted signal from the received signal [44]. SVM, SVR, neural networks, PW classifier, KNN, and k-means all belong to machine learning algorithms. SVM, SVR, neural networks, PW classifier, and k-means are supervised learning algorithms that require training sequences assistance. The EM algorithm and KNN pertain to unsupervised learning algorithms that have a simpler structure. The main target of machine learning algorithms used for optical communication systems is to generate improved nonlinear decision regions more adapted to nonlinear fiber transmission. Compared with Volterra-series transfer-function-based algorithms and DBP, machine learning algorithms do not require the parameters of the optical fiber transmission link, and have relatively better flexibility and adaptive capacity. In addition, unsupervised learning algorithms do not require training sequences assistance, and have low data redundancy and low computational complexity.

In this paper, we propose a method of a decision algorithm for a coherent optical communication system by using a mixture-of-Gaussian (MoG) clustering algorithm to mitigate the nonlinear effects. We construct a single-carrier polarization division multiplexed (PDM) 16-QAM transmission system experimental platform to verify the performances of the proposed decision algorithm. The experimental results show that we achieved 0.41 dB Q factor improvement in the 975-km 16-QAM transmission system. And the transmission distance can be further improved by about 175 km at a Q factor of 9.96 dB. Compared with the traditional minimum Euclidean distance (MED)-based decision, the MoG clustering-based decision algorithm has higher nonlinear tolerance and can generate optimum nonlinear decision boundaries to efficiently reduce the bit error rate (BER) and improve the performance of the system.

Furthermore, we propose an optimization method for decision-directed least mean square (DD-LMS) [56] using the MoG clustering-based decision algorithm, called MoG-LMS. In the MoG-LMS algorithm, we apply the MoG clusteringbased decision to replace the traditional MED-based decision and optimize the error computation formula. The performance improvement of the MoG-LMS algorithm is verified by experiments.

7.2.1 Theoretical Analysis and Operational Principle

In this section, we first represent mixture-of-Gaussian clustering algorithm. Then, We demonstrate the experimental system and analyze the results.

Mixture-of-Gaussian Clustering Algorithm

MoG, a superposition of Gaussian distributions, is one of the most widely used probabilistic approaches to clustering. A Gaussian mixture model is a probabilistic model that assumes all the data points are generated from a mixture of a finite number of Gaussian distributions with unknown parameters. For the MoG clustering algorithm, a Gaussian component can be considered as a cluster. Reference [57] has a more detailed explanation of theMoG clustering algorithm.

The idea behind the MoG clustering algorithm is to find the parameters of each Gaussian distribution that best explain the data. According to the definition of multivariate Gaussian distribution [57], for a vector $\boldsymbol{x}$ in n-dimensional space, the expression of probability density function $p(\boldsymbol{x}|\boldsymbol{\mu}, \boldsymbol{\Sigma})$ is shown in Eq. (7.6):

$$p(x|\mu, \Sigma) = \frac{1}{(2\pi)^{\frac{n}{2}}|\Sigma|^{\frac{1}{2}}} e^{-\frac{1}{2}(x-\mu)^T \Sigma^{-1}(x-\mu)} \tag{7.6}$$

where $\boldsymbol{\mu}$ is the n-dimensional mean vector, $\boldsymbol{\mu} = -[\mu_1, \mu_2, \ldots, \mu_n]^T$, 6 is the $n \times n$ dimensional covariance matrix, and $\sum = (ci, j)\ n \times n$, where $c_{i,j} = cov(x_i, x_j)$. MoG distribution can be represented using Eq. (7.7):

$$p_M(x) = \sum_{i=1}^{k} \alpha_i \cdot p(x|\mu_i, \Sigma_i). \tag{7.7}$$

The distribution consists of k mixed components. Each mixed component corresponds to a Gaussian distribution, where $\alpha_i (\alpha_i > 0)$ is the mixture coefficient of each Gaussian component, $\sum_{i=1}^{k} \alpha_i = 1$, μ_i are $\sum_i$ the mean vector and covariance of each Gaussian distribution.

The EM algorithm is a common method of estimating latent variables. The EM algorithm is usually applied for learning MoG model parameters [54]. The iteration of the MoG clustering algorithm consists of two steps: expectation (E) and maximization (M) steps [58].

The E step calculates the membership of each data point for each Gaussian distribution, i.e., the E step computes the a posteriori probability γ_{ji} of each data point belonging to each mixed component using the current means and covariance matrices of the Gaussian distribution. The a posteriori probability of data point $xj (1 \leq j \leq m)$ of data set $D = \{x_1, x_2, \ldots, x_m\}$ generated by the k mixed component can be computed by Eq. (7.8):

$$\begin{aligned}\gamma_{ji} = p_M(z_j = i|x_j) &= \frac{p(z_j = i)\cdot_M (x_j|z_j = i)}{p_M(x_j)} \\ &= \frac{\alpha_i \cdot (\mathbf{x}_j|\boldsymbol{\mu}_i, \boldsymbol{\Sigma}_i)}{\sum_{i=1}^{k} \alpha_l \cdot (\boldsymbol{x}_j|\boldsymbol{\mu}_l \boldsymbol{\Sigma}_l)} (1 \le i \le k).\end{aligned} \tag{7.8}$$

The M step updates the mixture coefficient, mean vectors, and covariance matrices $\{(\alpha_i', \mu_i', -\sum_i')|1 \le i \le k\}$ by Eq. (4) according to the updated a posteriori probability of the data points:

$$\begin{aligned}\alpha_i' &= \frac{\sum_{j=1}^{m} \gamma_{ji}}{m}, \\ \boldsymbol{\mu}_i' &= \frac{\sum_{j=1}^{m} \gamma_{ji} x_j}{\sum_{j=1}^{m} \gamma_{ji}}, \\ \boldsymbol{\Sigma}_i' &= \frac{\sum_{j=1}^{m} \gamma_{ji} (x_j - \mu_i')(x_j - \mu_i')^T}{\sum_{j=1}^{m} \gamma_{ji}}.\end{aligned} \tag{7.9}$$

The log-likelihood function $LL(D)$ expressed in Eq. (7.10) is used as an iteration-stopping criterion for the EM algorithm. The objective of the EM algorithm is to maximize the loglikelihood function. According to Chapter 9 in Ref. [21], in the process of maximizing the log-likelihood function, if we set the derivatives of the log-likelihood function to zero, we will not obtain a closed-form solution. Therefore, the lower bound of the log-likelihood function is constructed. Maximizing the log-likelihood function can be seen as the process of using the coordinate descent method to maximize the log-likelihood lower bound. When the log-likelihood function $LL(D)$ is maximized, the iteration of the EM algorithm can be stopped. The EM algorithm is close to convergence:

$$\begin{aligned}LL(D) &= \ln (\prod_{j=1}^{m} p(x_j)) \\ &= \sum_{j=1}^{m} \text{In} (\sum_{i=1}^{k} \alpha_i \cdot (\boldsymbol{x}_j|\boldsymbol{\mu}_i, \boldsymbol{\Sigma}_i)).\end{aligned} \tag{7.10}$$

Finally, the parameters of the MoG distribution in Eq. (2) can be determined. Then, data set $D = \{x_1, x_2, \ldots, x_m\}$ is divided into k clusters $C = \{C_1, C_2, \ldots, C_k\}$. The cluster label λ_j of each data point x_j can be determined by Eq. (7.11):

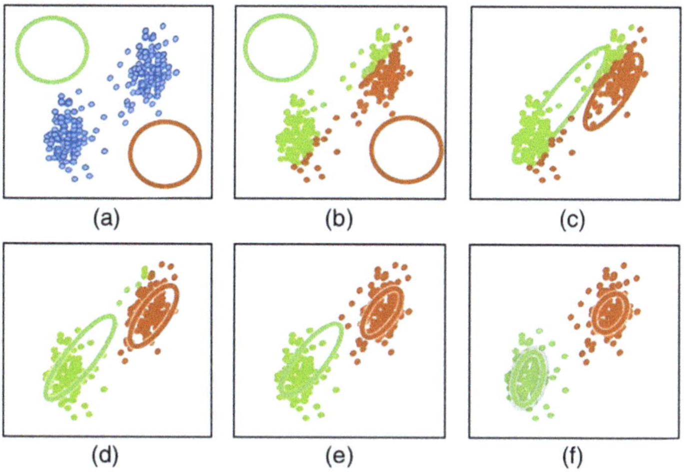

Fig. 7.10 Illustration of the iteration process of the MoG clustering algorithm. (**a**) Data points and parameter initialization. (**b**) Results after first E step; (**c**) results after first M step; (**d**) results after n_1 complete EM iterations; (**e**) results after n_2 complete EM iterations;(**f**) results after n_3 complete EM iterations. ($n_1 < n_2 < n_3$)

$$\lambda_j = \underset{i \in \{1,2,\ldots,k\}}{\operatorname{argmax}} \gamma_{ji}. \tag{7.11}$$

An example of the MoG clustering algorithm for a mixture of two Gaussian components is graphically shown in Fig. 7.10. As shown in Fig. 7.10a, the data points in blue are assigned into two clusters by using the MoG clustering algorithm. The two initialized Gaussian components are depicted by green and orange circles. Figure 7.10b, c show the process of the first iteration of the EM algorithm. Figure 7.10b shows the results of the initial first E step. The data points in green indicate that they belong to the green cluster with a higher a posteriori probability. Similarly, the data points in orange represent that they belong to the orange cluster with a higher a posteriori probability. The results after the first M step are shown in Fig. 7.10c. According to the result of the E step, the means and covariance matrices of the green and orange Gaussian components are updated by Eq. (7.9). Figure 7.10d, f show the results after n_1, n_2, and n_3 ($n_1 < n_2 < n_3$) complete iterations EM algorithm, respectively. Figure 7.10f shows that the data points are assigned into two clusters effectively. After several iterations, the algorithm is close to convergence.

Different from the k-means clustering algorithm, which distributes each data point uniquely to one cluster, the MoG clustering algorithm calculates the degree of membership of each data point to each cluster [59]. The difference between k-means and MoG clustering algorithms is shown in Fig. 7.11. In the MoG clustering algorithm, covariance information of data is also taken into account. In other

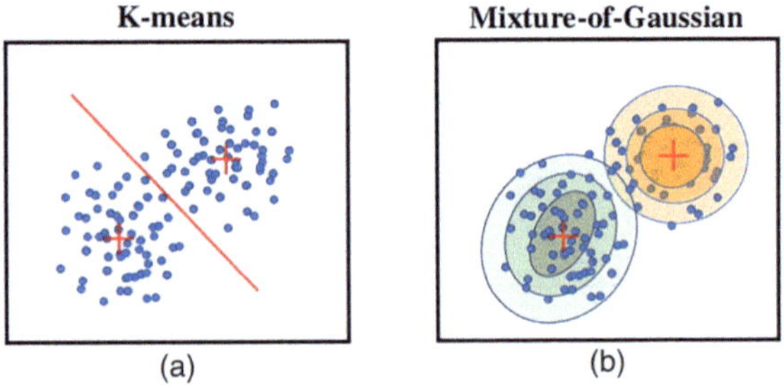

Fig. 7.11 Difference between k-means and MoG clustering algorithm for a set of data

words, MoG clustering can be considered as generalized k-means clustering by incorporating the information about the covariance structure of each cluster [60].

MoG Clustering Algorithm for M-QAM Symbol Decision

The k Gaussian components of MoG distribution correspond to k clusters $C = \{C_1, C_2, \ldots, C_k\}$. For M-QAM signals, corresponding to the number of constellation points, M is the number of clusters; the M-QAM signal can be expressed as a MoG distribution with M Gaussian components. The detailed principle of the MoG clustering-based decision algorithm for M-QAM is described as follows (16-QAM signals are used as an example).

(a) We assume that the received 16-QAM data set $D = \{x_1, x_2, \ldots, x_m\}$ can be expressed by MoG distribution [Eq. (7.7)], giving 16 as the value of k because there are 16 clusters in the 16-QAM signal constellation diagram.
(b) According the received 16-QAM data set, we initialize the model parameter of the MoG distribution $\{(\alpha_i, \mu_i, \sum_i)|1 \leq i \leq k\}$. The complexity of MoG clustering algorithms is determined by the number of iterations required for convergence and the complexity of each E and M step, so that for the initialization step, the mean of each Gaussian component can be initialized to the corresponding standard constellation point. The initialization of the covariances and the mixture coefficient of each Gaussian component can be obtained by using the k-means algorithm. Thus, the number of iterations can be decreased and computational complexity can be reduced.
(c) According to the current model parameters, we calculate the a posteriori probability $\gamma_{ji} = p_M(z_j = i|x_j)(1 \leq i \leq k)$ of $x_j(1 \leq j \leq m)$ generated by the mixed component by Eq. (3).
(d) After the a posteriori probability of all data points being calculated, we update the model parameter $\{(\alpha_i, \mu_i, \sum_i)|1 \leq i \leq k\}$ by Eq. (7.9).
(e) Repeat steps (c) and (d) until likelihood function $LL(D)$ expressed in Eq. (7.10) is maximized.
(f) Then, the parameters of theMoG distribution can be determined. Cluster label λ_j of each data point x_j can be determined by Eq. (7.11).

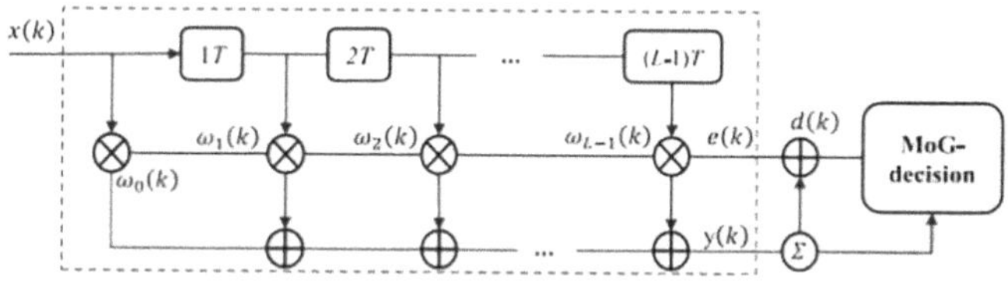

Fig. 7.12 Schematics of MoG-LMS algorithm

(g) Finally, each data point x j will be assigned into the corresponding cluster according to their max a posteriori probability. The optimum 16 clusters $C = \{C_1, C_2, \ldots, C_{16}\}$ for the received 16-QAM data set can be determined. The converged MoG model generates optimum nonlinear decision regions, then is used for M-QAM signal symbols decision. As a decision algorithm, in contrast to the traditional MED-based decision methods, the MoG-based algorithm is insensitive to the system's nonlinear impairments and achieves a more accurate classification of constellation points to improve the nonlinear tolerance of the coherent optical communication system.

The converged MoG model generates optimum nonlinear decision regions, then is used for M-QAM signal symbols decision. As a decision algorithm, in contrast to the traditional MED-based decision methods, the MoG-based algorithm is insensitive to the system's nonlinear impairments and achieves a more accurate classification of constellation points to improve the nonlinear tolerance of the coherent optical communication system.

MoG Least Mean Square

Due to the advantages of the MoG clustering algorithms, we optimize the DD-LMS [56] using the MoG clustering-based decision algorithm, called MoG-LMS. The schematics of the MoG-LMS algorithm is shown in Fig. 7.12. In the MoG-LMS algorithm, we use the MoG clustering-based decision to replace the traditional MED-based decision and optimize the error computation formula.

The relationship between input $x(k)$ and output $y(k)$ for a MoG-LMS equalizer at time k with L tap can be expressed by Eq. (7.12):

$$y(k) = \omega(k)^H x(k), \tag{7.12}$$

where $\omega(k)$ is the tap coefficient. $()^H$ is the conjugate transpose of the matrix. The expressions for $x(k)$ and $\omega(k)$ are represented, respectively, by Eqs. (7.13) and (7.14):

$$x(k) = [x(k), x(k-1), x(k-2), \ldots, x(k-L+1)]^T, \tag{7.13}$$

$$\omega(k) = [\omega_0(k), \omega_1(k), \omega_2(k), \ldots, \omega_{L-1}(k)]^T \tag{7.14}$$

The output $y(k)$ is sent to the decision module for the symbol decision. In the decision module, we use the MoG clustering-based decision to replace the traditional MED-based decision. Then, $d(k)$ is the mean vector of the Gaussian component that the output $y(k)$ belongs to. So the error computation formula can be optimized by Eq. (7.15):

$$e(k) = \| d(k) - y(k) \|^2 . \tag{7.15}$$

According to the error value $e(k)$, the tap coefficient can be updated at the next moment. The iteration expression is represented by Eq. (7.16):

$$\omega(k+1) = \omega(k) + \mu e(k) x^*(k), \tag{7.16}$$

where μ is the step size. After the update of the tap coefficient $\omega(k+1)$ at time $(k+1)$ is completed, a new sequence inputs, and the update of the output sequence iteratively continues to be completed.

For the MoG-LMS algorithm, we use the MoG clustering-based decision to replace the traditional MED-based decision in the decision module. The mean vector obtained by the MoG clustering-based decision algorithm replaces the standard constellation point for the error computation, so that the influence of the nonlinear distortion constellation can be mitigated, and the compensation effect of the algorithm can be improved.

7.2.2 *Experiment Setup and Results*

In this section, we first present the experimental system. Then, we analyze the results of MoG clustering algorithm for 16-QAM symbol decision.

Experiment Setup

This section presents the details of the experimental platform that is performed to demonstrate the usability of the MoG clustering-based decision algorithm for coherent optical transmission systems. The experimental setup for the single-carrier PDM 16-QAM coherent optical communication system is shown in Fig. 7.13. The total bit rate is 80G bit/s. At the transmitter, optical waves with linewidths of 100 kHz and wavelengths of 1550 nm are sent into the I/Q modulator. The symbol sequences of the 16-QAM signal are sent into a 25 GSa/s arbitrary waveform generator (AWG). Then, the in-phase (I) and quadrature (Q) component sequences of the 16-QAM symbols are converted into two electrical signals by AWG. The I/Q

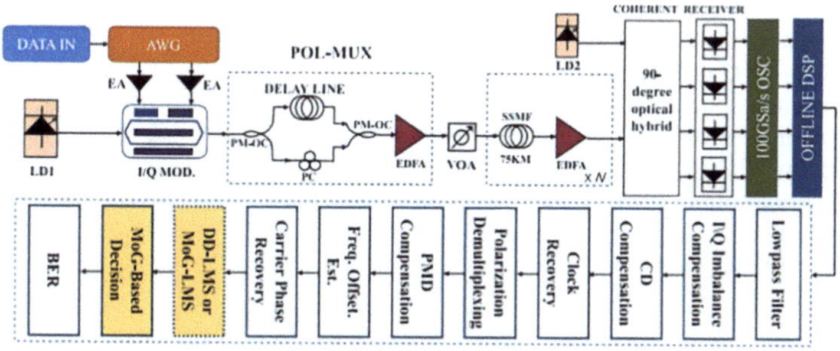

Fig. 7.13 Experimental architecture of single-carrier PDM 16-QAM coherent optical transmission system

modulator is driven by those two amplified electrical signals. Then, the modulated 16-QAM optical signal is generated by the I/Q modulator. The modulated 16-QAM optical signal is divided into two paths by a polarization-maintaining optical fiber coupler (PM-OC); one passes through the delay line, one passes through the polarization controller, and then they are combined by the PM-OC to form a polarization multiplexed signal. The transmission link consists of multi-span G.652D single-mode fiber (SMF). An erbium-doped fiber amplifier (EDFA) with a 20-dB gain is used at each span of 75 km. At the receiver, A 100-kHz local oscillator (LO) is used and superimposed with the polarization-multiplexed signal in the coherent receiver. A 100 GSa/s real-time oscilloscope is used as an analog-to-digital (A/D) converter to acquire data.

Offline DSP includes a lowpass filter, imbalance compensation, chromatic dispersion (CD) compensation, clock recovery,polarization demultiplexing, polarization mode dispersion (PMD) compensation, frequency offset estimation, carrier phase recovery, and DD-LMS or MoG-LMS (if used), and the MoG clustering-based decision is used before 16-QAM demapping and BER calculation.

Results of MoG Clustering Algorithm for 16-QAM Symbol Decision

In this section, we verify the performance of the MoG clustering algorithm for the 16-QAM symbol decision in combating the nonlinearity effects.

The PDM 16-QAM signal over 13-span 75-km SMF transmission is used to verify the feasibility of the proposed algorithms. Figure 7.14 shows the Q factor versus optical signal-to-noise ratio (OSNR) for 16-QAM after 975 km of transmission through SMF. The relationship between the Q factor and BER can be expressed as $Q = 20log_{10}[\sqrt{2}erfc^{-1}(2BER)]$, and the Q factor corresponding to the BER of 10^{-3} forward error correction (FEC) limit) is 9.8 dB [26]. We compare the performance of the MoG clustering-based algorithm with the MED-based algorithm [referred to as (w/o clustering algorithm) in the figure], and k-means-based algorithm. It can be seen from the trend of the curve that as the launched

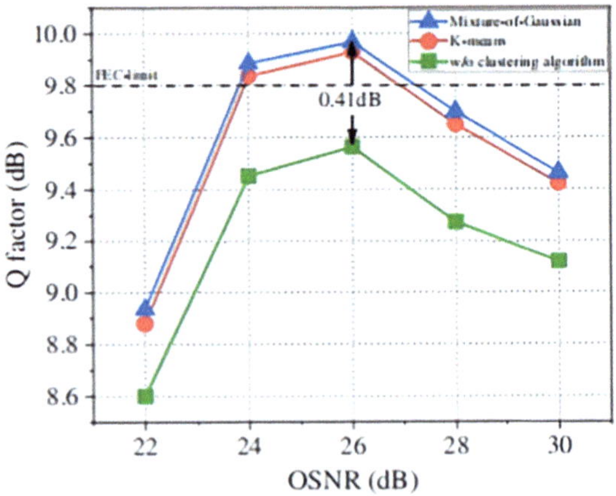

Fig. 7.14 Q factor curves versus OSNR after 975 km of transmission in 80-Gb/s PDM 16-QAM coherent optical communication system with MoG algorithm (triangle); with k-means algorithm (circle); without any clustering algorithm (square)

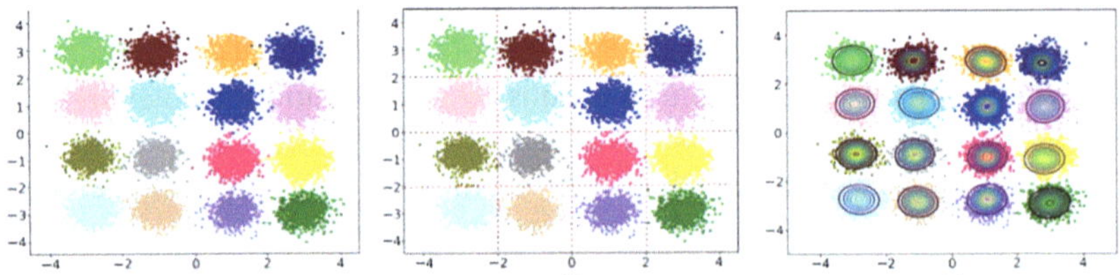

Fig. 7.15 Constellation diagrams of 16-QAM signals with the OSNR at 26 dB. (**a**) Correct cluster of the received data points; (**b**) results without any clustering algorithm and the decision region of MED-decision; (**c**) cluster result by using MoG clustering-based decision algorithm

signal power increases, the OSNR of the received signal becomes higher, and system performance improves. However, once the launched signal power exceeds a value, the effects of fiber nonlinearity increase, and system performance is deteriorated by the nonlinearity damage. In Fig. 7.14, it can be seen that after applying the MoG clustering-based algorithm and k-means-based algorithm, a better Q factor can be obtained, and the Q factors exceed the FEC threshold at the OSNR of 24 and 26 dB. The Q factor of 80-Gb/s 16-QAM signal after 975-km transmission is improved by about 0.41 dB at the OSNR of 26 dB.

Figure 7.15 shows the constellation diagrams of 16-QAM signals with the OSNR at 26 dB. Constellation points with the same color belong to the same cluster. Figure 7.15a shows the correct cluster color of the received data points. The color of each data point represents the correct cluster that each data point belongs to; Fig. 7.15b shows the decision results without any clustering algorithm and the decision region of the traditional MED-based decision (MED-decision); Fig. 7.15c shows the cluster result by using the MoG clustering-based decision algorithm. The contour lines of each Gaussian component are described. Each data point is assigned into the corresponding cluster according to its max a posteriori probability. So the MoG clustering-based decision algorithm generates nonlinearity-

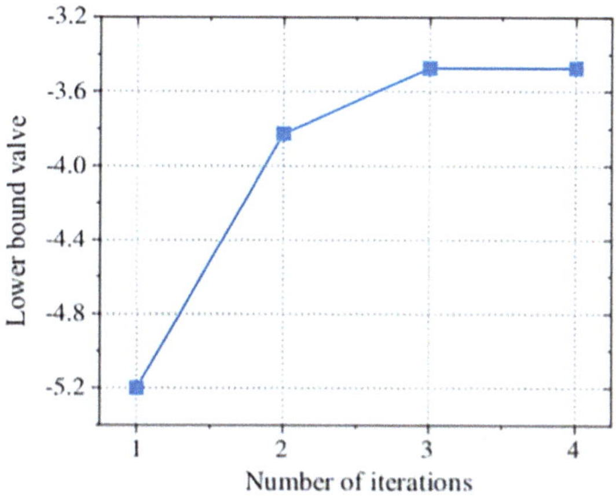

Fig. 7.16 Plot of the lower-bound valve of the log-likelihood function for the MoG clustering algorithm

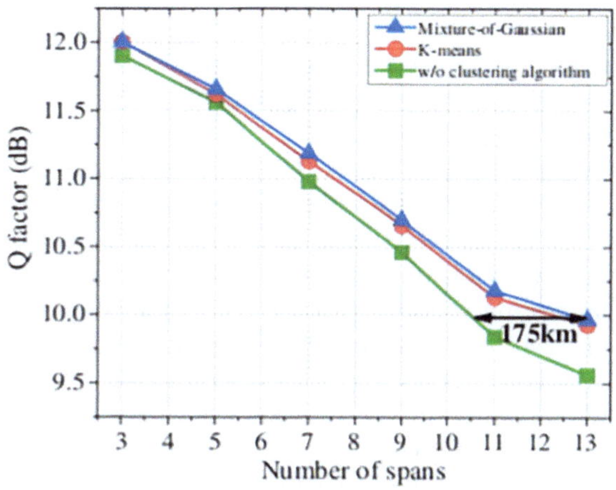

Fig. 7.17 *Q* factor curves versus different transmission distances in 80-Gb/s PDM 16-QAM coherent optical communication system

insensitive decision regions, contributes to achieving a more accurate classification of constellation points, and enhances the nonlinear system tolerance.

A plot of the lower-bound valve of the log-likelihood function for the MoG clustering algorithm is shown in Fig. 7.16. As in the initialization setup, we use the parameters of standard constellation points to initialize the MoG model; thus, the number of iterations is decreased, and then the computational complexity can be reduced. It can be seen that after four iterations, the lower-bound valve of the log-likelihood function is maximized. Correspondingly, the log-likelihood function is maximized. The algorithm has converged after the third iteration, and the final iteration cycle produces no changes in model parameters. Figure 7.17 shows the *Q* factor versus the transmission distance for an 80-Gb/s PDM 16-QAM signal at the OSNR of 26 dB. The transmission link is varied from three-span standard single-mode fiber (SSMF) to 13-span SSMF (two spans per change). It can be seen that as transmission distance increases, system performance decreases. The performance of the MoG clustering-based algorithm is compared with the MED-based algorithm [referred to as (w/o clustering algorithm) in the figure], and k-means-based algorithm. The transmission distance can be further improved by about 175 km at the Q factor of 9.96 dB by applying theMoG clustering-based decision.

7.3 Data-Defined Naïve Bayes (DNB) Based Decision Scheme for Nonlinear Mitigation

In recent years, various multiplexing techniques have been widely used in optical fiber communication systems in order to exceed the Shannon limit on the capacity of a single mode fiber(SMF), including the wavelength, amplitude, phase, and polarization multiplexing. Orbital angular momentum (OAM) is an approach that can increase the channel capacity in an optical fiber communication system and is attracting growing interest [61]. OAM provides an additional degree of freedom in multiplexing, which depends on a phase term with infinite integer values of topological charge index l [62, 63]. OAM transmission has been experimented over different distances [64–68]. It now can be achieved with multi-mode fiber (MMF) or few-mode fiber (FMF) over 8.8 km [64]. One of the main restrictions on OAM transmission in FMF or MMF is crosstalk. A common approach for the compensation of the crosstalk is the multi-input multi-output (MIMO) digital signal processing (DSP). However, it is impractical to apply MIMO in real-time FMF or MMF transmission due to the high complexity of MIMO DSP. For instance, a 10-mode channel over FMF transmission would require a 20 × 20 MIMO DSP [69]. Compared with conventional fibers, ring core fiber (RCF) is a special type of optical fiber with large effective refractive index ($\triangle neff$) difference between mode groups (MGs), which can separate the eigenmodes effectively [70]. RCF has attracted a great deal of attention due to the dramatic reduction in complexity: an eight-mode OAM mode-division multiplexed transmission only needs 4 × 4 MIMO DSP for RCF transmission over 100 km [71].

Although MIMO DSP could suppress crosstalk effectively, nonlinearity is still a key problem limiting the spectral efficiency. The nonlinear effect in communication system can be divided into two categories: optical fiber nonlinearity and device nonlinearity. Optical fiber nonlinearity, which include self-phase modulation (SPM), intra or inter cross-phase modulation (IXPM), and four wave mixing (FWM), often causes serious signals distortion over long distance transmission [72]. Digital back-propagation (DBP) algorithm and perturbation method are traditional approaches to compensate the nonlinearity of optical fiber, which are based on the analysis of the nonlinear Schrodinger equation (NLSE) of field propagation in optical fiber [44]. Besides, multiple optoelectronic devices are required to modulate signals for OAM mode-division multiplexed optical fiber communication, such as Mach-Zehnder modulator (MZM), photodiode (PD) and spatial light modulator (SLM) [73]. Device nonlinearity is the inherent characteristic of these opto-electronic devices, which

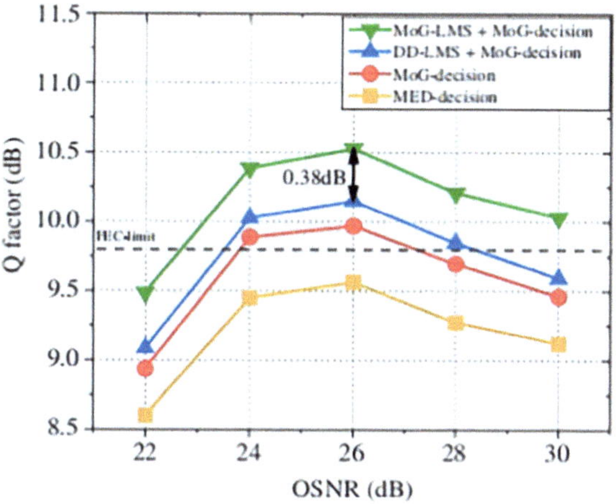

Fig. 7.18 Q factor curves versus OSNR after 975 km of transmission in 80-Gb/s PDM 16-QAM coherent optical communication system: with MED-decision (square); with MoG-decision (circle); with DD-LMS + MoG-decision (regular triangle); with DD-LMS + MoG-decision (inverted triangle)

will cause serious deterioration on the signal [74]. Traditional nonlinearity compensation approaches, such as Volterra equalizer, usually compensate the nonlinearity through the Volterra series to fit nonlinear model of devices [75]. However, it is very hard to build an accurate theoretical model for multiple opto-electronic devices by using Volterra series. More importantly, some special opto-electronic devices are still lack the related theoretical model. For example, the nonlinear effect of the SLM may be generated through several physical phenomena, such as self-phase modulation and memory effect, which is difficulty to determine an accurate model [76, 77]. Hence, few reports has been published for an accurate and comprehensive model of SLM nonlinearities. As a result, it is very hard to compensate the nonlinear impairment of the special opto-electronic devices in the OAM mode division multiplexed optical fiber communication through the conventional approaches, due to the uncertainty of the nonlinear model. Meanwhile, the computational complexity of conventional nonlinear equalizer, such as Volterra equalizer, is very high due to the complex multiplication operation (Fig. 7.18).

Machine learning algorithms of classification could classify the test data by using the mathematics method based on the attribute from the training data, which is very feasible to build an accurate nonlinear model for the complex and uncertain opto-electronic devices. Machine learning algorithms for classification have all shown good performances and tolerance on nonlinearity mitigation. For instance, k-nearest neighbors (KNN) algorithm for mitigating nonlinearity in 16/64-QAM coherent optical communication system has been demonstrated [52]. Besides, the support vector machines (SVM) for the compensation of nonlinear equalization in singe-carrier and multi-carrier coherent optical orthogonal frequency-division multiplexing (OFDM) have also been explored [78–80]. Moreover, the neural network is a powerful tool to build the nonlinear model for optical fiber communication noise, showing good performances on nonlinearity mitigation in many researches, such as deep neural network (DNN), artificial neural network (ANN) and long short-term memory (LSTM) neural network [55, 81, 82]. Especially, the naïve Bayes algorithm, which is a statistical classification algorithm that obtains the posterior probability of a data set through the prior and likelihood function, is ideal to mitigate the nonlinear impairments in OAM optical fiber communication with high accuracy and low complexity [83].

In this manuscript, a data-defined naïve Bayes (DNB)-based decision scheme for nonlinear mitigation is presented for an OAM mode-division multiplexed optical fiber communication system. An experiment is carried out to verify the effectiveness of the proposed scheme, and the results demonstrate that the accuracy and calculation complexity of the DNB algorithm is significantly improved compared with other machine learning algorithms.

7.3.1 Principle of the DNB Based Decision

The naïve Bayes algorithm can be described as a statistical classification algorithm that gives the posterior probability of a dataset based on a prior probability and likelihood probability. Consider a finite set $B = \{X_1, \ldots, X_i, \ldots, X_n\}$ of discrete random variables. Assuming that each variable X_i is attributed to a class $C = \{C_1, \ldots, C_i, \ldots, C_l\}$, the likelihood probability, $P(B|C_i)$ that describes the probability of each instance $B = \{X_1, \ldots, X_i, \ldots, X_n\}$ is attributed to a class C_i. And the variables Xi are features of instance can be given as [84]

$$P(B|C_i) = \frac{P(C_i|B)P(B)}{P(C_i)}. \tag{7.17}$$

$P(C_i)$ is the prior probability of set class $P(C_i|B)$ is the posterior probability defined as the probability of class C_i with the occurrence of set B, which is then given by

$$P(C_i|B) = \frac{P(C_i)P(B|C_i)}{P(B)} = \frac{P(C_i)P(B|C_i)}{\sum_{i=1}^{n} P(C_i)P(B|C_i)} \tag{7.18}$$

Equation (7.18) shows that the probability of class C_i can be predicted based on the prior probability and the likelihood probability [85]. From Eq. (7.18), it can be seen that both the posterior probability and the likelihood probability contain the probability distribution information for set B.

In an optical fiber communication system, the received signals always contain the features of the transmission channel. The DNB algorithm is therefore suitable for the mitigation of uncertain nonlinear impairments over OAM mode-division multiplexed transmission, since the distribution information of the uncertain nonlinear noise can be obtained based on the likelihood probability. The transmission signal can be described as set B, while the constellation point can be described as class C_i. The prior probability ($P(C_i)$) and likelihood probability $P(B|C_i)$ can be obtained from the training data, which contains the probability distribution of the uncertain nonlinear noise.

In Eq. (7.18), $\sum_{i=1}^{n} P(C_i)P(B|C_i) = P(B)$ can be ignored, since it is the same for all constellation points [86]. The posterior probability for each constellation point can then be computed based on the prior probability and likelihood probability, which can be simplified as [84], where a is the number of instance B:

$$f_i(B) = P(C_i)P(B|C_i) = P(C_i)\prod_{j=1}^{a} P(B_j|C_i). \tag{7.19}$$

For the received OAM signal, the posterior probability of each item of signal data $f_i(B)$ for each constellation point can be obtained by submitting the value of the I/Q axis into Eq. (7.19), and the decision process of the received OAM signal can then be executed directly based on the maximum posterior probability of each constellation point. The posterior probability also contains the distribution information of the uncertain nonlinear noise due to the likelihood probability according to Eq. (7.19). Hence, the DNB algorithm can effectively mitigate uncertain nonlinear impairments based on the posterior probability.

7.3.2 *Experimental Setup*

The experimental setup for 32 Gbaud OAM mode division multiplexed optical fiber communication system over 10 km RCF transmission is represented as Fig. 7.19.

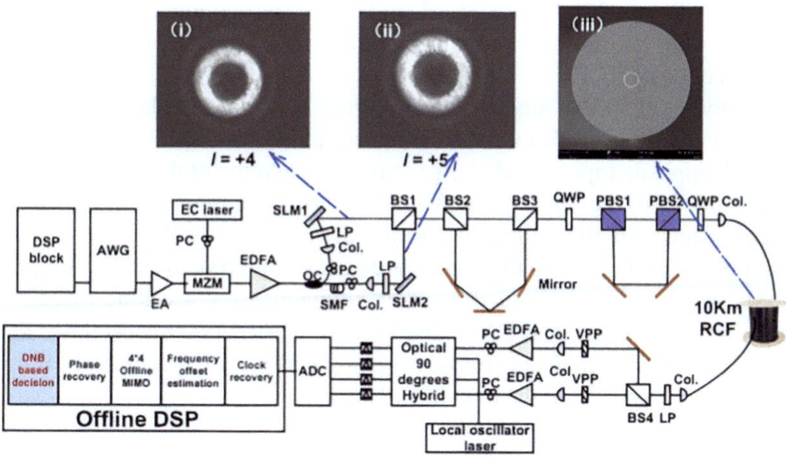

Fig. 7.19 Experimental setup for 32Gband OAM mode division multiplexed optical fiber communication system over 10 km RCF transmission

At transmitter, data passed digital signal process (DSP) block is encoded onto waveform generated by arbitrary waveform generator (AWG) with a sample rate of 64 GSa/s. The optical carrier is generated by using an external cavity (EC) laser with a wavelength of 1550.12 nm [87], followed by the modulation of the Mach-Zehnder modulator (MZM). The modulated electrical data sequence is pseudo-random binary sequence (PRBS) with pattern length of 218. Signal is amplified by using erbium-doped fiber amplifier (EDFA) for improving transmission distance. Then it is split into two components by using an optical coupler. One of the components is delayed through a 10 m single mode fiber (SMF) to reduce correlation. Signal beams are coupled from fiber to space through polarization controllers (PC), collimators (Col.), and linear polarization (LP) [88]. For modulating OAM modes with l = +4 and l = +5, two spatial light modulators (SLM) are used. Insets (i) and (ii) present the amplitudes of OAM mode with topological charge l = +4 and l = +5. Then two OAM beams are multiplexed with a beam splitter (BS1).

Then the multiplexed OAM beams are split into two beams by using BS2. One beam transmits through the BS2 with the OAM modes of l = +4 and l = +5, while the other beam is reflected with BS2, three reflected mirrors, and BS3, respectively. As a result, this beam contains two OAM modes with l = −4 and l = −5 due to five times reflections. Thus after the combination of two beams by using the BS3, the multiplexed OAM beams contain four modes of l = ±4 and l = ±5. Then the multiplexed OAM beams is converted into circular polarization from linear polarization by a quarter wave plate (QWP1). These four circular polarization OAM modes are again split into two linear polarizations (horizontal and vertically polarization) by using PBS1, and combined again with the PBS2 and two reflected mirrors. QWP2 is used to convert the linear polarization back into the circular polarization (CP), As a result, the multiplexed OAM beams contain eight modes (< + 4, left CP>, < + 4, right CP>,< - 4, left CP>, < - 4, right CP>; < + 5, left CP>, < + 5, right CP>,< - 5, left CP>, < - 5, right CP>) in two mode groups. The eight OAM modes are multiplexed and coupled into the 10 km RCF. Insets (iii) presents the cross section of RCF.

At receiver, all OAM modes are converted into linear polarization and split into two beams by using BS4. Two beams are passed through vortex phase plates (VPP) with opposite topological charge. After propagation of the VPP with opposite topological charge, the OAM beam with corresponding mode can be converted to Gaussian beams, which is coupled into the SMF with the collimator. The optical signals are received by using the coherent detection and processed by using the off-line DSP, which includes frequency offset estimation, 4 × 4 offline MIMO, clock recovery, phase recovery and DNB based decision [89]. Two uncorrelated datasets with the maximum normalized cross-correlation of 0.8% are transmitted for training and testing in order to keep the data independence.

7.3.3 Experimental Results and Analysis

An experiment is carried out using a 32 Gband eight-mode OAM optical fiber communication system with RCF transmission over 10 km to verify the effectiveness of the proposed method. Two OAM mode groups containing eight OAM modes with $|l|=4$, $|l|=5$ and < left, right > circular polarization are transmitted using QPSK modulation. Figure 7.20 illustrates the training process for the experimental data under an OSNR of 23 dB. The proposed DNB algorithm trains a probability density for each class corresponding to the constellation point to obtain the likelihood probability. The length of training dataset, validation dataset, and test dataset are 200, 10000, and 19807 constellation points, respectively. Two uncorrelated datasets with the maximum normalized cross-correlation of 0.8% are transmitted for training and testing in order to keep the data independence. In the training stage, the bandwidths are optimized as [0.0577, 0.0570, 0.0584, 0.0519], and the constellation diagram are divided into 256 × 256 spaces with the scale form along the x- and y-axes. Figure 7.20a shows the constellation diagram, while Fig. 7.20b shows the discrete probability density of four constellation points for each space

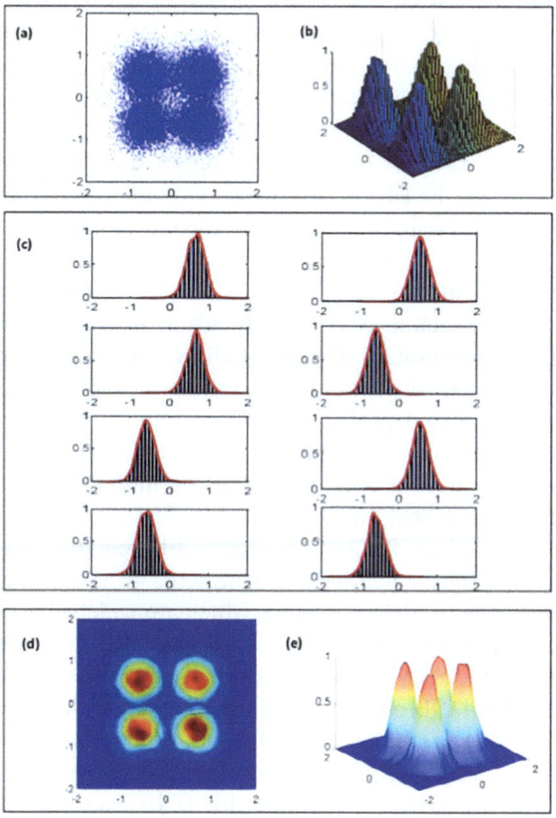

Fig. 7.20 Under the OSNR of 23 dB. Distributions in a training progress of one of experimental data constellation diagrams under the OSNR of 23 dB

using the Gaussian kernel function. The independent discrete probability density of the four constellation points along the x- and y-axes are shown in Fig. 7.20c. All probability densities for the different constellation points are distributed with an approximately normal distribution due to the amplified spontaneous emission noise in the RCF channel. However, the normal distribution of each constellation point is deformed differently under the uncertain nonlinear impairment. For example, the distributions of the constellation points at (+0.5, +0.5) and (−0.5, −0.5) in Fig. 7.20c are deformed much more severely than those of the other two constellation points. A two-dimensional discrete probability density at each small space is then fitted to obtain a continuous probability distribution, as shown in Fig. 7.20d. Figure 7.20e shows the contours (for the same probability density) of the four classes for four constellation points. The contour lines at the center of each class are denser, indicating a higher probability of constellation coordinate points in the centre. Conversely, the contour lines at the edge are sparser, indicating a lower probability. The probability density corresponding to the two constellation points of (+0.5, +0.5) and (+0.5, −0.5) are distributed widely compared with the other two constellation points, indicating a strong nonlinear impairment. Based on the training process mentioned above, the probability densities of all training data with the OSNR from 14 to 23 dB are calculated to obtain the corresponding likelihood probability at different OSNR.

Before the comparison between DNB algorithm and other machine learning algorithms, we firstly compensate the nonlinear noise using the conventional Volterra equalizer followed by the hard-decision. Simultaneously, the DNB based decision algorithm is also tested for all of the received OAM signals. The BER of the OAM signals by using the Volterra and DNB approaches are shown in Fig. 7.21. However, the BER by using the DNB algorithm is much lower than that by using the Volterra equalizer at each OSNR, which confirm the effectiveness of proposed DNB based decision algorithm for the mitigation of nonlinear impairments in OAM mode division multiplexed transmission. The nonlinear model of some opto-electronic devices, such as SLM, is difficult to fit precisely by using Volterra series. Hence the conventional Volterra equalizer cannot compensate nonlinear impairments in OAM mode division multiplexed transmission effectively. On the contrary, through the likelihood probability of the training data, the DNB algorithm could learn the probability distribution of the nonlinear noise, which mitigates the nonlinear impairment in OAM mode division multiplexed transmission effectively with the posterior probability.

To evaluate the performance of different machine learning algorithms in terms of mitigating the nonlinear impairment, we test two other machine learning decision algorithms, KNN and SVM, which are widely used in the DSP of optical communication systems. Figure 7.21 shows the BER for eight OAM modes using the DNB, KNN and SVM algorithms. A comparison with different decision algorithms shows that the DNB outperformed the KNN algorithm in terms of BER by a minimum of 11% at $l = +5$ for OSNR=18 with right polarization, and a maximum of 66% at $l = -4$ for OSNR=23 dB with right polarization using $k = 5$ and Euclidean distance nearest neighbors. The BER for DNB exceeded that of the SVM algorithm by a

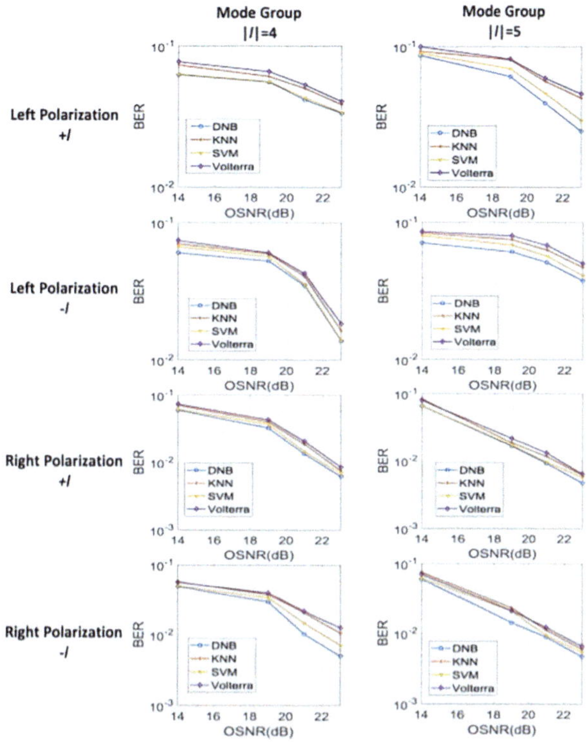

Fig. 7.21 Performances on BER of 8 OAM modes

minimum of 6% at l = +4 for OSNR=14 dB with right polarization, and a maximum of 35% at l = − 4 for OSNR=23 dB with right polarization, using an extension of the original nonlinear binary classification in SVM.

Table 7.1 gives the computational complexity of DNB, KNN and SVM algorithm in this experiment, including the training complexity and running complexity. For training complexity, the complexity of DNB algorithms is $O\{[d + 12d(d - 1)]n\}$, where d is dimensions (d = 2 for QPSK modulation), and n is the number of training sample [90]. As an unsupervised classification algorithm, the training complexity of KNN is zero. The training complexity of SVM is O[12dc(c-1)n] by using an extension of binary classification of SVM [91], where c and d are constants for QPSK modulation. Hence the training complexity of SVM is much higher than that of the DNB because the training complexity of SVM is accumulated with the square operation (n^2) compared with the DNB (n), especially with a large training sample.

Table 7.1 Computational complexity of DNB, KNN and SVM

	DNB	KNN	SVM
Training complexity	$O\{[\mathrm{d} + \frac{1}{2}d(d-1)]n\}$	none	$O[\frac{1}{2}\mathrm{dc}(c-1)n^2]$
Running complexity	$O(cd)$	$O(mnk)$	$O(\frac{1}{2}\mathrm{sdc}(c-1))$

The running complexity is more likely to cause computational redundancy and transmission delay. For the DNB, the complexity is a fixed value O(cd), that is independent of sample number since c and d are constants for QPSK modulation. However, the running complexity of KNN is proportional to training sample and test sample, which is very high with a large sample. For the SVM, the running complexity of SVM is proportional to the number of support vector s, which is very large with the complex of the fuzy boundaries in the sample datasets. Therefore, compared with O(mnk) of KNN (where m is the number of running sample, k is the number of k nearest points.) and $O(\frac{1}{2}sdc(c-1))$ of SVM (where s is the number of support vector which indicates the number of samples that determine the boundary), the running complexity of O(cd) for DNB is decreases significantly, as shown in Table 7.1 [90–92].

To verify the complexity of three machine learning algorithms, the computational time of three algorithms is recorded. The results are similar for each mode group at each OSNR. The computational time is represented as Fig. 7.22 with mode group l = |4| at OSNR=14, 19, 23 dB respectively. It can be seen that the computational time

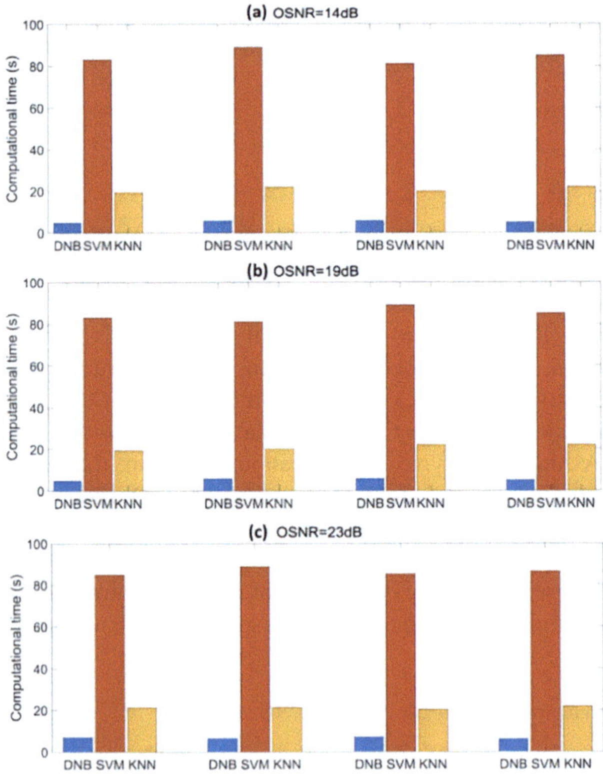

Fig. 7.22 Computational time with mode group l = |4| at different OSNR

Fig. 7.23 BERs under different bandwidth and band borders of a data set at OSNR=23 dB

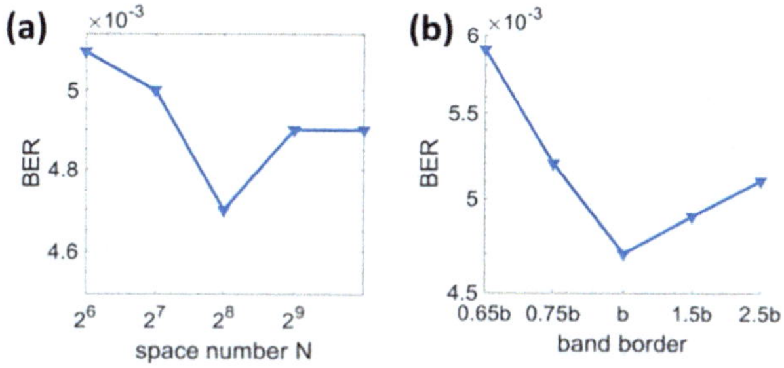

of DNB is significantly reduced compared with that of other two algorithms, which confirm the complexity comparison in Table 7.1.

In this experiment, the Gaussian function is applied as the kernel function. According to [86], the values of the bandwidth and band border have an impact on the estimation results, which should be optimized through cross-validation in order to avoid under/over-fitting. To illustrate the effectiveness of the bandwidth and band border, BERs for different bandwidths and band borders for a dataset with OSNR=23 dB are presented in Fig. 7.23. And space number N represents bandwidth which there is a monotone transformation given by Eq. (7.5), b represents the band border length calculated by the maximum minus the minimum in the training dataset. The performance in terms of the BER is optimized based on suitable values of the bandwidth and band border. For other values, the estimations would be under/over-fitted, which would reduce the BER performance.

7.4 Summary

In Chap. 7, the advanced nonlinear compensation technology using artificial intelligence in optical communication system is discussed. Section 7.1 introduces a robust weighted K-means clustering algorithm tailored for PS-64QAM coherent optical communication system, emphasizing the probabilistic shaping principle. The experimental setup validated these concepts, revealing significant improvements in signal integrity and error reduction. In Sect. 7.2, the decision technology based on hybrid Gaussian clustering is studied in depth, providing a theoretical framework and operational principles, and its effectiveness in improving system performance is demonstrated by experimental results. Finally, Sect. 7.3 introduces an innovative data-defined naive Bayes (DNB) based decision scheme that specifically targets nonlinear mitigation in complex communication environments. Experimental Settings and results show that DNB method is superior to traditional methods in both accuracy and computational efficiency, and has a remarkable prospect. Taken together, these sections highlight the potential of AI-driven approaches to revolutionize nonlinear compensation technology, setting a new benchmark for the future development of optical communication systems.

References

1. X. Wang, Q. Zhang, X. Xin, R. Gao, Q. Tian, F. Tian, C. Wang, X. Pan, Y. Wang, L. Yang, Robust weighted k-means clustering algorithm for a probabilistic-shaped 64QAM coherent optical communication system. Opt. Exp. **27**(26), 37601–37613 (2019)
2. X. Liu, Y. Wang, X. Wang, F. Tian, X. Xin, Q. Zhang, Q. Tian, L. Yang, Mixture-of-Gaussian clustering-based decision technique for a coherent optical communication system. Appl. Opt. **58**(33), 9201–9207 (2019)
3. S. Zhou, R. Gao, Q. Zhang, H. Chang, X. Xin, Y. Zhao, J. Liu, Z. Lin, Data-defined naïve bayes (DNB) based decision scheme for the nonlinear mitigation for oam mode division multiplexed optical fiber communication. Opt. Exp. **29**(4), 5901–5914 (2021)
4. M. Kong, J. Yu, H.C. Chien, K. Wang, X. Li, J. Shi, L. Zhao, B. Zhu, X. Pan, X. Xin, B. Ye, Y. Xia, T. Wang, Y. Chen, *WDM Transmission of 600G Carriers Over 5,600 km with Probabilistically Shaped 16QAM at 106 GBAUD* (Optica Publishing Group, Washington, 2019), p. M4I.3
5. J. Yu, K. Wang, H. Chien, B. Zhu, X. Li, J. Zhang, J. Shi, Y. Zhang, S. Shi, C. Ge, Y. Xia, Y. Chen, 8×528gb/s ps-64QAM transmission over 3000km in a terrestrial system with an amplifier span of 100km, in *2018 European Conference on Optical Communication (ECOC)* (2018), pp. 1–3
6. H.C. Chien, J. Yu, Y. Cai, B. Zhu, X. Xiao, Y. Xia, X. Wei, T. Wang, Y. Chen, Approaching terabits per carrier metro-regional transmission using beyond-100GBD coherent optics with probabilistically shaped DP-64QAM modulation. J. Lightw. Technol. **37**(8), 1751–1755 (2019)
7. M. Kong, X. Li, J. Zhang, K. Wang, X. Xin, F. Zhao, J. Yu, High spectral efficiency 400 Gb/s transmission by different modulation formats and advanced DSP. J. Lightw. Technol. **37**(20), 5317–5325 (2019)
8. X. Zhou, L.E. Nelson, P. Magill, Rate-adaptable optics for next generation long-haul transport networks. IEEE Commun. Mag. **51**(3), 41–49 (2013)
9. H.C. Chien, J. Zhang, J. Yu, Y. Cai, Single-carrier 400G PM-256QAM generation at 34 GBaud trading off bandwidth constraints and coding overheads, in *Optical Fiber Communication Conference* (Optica Publishing Group, Washington, 2017), pp. W1J–3
10. F. Li, J. Yu, Z. Cao, J. Zhang, M. Chen, X. Li, Experimental demonstration of four-channel WDM 560 Gbit/s 128QAM-DMT using IM/DD for 2-km optical interconnect. J. Lightw. Technol. **35**(4), 941–948 (2017)
11. K. Wang, Y. Wei, M. Zhao, W. Zhou, J. Yu, 140-Gb/s PS-256-QAM transmission in an OFDM system using kramers–kronig detection. IEEE Photon. Technol. Lett. **31**(17), 1405–1408 (2019)
12. H.C. Chien, J. Yu, B. Zhu, J. Shi, Y. Cai, X. Xiao, Y. Xia, X. Wei, Y. Chen, Probabilistically shaped DP-64QAM coherent optics at 105 GBd achieving 900 Gbps net bit rate per carrier over 800 km transmission, in *2018 European Conference on Optical Communication (ECOC)* (IEEE, Piscataway, 2018), pp. 1–3
13. J. Yu, M. Kong, H.C. Chien, K. Wang, J. Shi, X. Li, X. Pan, X. Xin, Y. Xia, B. Ye, et al., 400g/channel 50-GHZ WDM coherent transmission: PS 64QAM versus hybrid 32/64QAM, in *2019 Optical Fiber Communications Conference and Exhibition (OFC)* (IEEE, Piscataway, 2019), pp. 1–3
14. X. Li, J. Yu, L. Zhao, K. Wang, C. Wang, M. Zhao, W. Zhou, J. Xiao, 1-Tb/s millimeter-wave signal wireless delivery at D-band. J. Lightw. Technol. **37**(1), 196–204 (2019)
15. C. Xie, Z. Chen, S. Fu, W. Liu, Z. He, L. Deng, M. Tang, D. Liu, Achievable information rate enhancement of visible light communication using probabilistically shaped ofdm modulation. Opt. Exp. **26**(1), 367–375 (2018)
16. J. Zhang, W. Chen, M. Gao, G. Shen, K-means-clustering-based fiber nonlinearity equalization techniques for 64-QAM coherent optical communication system. Opt. Exp. **25**(22), 27570–27580 (2017)

17. N.G. Gonzalez, D. Zibar, A. Caballero, I.T. Monroy, Experimental 2.5-Gb/s QPSK WDM phase-modulated radio-over-fiber link with digital demodulation by a k-means algorithm. IEEE Photon. Technol. Lett. **22**(5), 335–337 (2010)
18. N.G. Gonzalez, D. Zibar, I.T. Monroy, Cognitive digital receiver for burst mode phase modulated radio over fiber links, in *36th European Conference and Exhibition on Optical Communication* (IEEE, Piscataway, 2010), pp. 1–3
19. T. Zhao, A. Nehorai, B. Porat, K-means clustering-based data detection and symbol-timing recovery for burst-mode optical receiver. IEEE Trans. Commun. **54**(8), 1492–1501 (2006)
20. Q. Zhang, C. Shu, Performance evaluation of k-means clustering assisted common phase error estimation, in *2019 Conference on Lasers and Electro-Optics (CLEO)* (IEEE, Piscataway, 2019), pp. 1–2
21. L. Sun, J. Du, K. Xu, B. Liu, Z. He, K-means assisted soft decision of PAM4 to mitigate level nonlinearity and level-dependent noise for VCSEL-based 100-Gbps 100-m MMF optical interconnection, in *Optical Fiber Communication Conference* (Optica Publishing Group, Washington, 2019), pp. M1F–5
22. K. Wang, J. Zhang, M. Zhao, G. Li, C. Wang, J. Yu, A joint algorithm for photonics-aided microwave-communication system at K-band. IEEE Photon. Technol. Lett. **30**(20), 1807–1810 (2018)
23. C. Pan, F.R. Kschischang, Probabilistic 16-QAM shaping in WDM systems. J. Lightw. Technol. **34**(18), 4285–4292 (2016)
24. F. Buchali, F. Steiner, G. Böcherer, L. Schmalen, P. Schulte, W. Idler, Rate adaptation and reach increase by probabilistically shaped 64-QAM: an experimental demonstration. J. Lightw. Technol. **34**(7), 1599–1609 (2015)
25. B. Liu, Y. Zhang, K. Wang, M. Kong, L. Zhang, Q. Zhang, Q. Tian, X. Xin, Performance comparison of PS star-16QAM and PS square-shaped 16QAM (square-16QAM). IEEE Photon. J. **9**(6), 1–8 (2017)
26. J. Yu, K. Wang, J. Zhang, B. Zhu, S. Dzioba, X. Li, H.C. Chien, X. Xiao, Y. Cai, J. Shi, et al., 8 $\times$ 506-gb/s 16QAM WDM signal coherent transmission over 6000-km enabled by PS and HB-CDM, in *Optical Fiber Communication Conference* (Optica Publishing Group, Washington, 2018), pp. M2C–3
27. S. Lloyd, Least squares quantization in PCM. IEEE Trans. Inf. Theory **28**(2), 129–137 (1982)
28. G. Biondini, W.L. Kath, C.R. Menyuk, Importance sampling for polarization-mode dispersion. IEEE Photon. Technol. Lett. **14**(3), 310–312 (2002)
29. P. Schulte, G. Böcherer, Constant composition distribution matching. IEEE Trans. Inf. Theory **62**(1), 430–434 (2015)
30. L. Schmalen, Probabilistic constellation shaping: challenges and opportunities for forward error correction, in *2018 Optical Fiber Communications Conference and Exposition (OFC)* (IEEE, Piscataway, 2018), pp. 1–3
31. A. Alvarado, E. Agrell, D. Lavery, R. Maher, P. Bayvel, Replacing the soft-decision FEC limit paradigm in the design of optical communication systems. J. Lightw. Technol. **33**(20), 4338–4352 (2015)
32. D. Che, C. Sun, W. Shieh, Maximizing the spectral efficiency of stokes vector receiver with optical field recovery. Opt. Exp. **26**(22), 28976–28981 (2018)
33. J. Ren, B. Liu, X. Xu, L. Zhang, Y. Mao, X. Wu, Y. Zhang, L. Jiang, X. Xin, A probabilistically shaped star-cap-16/32 modulation based on constellation design with honeycomb-like decision regions. Opt. Exp. **27**(3), 2732–2746 (2019)
34. H. Zhou, Y. Li, D. Lu, L. Yue, C. Gao, Y. Liu, R. Hao, Z. Zhao, W. Li, J. Qiu, et al., Joint clock recovery and feed-forward equalization for PAM4 transmission. Opt. Exp. **27**(8), 11385–11395 (2019)
35. X. Li, J. Xiao, J. Yu, Heterodyne detection and transmission of 60-Gbaud PDM-QPSK signal with se of 4b/s/Hz. Opt. Exp. **22**(8), 9307–9313 (2014)
36. I.T. Union, Telecommunication standardization section, “Forward error correction for high bit-rate DWDM submarine system”. ITU-T G. 975.1 (2004)

37. J. Yu, Photonics-assisted millimeter-wave wireless communication. IEEE J. Quant. Electron. **53**(6), 1–17 (2017)
38. J. Lin, H. Sepehrian, L.A. Rusch, W. Shi, Single-carrier 72 GBaud 32QAM and 84 GBaud 16QAM transmission using a SIP IQ modulator with joint digital-optical pre-compensation. Opt. Exp. **27**(4), 5610–5619 (2019)
39. S.J. Savory, Digital coherent optical receivers: algorithms and subsystems. IEEE J. Sel. Top. Quant. Electron. **16**(5), 1164–1179 (2010)
40. A. Amari, X. Lin, O.A. Dobre, R. Venkatesan, A. Alvarado, A machine learning-based detection technique for optical fiber nonlinearity mitigation. IEEE Photon. Technol. Lett. **31**(8), 627–630 (2019)
41. L. Liu, L. Li, Y. Huang, K. Cui, Q. Xiong, F.N. Hauske, C. Xie, Y. Cai, Intrachannel nonlinearity compensation by inverse volterra series transfer function. J. Lightw. Technol. **30**(3), 310–316 (2011)
42. E. Giacoumidis, I. Aldaya, M.A. Jarajreh, A. Tsokanos, S.T. Le, F. Farjady, Y. Jaouën, A.D. Ellis, N.J. Doran, Volterra-based reconfigurable nonlinear equalizer for coherent OFDM. IEEE Photon. Technol. Lett. **26**(14), 1383–1386 (2014)
43. F.P. Guiomar, S.B. Amado, C.S. Martins, A.N. Pinto, Time-domain volterra-based digital backpropagation for coherent optical systems. J. Lightw. Technol. **33**(15), 3170–3181 (2015)
44. E. Ip, J.M. Kahn, Compensation of dispersion and nonlinear impairments using digital backpropagation. J. Lightw. Technol. **26**(20), 3416–3425 (2008)
45. H. Nakashima, T. Oyama, C. Ohshima, Y. Akiyama, Z. Tao, T. Hoshida, Digital nonlinear compensation technologies in coherent optical communication systems, in *Optical Fiber Communication Conference* (Optica Publishing Group, Washington, 2017), pp. W1G–5
46. F. Zhang, Q. Zhuge, M. Qiu, X. Zhou, M.Y. Sowailem, T.M. Hoang, M. Xiang, D.V. Plant, Blind adaptive digital backpropagation for fiber nonlinearity compensation. J. Lightw. Technol. **36**(9), 1746–1756 (2018)
47. M. Li, S. Yu, J. Yang, Z. Chen, Y. Han, W. Gu, Nonparameter nonlinear phase noise mitigation by using M-ary support vector machine for coherent optical systems. IEEE Photon. J. **5**(6), 7800312–7800312 (2013)
48. D. Wang, M. Zhang, Z. Cai, Y. Cui, Z. Li, H. Han, M. Fu, B. Luo, Combatting nonlinear phase noise in coherent optical systems with an optimized decision processor based on machine learning. Opt. Commun. **369**, 199–208 (2016)
49. E. Giacoumidis, S. Mhatli, T. Nguyen, S. T. Le, I. Aldaya, M.E. McCarthy, A.D. Ellis, B.J. Eggleton, Comparison of DSP-based nonlinear equalizers for intra-channel nonlinearity compensation in coherent optical OFDM. Opt. Lett. **41**(11), 2509–2512 (2016)
50. D. Wang, M. Zhang, Z. Li, C. Song, M. Fu, J. Li, X. Chen, System impairment compensation in coherent optical communications by using a bio-inspired detector based on artificial neural network and genetic algorithm. Opt. Commun. **399**, 1–12 (2017)
51. M.A. Jarajreh, E. Giacoumidis, I. Aldaya, S.T. Le, A. Tsokanos, Z. Ghassemlooy, N.J. Doran, Artificial neural network nonlinear equalizer for coherent optical OFDM. IEEE Photon. Technol. Lett. **27**(4), 387–390 (2014)
52. J. Zhang, M. Gao, W. Chen, G. Shen, Non-data-aided k-nearest neighbors technique for optical fiber nonlinearity mitigation. J. Lightw. Technol. **36**(17), 3564–3572 (2018)
53. D. Wang, M. Zhang, M. Fu, Z. Cai, Z. Li, H. Han, Y. Cui, B. Luo, Nonlinearity mitigation using a machine learning detector based on *k*-nearest neighbors. IEEE Photon. Technol. Lett. **28**(19), 2102–2105 (2016)
54. D. Zibar, O. Winther, N. Franceschi, R. Borkowski, A. Caballero, V. Arlunno, M.N. Schmidt, N.G. Gonzales, B. Mao, Y. Ye et al., Nonlinear impairment compensation using expectation maximization for dispersion managed and unmanaged PDM 16-QAM transmission. Opt. Exp. **20**(26), B181–B196 (2012)
55. X. Lu, C. Lu, W. Yu, L. Qiao, S. Liang, A.P.T. Lau, N. Chi, Memory-controlled deep LSTM neural network post-equalizer used in high-speed PAM VLC system. Opt. Exp. **27**(5), 7822–7833 (2019)

56. L. Tao, Y. Wang, Y. Gao, A.P.T. Lau, N. Chi, C. Lu, 40 Gb/s CAP32 system with DD-LMS equalizer for short reach optical transmissions. IEEE Photon. Technol. Lett. **25**(23), 2346–2349 (2013)
57. C.M. Bishop, N.M. Nasrabadi, *Pattern recognition and Machine Learning*, vol. 4, no. 4 (Springer, Berlin, 2006)
58. F.N. Khan, Q. Fan, C. Lu, A.P.T. Lau, An optical communication's perspective on machine learning and its applications. J. Lightw. Technol. **37**(2), 493–516 (2019)
59. F. Musumeci, C. Rottondi, A. Nag, I. Macaluso, D. Zibar, M. Ruffini, M. Tornatore, An overview on application of machine learning techniques in optical networks. IEEE Commun. Surveys Tuts. **21**(2), 1383–1408 (2018)
60. K. Tian, S. Zhou, J. Guan, Deepcluster: a general clustering framework based on deep learning, in *Machine Learning and Knowledge Discovery in Databases: European Conference, ECML PKDD 2017, Skopje, Macedonia, September 18–22, 2017, Proceedings, Part II 17* (Springer, Berlin, 2017), pp. 809–825
61. J. Wätzel, A. Blättermann, D. Schulz, C.T. Chiang, J. Berakdar, Imprinting photon orbital angular momentum during laser-assisted photoemission from quantum wells. Opt. Lett. **45**(21), 5970–5973 (2020)
62. L. Allen, M.W. Beijersbergen, R. Spreeuw, J. Woerdman, Orbital angular momentum of light and the transformation of Laguerre-Gaussian laser modes. Phys. Rev. A **45**(11), 8185 (1992)
63. S. Zhao, S. Chen, X. Wang, R. Liu, P. Zhang, H. Li, H. Gao, F. Li, Measuring the complex spectrum of orbital angular momentum and radial index with a single-pixel detector. Opt. Lett. **45**(21), 5990–5993 (2020)
64. A. Wang, L. Zhu, L. Wang, J. Ai, S. Chen, J. Wang, Directly using 8.8-km conventional multi-mode fiber for 6-mode orbital angular momentum multiplexing transmission. Opt. Exp. **26**(8), 10038–10047 (2018)
65. R. Zhang, H. Tan, J. Zhang, L. Shen, J. Liu, Y. Liu, L. Zhang, S. Yu, A novel ring-core fiber supporting mimo-free 50km transmission over high-order oam modes, in *2019 Optical Fiber Communications Conference and Exhibition (OFC)* (IEEE, Piscataway, 2019), pp. 1–3
66. J. Zhang, G. Zhu, J. Liu, X. Wu, J. Zhu, C. Du, W. Luo, Y. Chen, S. Yu, Orbital-angular-momentum mode-group multiplexed transmission over a graded-index ring-core fiber based on receive diversity and maximal ratio combining. Opt. Exp. **26**(4), 4243–4257 (2018)
67. N. Bozinovic, Y. Yue, Y. Ren, M. Tur, P. Kristensen, H. Huang, A.E. Willner, S. Ramachandran, Terabit-scale orbital angular momentum mode division multiplexing in fibers. Science **340**(6140), 1545–1548 (2013)
68. L. Wang, R.M. Nejad, A. Corsi, J. Lin, Y. Messaddeq, L. Rusch, S. LaRochelle, Linearly polarized vector modes: enabling mimo-free mode-division multiplexing, Opt. Exp. **25**(10), 11736–11749 (2017)
69. G. Zhu, Z. Hu, X. Wu, C. Du, W. Luo, Y. Chen, X. Cai, J. Liu, J. Zhu, S. Yu, Scalable mode division multiplexed transmission over a 10-km ring-core fiber using high-order orbital angular momentum modes. Opt. Exp. **26**(2), 594–604 (2018)
70. C. Brunet, P. Vaity, B. Ung, Y. Messaddeq, S. LaRochelle, L.A. Rusch, Design of a family of ring-core fiber for OAM, in *2015 Optical Fiber Communications Conference and Exhibition (OFC)* (IEEE, Piscataway, 2015), pp. 1–3
71. J. Zhang, J. Liu, L. Shen, L. Zhang, J. Luo, J. Liu, S. Yu, Mode-division multiplexed transmission of wavelength-division multiplexing signals over a 100-km single-span orbital angular momentum fiber. Photon. Res. **8**(7), 1236–1242 (2020)
72. E. Bidaki, S. Kumar, A raman-pumped dispersion and nonlinearity compensating fiber for fiber optic communications. IEEE Photon. J. **12**(1), 1–17 (2019)
73. Y. Bao, Z. Li, J. Li, X. Feng, B.O. Guan, G. Li, Nonlinearity mitigation for high-speed optical OFDM transmitters using digital pre-distortion. Opt. Exp. **21**(6), 7354–7361 (2013)
74. N. Yokota, T. Miki, K. Abe, H. Yasaka, Nonlinearity of semiconductor mach-zehnder modulator for flat optical frequency comb. IEEE Photon. Technol. Lett. **27**(21), 2219–2221 (2015)

75. G. Saavedra, G. Liga, P. Bayvel, Volterra-assisted optical phase conjugation: a hybrid optical-digital scheme for fiber nonlinearity compensation. J. Lightw. Technol. **37**(10), 2467–2479 (2019)
76. A. Miniewicz, J. Mysliwiec, P. Pawlaczyk, M. Zielinski, Photorefractive-like all-optical switching in nematic-photoconducting polymer liquid crystal cell. Mol. Cryst. Liq. Cryst. **489**(1), 119–445 (2008)
77. M.A. Cibula, D.H. McIntyre, General algorithm to optimize the diffraction efficiency of a phase-type spatial light modulator. Opt. Lett. **38**(15), 2767–2769 (2013)
78. E. Giacoumidis, J. Wei, S. Mhatli, M.F. Stephens, N.J. Doran, A.D. Ellis, B.J. Eggleton, Nonlinear inter-subcarrier intermixing reduction in coherent optical OFDM using fast machine learning equalization, in *2017 Optical Fiber Communications Conference and Exhibition (OFC)* (IEEE, Piscataway, 2017), pp. 1–3
79. T. Nguyen, S. Mhatli, E. Giacoumidis, L. Van Compernolle, M. Wuilpart, P. Mégret, Fiber nonlinearity equalizer based on support vector classification for coherent optical OFDM. IEEE Photon. J. **8**(2), 1–9 (2016)
80. H. Mrabet, A. Sangeetha, S. Mhatli, R. Attia, et al., 40 & 100 Gb/s optical communications systems based on blind support vector machine with electrical equalizer. J. Optoelectron. Adv. Mater. **19**(3), 146–152 (2017)
81. Q. Fan, G. Zhou, T. Gui, C. Lu, A.P.T. Lau, Advancing theoretical understanding and practical performance of signal processing for nonlinear optical communications through machine learning. Nat. Commun. **11**(1), 3694 (2020)
82. M.A.K. Jarajreh, S. Rajbhandari, E. Giacoumidis, N.J. Doran, Z. Ghassemlooy, Fibre impairment compensation using artificial neural network equalizer for high-capacity coherent optical OFDM signals, in *2014 9th International Symposium on Communication Systems, Networks & Digital Sign (CSNDSP)* (IEEE, Piscataway, 2014), pp. 1112–1117
83. P. Langley, W. Iba, K. Thompson, et al., An analysis of Bayesian classifiers, in *AAAI'92: Proceedings of the Tenth National Conference on Artificial Intelligence*, vol. 90 (Citeseer, 1992), pp. 223–228
84. N. Friedman, D. Geiger, M. Goldszmidt, Bayesian network classifiers. Mach. Learn. **29**, 131–163 (1997)
85. M. Dunlop-Gray, P.K. Poon, D. Golish, E. Vera, M.E. Gehm, Experimental demonstration of an adaptive architecture for direct spectral imaging classification. Opt. Exp. **24**(16), 18307–18321 (2016)
86. P. Domingos, M. Pazzani, On the optimality of the simple bayesian classifier under zero-one loss. Mach. Learn. **29**, 103–130 (1997)
87. J. Zhang, J. Liu, Z. Lin, J. Liu, L. Shen, S. Yu, Nonlinearity-aware adaptive bit and power loading DMT transmission over low-crosstalk ring-core fiber with mode group multiplexing. J. Lightw. Technol. **38**(21), 5875–5882 (2020)
88. E. Ip, A.P.T. Lau, D.J. Barros, J.M. Kahn, Coherent detection in optical fiber systems. Opt. Exp. 16(2), 753–791 (2008)
89. J. Li, C. Cai, J. Du, S. Jiang, L. Ma, L. Wang, L. Zhu, A. Wang, M.J. Li, H. Chen, et al., Ultra-low-noise mode-division multiplexed wdm transmission over 100-km FMF based on a second-order few-mode raman amplifier. J. Lightw. Technol. **36**(16), 3254–3260 (2018)
90. M.P. Wand, M.C. Jones, *Kernel Smoothing* (CRC Press, Boca Raton, 1994)
91. L. Guo, S. Boukir, Fast data selection for SVM training using ensemble margin. Pattern Recognit. Lett. **51**, 112–119 (2015)
92. D. Cannisi, B. Yuan, Design space exploration for k-nearest neighbors classification using stochastic computing, in *2016 IEEE International Workshop on Signal Processing Systems (SiPS)* (IEEE, Piscataway, 2016), pp. 321–326

MIX
Papier aus verantwortungsvollen Quellen
Paper from responsible sources
FSC® C105338

If you have any concerns about our products,
you can contact us on
ProductSafety@springernature.com

In case Publisher is established outside the EU,
the EU authorized representative is:
Springer Nature Customer Service Center GmbH
Europaplatz 3, 69115 Heidelberg, Germany

Printed by Libri Plureos GmbH
in Hamburg, Germany